Ecosystem Management and Non-Conventional Energy Sources

Ecosystem Management and Non-Conventional Energy Sources

Edited by **Craig Zodikoff**

SYRAWOOD
PUBLISHING HOUSE

New York

Published by Syrawood Publishing House,
750 Third Avenue, 9th Floor,
New York, NY 10017, USA
www.syrawoodpublishinghouse.com

Ecosystem Management and Non-Conventional Energy Sources
Edited by Craig Zodikoff

International Standard Book Number: 978-1-68286-173-8 (Hardback)

Printed in the United States of America.

Contents

ADVANCED OXIDATION PROCESSES FOR FOOD INDUSTRIAL WASTEWATER DECONTAMINATION

Dorota Krzemińska[1], Ewa Neczaj[1], Gabriel Borowski[2]

[1] Institute of Environmental Engineering, Czestochowa University of Technology, Brzeznicka 60a, 42-200 Czestochowa, Poland, e-mail: dkrzeminska@is.pcz.czest.pl

[2] Faculty of Fundamentals of Technology, Lublin University of Technology, Nadbystrzycka 38, 20-618 Lublin, Poland, e-mail: g.borowski@pollub.pl

ABSTRACT

High organic matter content is a basic problem in food industry wastewaters. Typically, the amount and composition of the effluent varies considerably. In the article four groups of advanced processes and their combination of food industry wastewater treatment have been reviewed: electrochemical oxidation (EC), Fenton's process, ozonation of water and photocatalytic processes. All advanced oxidation processes (AOP`s) are characterized by a common chemical feature: the capability of exploiting high reactivity of HO⋅ radicals in driving oxidation processes which are suitable for achieving decolonization and odour reduction, and the complete mineralization or increase of bioavailability of recalcitrant organic pollutants.

Keywords: advanced oxidation process, wastewater, food industry.

INTRODUCTION

Industrial wastewater characteristics vary not only between the industries that generate them, but also within each industry. These characteristics are also much more diverse than domestic wastewater, which is usually qualitatively and quantitatively similar in its composition. On the contrary, industry produces large quantities of highly polluted wastewater containing toxic substances, organic and inorganic compounds such as: heavy metals, pesticides, phenols and derivatives thereof, aromatic and aliphatic hydrocarbons, halogenated compounds, etc., which are generally resistant to destruction by biological treatment methods [Meriç et al. 2005, Shu 2006, Ledakowicz et al. 2001, Gogate & Pandit 2004].

Compared to other industrial sectors, food industry requires great amounts of water, since it is used throughout most of plant operations, such as production, cleaning, sanitizing, cooling and materials transport, among others [Mavrov & Belieres 2000, Cicek 2003, Álvarez et al. 2011].

As a result, meat, poultry, dairy, olive mill etc. processing plants are the facilities producing "difficult" wastewater with large total load of organic pollutants like proteins or fats and chemicals used for cleaning and sanitizing processing equipment [Álvarez et al. 2011, Zhukova et al. 2011].

The wastewater streams with different levels of pollution load (low, medium and high contamination) are collected and treated in an on-site installation or in a municipal sewage treatment plant. Increasing food production will increase the volume of sewage and the cost of disposal for food processing plants and present difficult challenges for municipal wastewater treatment plant operators [Mavrov & Belieres 2000, Cicek 2003]. Currently, in accordance with the legislation of the European Union introduced more stringent controls and rules concerning pollution of industrial wastewater [Marcucci et al. 2002, Mason 2000]. Due to the adverse impact of even small quantities of these compounds on the organoleptic characteristics of the discharge of waste waters such as colour, odour, taste, etc., as well as a threat to living organisms began to

Preface

Ecosystem management revolves around the conservation and efficient use of natural resources. This book includes contributions and latest researches by international experts in the field of ecosystem management and alternative sources of energy which will further provide an insight into the sustainable management of the ecosystem. The objective of this book is to give a general view of the different topics associated with this field such as waste disposal techniques, water purification and management, alternative energy production, organic waste management and treatment, etc. Those seeking further information in this field will be greatly assisted by this book.

The information shared in this book is based on empirical researches made by veterans in this field of study. The elaborative information provided in this book will help the readers further their scope of knowledge leading to advancements in this field.

Finally, I would like to thank my fellow researchers who gave constructive feedback and my family members who supported me at every step of my research.

Editor

look for effective, efficient and economically viable methods to remove them.

In order to achieve these aims the potential and promising methods need to include advanced oxidation methods (AOP's), which include the Fenton reaction, UV photolysis, sonication, ozonation, electrochemical oxidation, etc. These processes involve the generation of highly free radicals, mainly hydroxyl radical (HO$^\bullet$) via chemical, photochemical and photocatalytic reactions. Their application is unavoidable for the treatment of refractory organic pollutants. Numerous researches have evaluated on the treatment of refractory compounds by different AOP`s [Ledakowicz et al. 2001, Oller et al. 2011, Tarek et al. 2011, Pera-Titus et al. 2004].

The application of AOP`s in wastewater treatment leads to the degradation of the pollutant rather than transferring it to another phase, making the relevant technologies effective in the removal of organic pollutants in solution. In recent years, one of the main objectives of these processes mainly with highly polluted effluents has been not to mineralise the pollutant totally (AOP`s for complete mineralization are very expensive), but to improve the biodegradability for a possible coupling of the AOP`s with a conventional biological treatment process [Sanz et al. 2003, Muñoz et al. 2005].

PROCESSES FOR FOOD INDUSTRY WASTEWATER TREATMENT

The AOP`s has been used to reduce the organic load or toxicity of wastewaters from different industries. They are based on the generation of hydroxyl free radicals, which have a high electrochemical oxidant potential (Table 1). The generation of hydroxyl radicals involves the combination of classical oxidants, such as H_2O_2 or O_3 with UV radiation or a catalyst. The formed radicals react with organic materials breaking them down gradually in a stepwise process. The generation of hydroxyl radicals can be achieved by a variety of reactions, such as O_3/UV, H_2O_2/UV, Fenton reaction, photo-Fenton, $TiO_2/ H_2O_2/UV$ etc. Such integrated solutions can lead to more efficient use of chemical oxidants while reducing the effect of toxic or inhibitory compounds in bioreactors, leading to more robust and stable biological treatment [Gogate & Pandit 2004, Pera-Titus et al. 2004, Turhan & Turgut 2009].

Table 1. Oxidizing potential for several chemical oxidizers [Pera-Titus et al. 2004, Turhan & Turgut 2009]

Oxidant	Electrochemical oxidation potential [V]
Fluorine (F_2)	3.03
Hydroxyl radical (HO$^\bullet$)	2.80
Atomic oxygen	2.42
Ozone (O_3)	2.07
Hydrogen peroxide (H_2O_2)	1.77
Hypobromous acid ($H_{br}O$)	1.59
Chlorine dioxide (ClO_2)	1.50
Chlorine (Cl_2)	1.36
Oxygen (molecular)	1.23
Bromine (Br_2)	1.09

Elektrochemical oxidation

It was observed that electrochemical oxidation (EC) has the potential to be a distinct economic and environmental choice for treatment of wastewater and due to strict environmental regulations [Kobya & Delipinar 2008, Gengec et al. 2012]. EC involves the formation of hydroxyl radicals at the active sites of anode and has been used for the decontamination of various inorganic and organic pollutants [Rizzo 2011]. One of the major advantages of electrochemistry is that on the surface of the electrodes only electrons are produced and consumed, thereby "pure reagents" and do not contribute to a further increase in the number of chemical compounds in the environment, which often takes place in other processes [Zaleska & Grabowska 2008]. Moreover, advantages of the EC compared to conventional chemical coagulation include reduced wastewater acidification and salinity, low dosage of coagulant, superior coagulant dispersion and intrinsic electroflotation separation capability [Gengec et al. 2012].

The efficiency of electrochemical process depends on electrode and supporting electrolyte types, applied current, solution pH, nature of target contaminant/water matrix and initial concentration of the pollutants [Rizzo 2011]. In the EC process different anodes have been investigated: graphite, Pt, TiO_2, IrO_2, PbO_2, several Ti-based alloys and boron-doped diamond electrodes but the most generally employed as a electrode material is iron (Fe) or aluminium (Al) [Gengec et al. 2012, Rizzo 2011].

Generally, six main processes occur during EC: (1) migration to an oppositely charged electrode (electrophoresis) and aggregation due to charge neutralization; (2) the cation or hydroxyl ion (OH) forms a precipitate with the pollutant; (3) the metallic cation interacts with OH⁻ to form a hydroxide, which has high adsorption properties thus bonding to the pollutant (bridge coagulation); (4) the hydroxides form larger lattice-like structures and sweeps through the water (sweep coagulation); (5) oxidation of pollutants to less toxic species; (6) removal by electroflotation or sedimentation and adhesion to bubbles [Kobya & Delipinar 2008, Roa-Morales et al. 2007].

Details of some food industry wastewater studies with EC performed in recent years are provided in Table 2. EC has been proved to be an efficient method for the treatment of food process wastewater such as distillery and fermentation wastewater [Kannan et al. 2006], yeast wastewater [Kobya & Delipinar 2008], potato chips manufacturing wastewater [Kobya et al. 2006], restaurant wastewater [Chen et al. 2000], egg process wastewater [Xu et al. 2002], oily wastewater [Calvo et al. 2003], poultry slaughterhouse wastewater [Bayramoglu et al. 2006] and olive oil wastewater [Rodrigo et al. 2010].

Fenton process

Many AOPs use hydrogen peroxide as the main oxidizing agent, which is a more efficient reagent than gaseous oxygen concerning the contaminants mineralization. Fenton's process has its origin in the discovery reported in 1894 that ferrous ion strongly promotes the oxidation of tartaric acid by hydrogen peroxide. However, only much later the oxidation activity has been ascribed to the hydroxyl radical [Herney-Ramirez et al. 2010].

Oxidation with Fenton's reagent is accomplished with a mixture of ferrous ions and hydrogen peroxide, and it takes advantage of the reactivity of the free hydroxyl radicals produced in acidic solution by the catalytic decomposition of hydrogen peroxide and of the coagulation produced by the ferric hydroxide precipitates [Gogate

Table 2. A brief summary of research studies in which EC were used for treating food industry wastewater

Biological and chemical treatment	Target wastewater	Concluding remarks	References
Aluminium electrocoagulation /H₂O₂	Pasta and cookie processing industrial wastewater	Under optimal conditions of pH 4 and 18.2 mA/m² current density, Treatment reduced chemical oxygen demand (COD) by 90%, biochemical oxygen demand (BOD₅) by 96%, total solids by 95% and fecal coliforms by 99.9%.	[Roa-Morales et al. 2007]
Electrocoagulation (electrode material Fe and Al)	Cattle-slaughterhouse wastewater	In the case of aluminium electrode, polyaluminum chloride (PAC) was used as the coagulant aid for the aforesaid purpose. COD removal of 94.4% was obtained by adding 0.75 g L/PAC. In the case of iron electrode, EC was conducted concurrent with the Fenton process. As a result, 81.1% COD removal was achieved by adding 9% H₂O₂.	[Ün et al. 2009]
Anaerobic electrocoagulation (AE) and (AAE) anaerobic-aerobic electrocoagulation	Baker's yeast wastewater	The maximum color, COD and TOC were 88%, 48% and 49% at 80 A/m², pH 4 and 30 min for AE and 86%, 49% and 43% at 12.5 A/m², pH 5 and 30 min for AAE, respectively.	[Gengec et al. 2012]
Conductive-diamond electrochemical oxidation	Synthetic melanoidins	The electrolysis was carried out the in lack of additional electrolyte (except for carbonates) and in the presence of NaCl (ranging from 17.5 to 85 mM). As it can be observed, the electrolysis without addition of NaCl (it remains around 20% of initial COD) while electrolyses with NaCl electrolyte obtain the complete removal of COD.	[Canizares et al. 2009]
Electrocoagulation (electrode material Fe and Al)	Baker's yeast wastewater	The maximum removal efficiencies of COD, TOC and turbidity under optimal operating conditions, i.e., pH 6.5 for (Al) and pH 7 for (Fe), current density of 70 A/m² and operating time of 50 min were 71, 53 and 90% for (Al) and 69, 52 and 56% for (Fe), respectively.	[Kobya & Delipinar 2008]
Aluminium electrocoagulation	Moroccan olive mill wastewater	Electrolysis time 15 min, NaCl concentration 2 g/L, initial pH 4.2 and current density 250 A/m²,the discoloration of the effluent, the reduction of the COD and the reduction of polyphenols exceeded 70%, the electrodes consumption was 0.085 kg Al/kg COD_removed.	[Hanafi et al. 2010]
Electrochemical oxidation (boron-doped diamond (BDD) as anode)	Sinapinic acid (representative polyphenolic type compounds present in olive oil mill wastewater)	Under optimal experimental conditions of flow rates (i.e. 300 L/h), temperature (T=50 °C) and current density (i.e. 10 mA/cm²), 97% of COD was removed in 3h electrolysis, with 17 kWh/m³energy consumption.	[Elaoud et al. 2011]

& Pandit 2004, Pera-Titus et al. 2004, Canizares et al. 2009].

The mechanism of the Fenton's process is quite complex, and some papers can be found in the literature where tens of equations are used for its description (Eq. 1) [Mert et al. 2010].

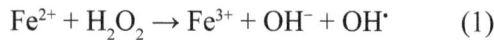

$$Fe^{2+} + H_2O_2 \rightarrow Fe^{3+} + OH^- + OH^\cdot \qquad (1)$$

As iron(II) acts as a catalyst, it has to be regenerated, which seems to occur through the following scheme [Pérez et al. 2002, Rivas et al. 2003]:

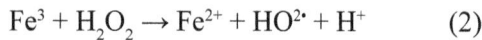

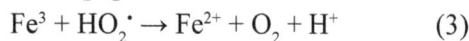

$$Fe^3 + H_2O_2 \rightarrow Fe^{2+} + HO^{2\cdot} + H^+ \qquad (2)$$

$$Fe^3 + HO_2^\cdot \rightarrow Fe^{2+} + O_2 + H^+ \qquad (3)$$

The photo-Fenton process, as its name suggests, is rather similar to the Fenton one, but employing also radiation [Tokumura et al. 2006].

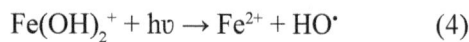

$$Fe(OH)_2^+ + h\upsilon \rightarrow Fe^{2+} + HO^\cdot \qquad (4)$$

Its effectiveness is attributed to the photolysis of Fe(III) cations in acidic media yielding Fe(II) cations (Eq. 4). In this process, the regeneration of Fe^{2+} by photo-reduction of Fe^{3+} is accelerated, this photo-reduction being an additional source of highly oxidative hydroxyl radicals, as compared with the "simple" Fenton's process [Herney-Ramirez et al. 2010, Stasinakis 2008].

The major parameters affecting Fenton process are solution's pH, amount of ferrous ions, concentration of H_2O_2, initial concentration of the pollutant and presence of other ions [Mert et al. 2010, Stasinakis 2008, de Sena et al. 2009].

The main advantage of the process is degradation organic as well as inorganic pollutants that will leading to high mineralization levels [Sanz et al. 2003, Canizares et al. 2009, Stasinakis 2008, de Sena et al. 2009, Martins et al. 2011]. Among the advanced oxidation processes, the easy-to-handle Fenton's reaction has proven to be more effective in terms of removal rate as well as operating expenses for the treatment of toxic and/or refractory food industrial wastewater (Table 3) such as baker's yeast industry effluents [Altinbas 2003], juice wastewater [Amora et al. 2012], meat industry wastewater [de Sena et al. 2009], coffee effluent [Tokumura et al. 2008], winery and distillery wastewater [Oller et al. 2011], olive mill wastewater [Zorpas & Costa 2010] etc.

Table 3. A brief summary of research studies in which Fenton process were used for treating food industry wastewater

Biological and chemical treatment	Target wastewater	Concluding remarks	References
Fenton's oxidation	Baker's yeast industry effluents	The best dosage was 1200 mg/L Fe^{2+}/800 mg/L H_2O_2 at pH 4 and in reaction time of 20 min for mineralization of DOC and COD. For these conditions, The maximum color, DOC and COD were 99%, 90% and 88%.	[Pala & Erden 2005]
Photo-Fenton oxidation	Synthetic apple-juice wastewater	At the selected operation conditions (5800 ppm H_2O_2, 40 ppm Fe^{2+}, pH 4.2 and 26 °C), 91% mineralization was achieved in 40 min when treating a synthetic wastewater with 700 ppm of the total organic carbon (TOC).	[Durán et al. 2011]
Fenton's oxidation	Livestock wastewater	The optimum ratio of H_2O_2 (mg/L) to the initial COD_{cr} was 1.05. The optimum initial pH were 3.5–4, H_2O_2/Fe^{2+} was 2 and the optimum reaction time 30 min. The removal ratios of COD_{cr} and color of the supernatant after static precipitation of the produced sludge were 88 and 95.4%, respectively.	[Lee & Shoda 2003]
Photo-Fenton oxidation	Coffee effluent	The results suggest that the $UV/H_2O_2/Fe^{2+}$ system is very efficient for the treatment of coffee effluents. The mineralization of 250 mg/L model coffee effluent was not complete but about 90% mineralization was found after 200 min of UV irradiation.	[Tokumura et al. 2006]
Solar photo-Fenton oxidation	Winery wastewaters	Under optimal conditions, 61% TOC was achieved in 360 min. Temperature and initial concentrations of H_2O_2 and oxalic acid were the most significant factors affecting the wastewater mineralization. The addition of H_2O_2 can be used to control the mineralization degree of this type of wastewater.	[Monteagudo et al. 2012]
Fenton's oxidation	Synthetic melanoidins	These high concentrations of refractory carbonis is not able to achieve the complete mineralization of the waste, and a high TOC remains at the end of the treatment (>40%).	[Canizares et al. 2009]
Fenton, Fenton-like process	Olive oil mill wastewater	Processes showed high COD (>80%) and total-phenol (>85%) removal performance on evaluated effluents	[Mert et al. 2010]
Electro-Fenton	Olive mill wastewater	Considered as a pre-treatment (COD removal 53%) before anaerobic digestion and ultrafiltration resulting in a complete detoxify of effluent.	[Khoufi et al. 2009]

Ozonation

Ozonation of water is a well-known technology and the strong oxidative properties of O_3 and its ability to effectively oxidize many organic compounds in aqueous solution have been well documented. Unlike other oxidizing agents such as Cl_2, oxidation with O_3 leaves no toxic residues that have to be removed or disposed [Sarayu et al. 2007, Ulson de Souza et al. 2010]. Ozonation is one of the AOP processes used to food wastewater treatment, which is versatile and environmentally sound [Sarayu et al. 2007]. Although, the cost of ozone production still is high interest in the use of ozone in wastewater treatment has increased considerably in last few years due to the numerous advantages of this process [Meriç et al. 2005, Ulson de Souza et al. 2010]. Ozonation can eliminate toxic substances, increase the biodegradability of organic pollutants and has high potential in decolourization [Zayas et al. 2007].

Ozone is very reactive towards compounds incorporating conjugated double bonds(such as C=C, C=N, N=N), often associated with colour [Gogate & Pandit 2004, Rizzo 2011, Coca et al. 2007]. Ozone can react with solutes either by direct oxidation between organic contaminants and molecular ozone (Eq. 5) or by indirect reactions with hydroxyl radicals resulting from ozone decomposition (Eq. 6, 7) [Meriç et al. 2005, Rizzo 2011, Coca et al. 2007].

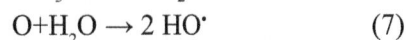

$$O_3 + R \rightarrow RO + O_2 \tag{5}$$

$$O_3 \leftrightarrow O + O_2 \tag{6}$$

$$O + H_2O \rightarrow 2\ HO^{\cdot} \tag{7}$$

This arises from the fact that pH affects the double action of ozone on the organic matter which may be a direct or an indirect (free radical) oxidation pathway. At low pH, ozone exclusively reacts with compounds with specific functional groups through selective reactions, such as electrophilic, nucleophilic or dipolar addition reactions (i.e. direct pathway). At basic pH, ozone decomposes yielding hydroxyl radicals, a highly oxidizing species which reacts nonselectively with a wide range of organic and inorganic com-

Table 4. A brief summary of research studies in which ozonation were used for treating food industry wastewater

Biological and chemical treatment	Target wastewater	Concluding remarks	References
Ozonation and aerobic biological degradation	Table olive wastewater	In this combined process, the ozonation of the biologically pretreated table black-olive wastewater yielded a conversion of 62%, considerably higher than the 52% attained in the ozonation process.	[Beltran-Heredia et al. 2000]
Ozonation	Six phenolic acids typically found in olive mill wastewater	(O_3 production was varied 10 g/L, flow rate was 30 l/h). Process were effective in degradation of phenolic compounds 91%. The COD abatement (42%) proves that ozonation pathways at those conditions are mainly through partial oxidation rather than directly to end products (CO_2, H_2O).	[Martins & Quinta-Ferreira 2011]
Ozonation	Red-meat-processing wastewater	The decrease in COD and BOD of the wastewater were 10.7% and 23.6%, respectively, decolorization of the wastewater after ozonation.	[Wu & Doan 2005]
Ozonation	Olive mill wastewater	The ozonation of OMW carried out at three different times (60, 90 and 120 min) The highest efficiencies were achieved at 120 min). Polyphenols and COD reductions were about 82,4% and 59,8%.	[Andreozzi et al. 2008]
Ozonation	Winery wastewaters	The ozone dose rate 0.68 g/h/L, pH 4. Ozonation reduced the initial COD by 12% At pH 4, ozone does not decompose to more reactive radical species, therefore the COD removal was limited.	[Lucas et al. 2010]
Ozonation with conventional aerobic oxidation	Distillery wastewater	Ozone was found to be effective in bringing down the COD (up to 27%) during the pretreatment step itself. The integrated process achieved ~79% COD reduction along with decoloration of the effluent sample as compared to 34.9% COD reduction for non-ozonated sample, over a similar treatment period.	[Sangave et al. 2007]
Ozonation	Synthetic melanoidins	As it can be observed, it achieves very significant removals of COD and TOC (>80% and >70%, respectively) and the almost complete disappearance of the color (ozone production: 1 g/h, 25°C, pH 12, time change)	[Canizares et al. 2006]
Ozonation	Molasses fermentation wastewater	The highest efficiencies were achieved at 40 °C (the ozone dose rate constants = 2.3 g/h/L). Color and COD reductions were about 90% and 37%, respectively. In no case, the percentage of TOC removed was higher than 10–15%.	[Coca et al. 2005]

pounds in water (i.e., indirect ozonation pathway) [Meriç et al. 2005, Pera-Titus et al. 2004, Turhan & Turgut 2009].

Ozone has many properties desirable for the treatment of the wastewater [Gogate & Pandit 2004, Rizzo 2011]:

- no sludge remains,
- danger is minimal,
- decolorization and degradation occur in one step,
- it is easily performed,
- little space is required,
- all residual ozone can be easily decomposed to oxygen and water.

Ozonation has been successfully applied to the treatment of winery and distillery wastewater [Zaleska & Grabowska 2008], olive mile wastewater [Oller et al. 2011], meat industry wastewater [Millamena 1992], molasses wastewater [Coca et al. 2007] etc.

Photocatalytic process

The photocatalytic or photochemical degradation processes are gaining importance in the area of wastewater treatment, since these processes result in complete mineralization with operation at mild conditions of temperature and pressure [Tarek et al. 2011, Ugurlu & Karaoglu 2011]. In the process, semiconductor material is excited by electromagnetic radiation possessing energy of sufficient magnitude, to produce conduction band electrons and valence band holes [Gogate & Pandit 2004, Stasinakis 2008, Żmudziński 2010].

The selection of the adequate catalyst must consider the following properties: chemical activity, stability, availability and handiness, cost and lack of toxicity [Navarro et al. 2005]. The surface area and the number of active sites offered by the catalyst (thus nature of catalyst, i.e. crystalline or amorphous is important) for the adsorption of pollutants plays an important role in deciding [Gogate & Pandit 2004]. Several catalytic materials have been studied in photocatalysis (various oxides such as TiO_2, ZnO, SnO_2, WO_3, ZrO_2, CeO etc. or sulfides such as CdS, ZnS etc.) [Gogate & Pandit 2004, Sakthivel et al. 2003, Habibi et al. 2001]. Among the semiconductors reported so far, outstanding stability and oxidative power makes TiO_2 the best semiconductor photocatalyst for environmental remediation and energy conversion processes [Gogate & Pandit 2004, Ugurlu & Karaoglu 2011, Navarro et al. 2005, Chatzi-

symeon et al. 2008, Banu et al. 2008]. However, there are two potential drawbacks associated with the use of TiO_2, namely: (a) its possible toxic effects on human health and (b) reduced activity due to the complexity of water matrix (i.e., presence of solids or inorganic ions) [Chatzisymeon et al. 2008].

TiO_2 in its anatase form has an energy bandgap of 3.2 eV and can be activated by UV radiation with a wavelength up to 387.5 nm. It only requires 1 W/m^2 of light [Navarro et al. 2005, Navgire 2012]. Photocatalytic degradation occurs through a multistep process that involves the formation of reactive species on the surface of the photocatalyst and the subsequent generation of hydroxyl radicals that result in the mineralization of most organic compounds [Navarro et al. 2005, Chatzisymeon et al. 2008, Banu et al. 2008, Navgire et al. 2012]. The photocatalysis can be explained by the following simplified reaction (with TiO_2) [Stasinakis 2008, Navarro et al. 2005]:

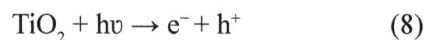

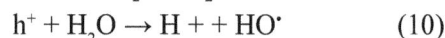

$$TiO_2 + h\upsilon \rightarrow e^- + h^+ \qquad (8)$$
$$e^- + O_2 \rightarrow O_2^- {}^\bullet \qquad (9)$$
$$h^+ + H_2O \rightarrow H + + HO^\bullet \qquad (10)$$
$$h^+ + OH^- \rightarrow HO^\bullet \qquad (11)$$
$$O_2^- {}^\bullet + H^+ \rightarrow HO_2^\bullet \qquad (12)$$

The UV radiation required for the photocatalytic processes can be obtained from artificial sources or the sun. There is a significant economic incentive for solar light based photocatalytic degradations [Banu et al. 2008].

The major factors affecting TiO_2/UV light process are: initial organic load, amount of catalyst, reactor's design, UV irradiation time, temperature, solution's pH, light intensity and the presence of ionic species. The use of excessive amounts of catalyst may reduce the amount of energy being transferred into the medium due to the opacity offered by the catalyst particles [Stasinakis 2008]. Organic compounds can then undergo both oxidative degradation through their reactions with valence band holes, hydroxyl and peroxide radicals and reductive cleavage through their reactions with electrons yielding various by-products and eventually mineral end-products [Gogate & Pandit 2004, Ugurlu & Karaoglu 2011, Chatzisymeon et al. 2008]. In several studies, photocatalytic process has been used to the treatment of winery and distillery wastewater [Lucas et al. 2009], olive mile wastewater [Chatzisymeon et al. 2008], dairy industry wastewater [Banu et al. 2008], molasses wastewater [Satyawali &

Table 5. A brief summary of research studies in which photocatalytic process were used for treating food industry wastewater

Biological and chemical treatment	Target wastewater	Concluding remarks	References
$UV/H_2O_2/TiO_2/Sep$	Olive mill wastewater	Optimum values of catalyst dose, temperature and H_2O_2 were found to be 318 K and 0.25 and 0.50 g/L, 30 ml/L H_2O_2, respectively. The degradation of lignin and phenol was favourable at pH 9–11.0 and natural sunlight. All pollutants could be removed under 24 h in 80–100% rates.	[Ugurlu & Karaoglu 2011]
Calcined $InYO_3$	Molasses fermentation wastewater	Specifically, $InYO_3$ calcined at 700 °C had a considerably larger surface area and lower reflectance intensity and showed higher photocatalytic activity. After 150-min reaction, the decolorization and COD removal were 98.23% and 92.98%, respectively.	[Qin et al. 2011]
Vacuum ultraviolet/TiO_2	Oily wastewater from the restaurant	Under the optimum conditions of irradiation 10 min, initial COD 3981 mg/L, TiO_2 150 mg/L, pH 7.0 and flow rate of air 40 L/h, the process TiO_2/VUV achieved removal efficiencies of COD, BOD_5 and oil about 63%, 43%, 70%.	[Kang et al. 2011]
TiO_2/H_2O_2	Black table olive processing wastewater	TiO_2 and H_2O_2 concentrations in the range 0.25–2 g/L and 0.025–0.15 g/L. Depending on the conditions employed, nearly complete decoloration (>90%) could be achieved, while mineralization never exceeded 50%.	[Chatzisymeon et al. 2008]
TiO_2/H_2O_2	Winery wastewater	The maximum COD removal were achieved at zero catalyst loading with COD removal of about 84%. Lower rates of chemical reaction in photocatalysis compared to photolysis were possibly because of the shielding of UV light by titania particles.	[Agustina et al. 2008]
TiO_2/H_2O_2, TiO_2/Fe-clays	Winery wastewater	The TiO_2/H_2O_2 treatment produces the highest efficiency, reaching to 52–58% of COD. The optimum dosage is: 2.5 mL/L H_2O_2 and 1.0 g/L TiO_2. Secondly, although the H_2O_2/clays system produces lower COD removal 34–45%, it requires a H_2O_2 dosage between three and six times lower than the TiO_2/H_2O_2 treatment.	[Navarro et al. 2005]
TiO_2/UV	Fresh-Cut Vegetable Industry wastewater	Photocatalysis was an effective disinfection method, reducing counts of bacteria, molds, and yeasts. Most of the treated wash waters had total bacteria reductions of 4.1±1.3 to 4.8±0.4 log CFU/ml after 10 min of treatment when compared with untreated water. That implementation of TiO_2/UV in the wash waters could allow the reuse of wash water.	[[Selma et al. 2008]
MoO_3/TiO_2/UV	Molasses wastewater	The color of the molasses solution decreases to around 70%, The maximum decreases in COD, BOD, and TDS about 90%, 90%, 50%, respectively. This system rate increases with an increase in temperature from 40 to 80 °C.	[Navgire et al. 2012]

Balakrishnan 2008], candy and sugar industry wastewater [Żmudziński 2010], fresh-cut vegetable industry wastewater [Selma et al. 2007] etc. Several applications are presented below (Table 5).

THE COMBINATION OF AOP`S FOR WASTEWATER TREATMENT

Differently combined AOP's have been developed and investigated by several research groups as alternatives for treating food industrial wastewater containing organic pollutants. This may commonly cause a reduction of toxicity or elimination of a specific pollutant or reduce the reaction time and economic cost [Oller et al. 2011, Stasinakis 2008, de Sena et al. 2009, Zayas et al. 2007].

Determining the target is an essential step in combination studies since it helps define process efficiency and provides a basis for comparing different operating conditions and optimizing the process [Oller et al. 2011, Stasinakis 2008].

CONCLUSION

Food industry uses large amounts of water for many different purposes including cooling and cleaning, as a raw material, as sanitary water for food processing, for transportation, cooking and dissolving, as auxiliary water etc. In principle, the water used in the food industry may be used as process and cooling water or boiler feed water. As a consequence of diverse consumption, the amount and composition of

Table 6. A brief summary of research studies in which combining various AOP's were used for treating food industry wastewater

Biological and chemical treatment	Target wastewater	Concluding remarks	References
Diamond electrochemical oxidation Fenton-ozonation	Synthetic melanoidins solution	Operation conditions – CDEO: natural pH; T: 25 °C; j: 300 $A \cdot cm^{-2}$. Ozone production: 1 g/h; T: 25 °C; pH 12. Fenton process: pH 3; T: 25°C; Fe^{2+}: 667 mg/dm^{-3}. Complete mineralization of the waste is obtained and no refractory compounds remain at the end of the process in both cases.	[Canizares et al. 2009]
sonication/ anaerobic fermentation	Cassava wastewater	Hydrogen yield for cassava wastewater pretreated at pH 7.0 with sonication for 45 min using anaerobic seed sludge was 0.913 mol H_2/g COD. In wastewater COD removal was 40%.	[Leaño et al. 2012]
UD,ozonation	Thermally pretreated distillery wastewater	The study clearly shows the suitability of ozone and ultrasound as a pre-treatment step for the thermally pretreated wastewater for aerobic treatment by increasing the COD removal efficiency. Ozone was more efficient in COD removal with a 25-times increase in the rate of biodegradation of ozonated sample along with discoloration of the effluent sample.	[Sangave et al. 2007]
$UV/H_2O_2/O_3$	Coffee wastewater	The $UV/H_2O_2/O_3$ process is capable of reducing the COD content of the wastewater by 87% in 35 min at pH 2.0. By comparison, the UV/H_2O_2 and UV/O_3 treatments under the same conditions reduced the COD by approximately 84%.	[Zayas et al. 2007]
$UV/H_2O_2/O_3$	Winery wastewater	The COD removal efficiency is slightly higher at pH10 (57%), than pH4 (49%) and pH7 (40%) after 300 min.	[Lucas et al. 2010]
Dissolved air flotation (DAF) with UV/H_2O_2 or photo-Fenton	Meat industry wastewater	DAF/UV/H_2O_2 results for BOD_5 and COD reduction was 82.9% and 91.1% respectively. For TS and VS, reductions of up to 72.5% and 77.0% were achieved, respectively. DAF/photo-Fenton results for BOD_5 and COD reduction was 95.7% and 97.6\% respectively. For TS and VS, reductions of up to 61.5% and 90.8% were achieved, respectively.	[de Sena et al. 2009]
UV/H_2O_2	Synthetic melanoidin	The oxidation process was much more capable of removing color 99%, dissolved organic carbon (DOC) 50% and dissolved organic nitrogen (DON) 25% at the optimal applied dose of H_2O_2 for the system (3.3 g/L)	[Dwyer et al. 2008]
UV/O_3	Olive mill wastewater	In particular, biodegradation of UV/O_3 pretreated OMW found to have the highest removal levels; the percent of COD removal reaches about 91%.	[Lafi et al. 2009]
Modified photo-Fenton /ozonation	Olive mill wastewater	For Fe(III)/air/solar light and ozonation the highest efficiencies Polyphenols and COD reductions were about 87.9% and 64.9%.	[Andreozzi et al. 2008]

food industry wastewaters varies considerably. Characteristics of the effluent consist of large amounts of suspended solids, nitrogen in several chemical forms, fats and oils, phosphorus, chlorides and organic matter.

AOP's constitute a promising technology for the treatment of food industry wastewaters containing difficult to biodegradable organic contaminants. It involves the generation of free hydroxyl radical (HO·), a powerful, nonselective chemical oxidant to change organic compounds to a more biodegradable form or of carbon dioxide and water. These processes can reduce a broad spectrum of chemical and biological contaminants which are otherwise difficult to remove with conventional treatment processes of food industry wastewater.

Acknowledgments

This work was supported by the Faculty of Environmental Protection and Engineering, Czestochowa University of Technology (project BS/MN-401-315/11).

REFERENCES

1. Agustina T.E., Ang H.M., Pareek V.K. 2008. Treatment of winery wastewater using a photocatalytic/photolytic reactor. Chem. Eng. J., 135 (1–2), 151–156.

2. Altinbas M., Aydin A.F., Sevimli M.F., Ozturk I. 2003. Advanced oxidation of biologically pretreated Baker's Yeast Industry effluents for high recalcitrant COD and color removal. J. Environ. Sci. Health A, 38 (10), 2229–2240.

3. Álvarez P.M., Pocostales J.P., Beltrán F.J. 2011. Granular activated carbon promoted ozonation of a food-processing secondary effluent. J. Hazard. Mater., 185, 776–783.

4. Amora C., Lucasa M.S., Pirraa A.J., Peresa J.A. 2012. Treatment of concentrated fruit juice wastewater by the combination of biological and chemical processes. J. Environ. Sci. Health A, 47 (12), 1809–1817.

5. Andreozzi R., Canterino M., Di Somma I., Lo Giudice R., Marotta R., Pinto G., Pollio A. 2008. Effect of combined physico-chemical processes on the phytotoxicity of olive mill wastewaters. Wat. Res., 42, 1684–1692.

6. Banu J.R., Anandan S., Kaliappan S., Yeom I-Y. 2008. Treatment of dairy wastewater using anaerobic and solar photocatalytic methods. Sol. Energy, 82, 812–819.

7. Bayramoglu M., Kobya M., Eyvaz M., Senturk E. 2006. Technical and economic analysis of electrocoagulation for the treatment of poultry slaughterhouse wastewater. Sep. Purif. Technol., 51, 404–408.

8. Beltran-Heredia J., Torregrosa J.,Dominguez J.R., Garcia J. 2000. Treatment of black-olive wastewaters by ozonation and aerobic biological degradation. Wat. Res., 34 (14), 3515–3522.

9. Calvo L.S., Leclerc J.P., Tanguy G., Cames M.C., Paternotte G., Valentin G., Rostan A., Lapicque F. 2003. An electrocoaguhtion unit for the purification of soluble oil wastes of high COD. Environ. Prog., 22, 57–65.

10. Canizares P., Hernández-Ortega M., Rodrigo M.A., Barrera-Díaz C.E., Roa-Morales G., Sáez C. 2009. A comparison between conductive-diamond electrochemical oxidation and other advanced oxidation processes for the treatment of synthetic melanoidins. J. Hazard. Mater., 164, 120–125.

11. Chatzisymeon E., Stypas E., Bousios S., Xekoukoulotakis N.P., Mantzavinos D. 2008. Photocatalytic treatment of black table olive processing wastewater. J. Hazard. Mater., 154, 1090–1097.

12. Chen X., Chen G., Yue P.L. 2000. Separation of pollutants from restaurant wastewater by electrocoagulation. Sep. Purif. Technol., 19, 65–76.

13. Cicek N. 2003. A review of membrane bioreactors and their potential application in the treatment of agricultural wastewater. CSBE, 43, 37–49.

14. Coca M., Pena M., Gonzalez G. 2007. Kinetic study of ozonation of molasses fermentation wastewater. J. Hazard. Mater., 149, 364–370.

15. Coca M., Pena M., Gonzalez G. 2005. Variables affecting efficiency of molasses fermentation wastewater ozonation. Chemosphere, 60, 1408–1415.

16. de Sena F.R., Tambosi J.L., Genena A.K., Moreira R., Schröder H.F., José H.J. 2009. Treatment of meat industry wastewater using dissolved air flotation and advanced oxidation processes monitored by GC–MS and LC–MS. Chem. Eng. J., 152, 151–157.

17. Durán A., Monteagudo J.M., Carnicer A. 2011. Photo-Fenton mineralization of synthetic apple-juice wastewater. Chem. Eng. J., 168, 102–107.

18. Elaoud S.Ch., Panizza M., Cerisola G., Mhiri T. 2011. Electrochemical degradation of sinapinic acid on a BDD anode. Desalination, 272, 148–153.

19. Gengec E., Kobya M., Demirbas E., Akyol A., Oktor K. 2012. Optimization of baker's yeast wastewater using response surface methodology by electrocoagulation. Desalination, 286, 200–209.

20. Gogate P.R., Pandit A.B. 2004. A review of imperative technologies for wastewater treatment: Oxidation technologies at ambient conditions. Advances in Environmental Research, 8 (3–4), 501–551.

21. Habibi, M.H., Tangestaninejad, S., Yadollahi, B. 2001. Photocatalytic mineralisation of mercaptans as environmental pollutants in aquatic system using TiO_2 suspension. Appl. Catal. B-Environ., 33 (1), 57.

22. Hanafi F., Assobhei O., Mountadar M., Detoxification and discoloration of Moroccan olive mill wastewater by electrocoagulation. J. Hazard. Mater., 2010, 174, 807–812.

23. Herney-Ramirez J., Vicente M.A., Madeira L.M. 2010. Heterogeneous photo-Fenton oxidation with pillared clay-based catalysts for wastewater treatment: A review. Appl. Catal. B-Environ., 98, 10–26.

24. Kang J-X., Lu L., Zhan W., Li B., Li D-S., Ren Y-Z., Liu D-Q. 2011. Photocatalytic pretreatment of oily wastewater from the restaurant by a vacuum ultraviolet/TiO_2 system. J. Hazard. Mater., 186, 849–854.

25. Kannan N., Karthikeyan G., Tamilselvan N. 2006. Comparison of treatment potential of electrocoagulation of distillery effluent with and without activated Areca catechu nut carbon. J. Hazard. Mater., 137, 1803–1809.

26. Khoufi, S., Aloui, F., Sayadi, S., Pilot scale hybrid process for olive mill wastewater treatment and reuse. Chem. Eng. Process., 2009, 48 (2), 643–650.

27. Kobya M., Delipinar S. 2008. Treatment of the baker's yeast wastewater by electrocoagulation. J. Hazard. Mater., 154, 1133–1140.

28. Kobya M., Hiz H., Senturk E., Aydiner C., Demirbas E. 2006. Treatment of potato chips manufacturing wastewater by electrocoagulation. Desalination., 190, 201–211.

29. Lafi W.K., Shannak B., Al-Shannag M., Al-Anber Z., Al-Hasan M. 2009. Treatment of olive mill wastewater by combined advanced oxidation and biodegradation. Sep. Purif. Technol., 70, 141–146.

30. Leaño E.P., Babel S. 2012. Effects of pretreatment methods on cassava wastewater for biohydrogen production optimization. Renew. Energ., 39, 339–346.

31. Ledakowicz S., Solecka M., Zylla R. 2001. Biodegradation, decolourisation and detoxification of textile wastewater enhanced by advanced oxidation processes. J. Biotechnol., 89, 175–184.

32. Lee H., Shoda M. 2008. Removal of COD and colour from livestock wastewater by the Fenton method. J. Hazard. Mater., 153, 1314–1319.

33. Lucas M.S., Peres J.A., Puma G.L. 2010. Treatment of winery wastewater by ozone-based advanced oxidation processes (O_3, O_3/UV and O_3/UV/H_2O_2) in a pilot-scale bubble column reactor and process economics. Sep. Purif. Technol., 72, 235–241.

34. Lucas M.S., Mosteo R., Maldonado M.I., Malato S., Peres J.A. 2009. Solar photochemical treatment of winery wastewater in a CPC reactor. J. Agric. Food Chem., 57 (23), 11242–11248.

35. Marcucci M., Ciardelli G., Matteucci A., Ranieri L., Russo M. 2002. Experimental campaigns on textile wastewater for reuse by means of different membrane processes. Desalination, 149 (1-3), 137–143.

36. Martins R.C., Quinta-Ferreira R.M. 2011. Remediation of phenolic wastewaters by advanced oxidation processes (AOPs) at ambient conditions: Comparative studies. Chem. Eng. Sci., 66, 3243–3250.

37. Mason T.J. 2000. Large scale sonochemical processing: aspiration and actuality. Ultrason. Sonochem., 7(4), 145–149.

38. Mavrov V., Belieres E. 2000. Reduction of water consumption and wastewater quantities in the food industry by water recycling using membrane processes. Desalination, 131, 75–86.

39. Meriç S., Selçuk H., Belgiorno V. 2005. Acute toxicity removal in textile finishing wastewater by Fenton's oxidation, ozone and coagulation-flocculation processes. Wat. Res., 39 (6), 1147–1153.

40. Mert B.K., Yonar T., Kilic M.Y., Kestioglu K. 2010. Pre-treatment studies on olive oil mill effluent using physicochemical Fenton and Fenton-like oxidations processes. J. Hazard. Mater., 174, 122–128.

41. Millamena O.M. 1992. Ozone treatment of slaughterhouse and laboratory wastewaters. Aquacult. Eng., 11 (1), 23–31.

42. Monteagudo J.M., Durán A., Corral J.M., Carnicer A., Frades J.M., Alonso M.A. 2012. Ferrioxalate-induced solar photo-Fenton system for the treatment of winery wastewaters. Chem. Eng. J., 181–182, 281–288.

43. Muñoz I., Rieradevall J., Torrades F., Peral J., Doménech X. 2005. Environmental assessment of different solar driven advanced oxidation processes. Sol. Energy, 79 (4), 369–375.

44. Navarro P., Sarasa J., Sierra D., Esteban S., Ovelleiro J.L. 2005. Degradation of wine industry wastewaters by photocatalytic advanced oxidation. Water Sci. Technol., 51 (1), 113–120.

45. Navgire M., Yelwande A., Tayde D., Arbad B., Lande M. 2012. Photodegradation of Molasses by a MoO_3-TiO_2 Nanocrystalline Composite Material. Chin. J. Catal., 33, 261–266.

46. Oller I., Malato S., Sánchez-Pérez J.A. 2011. Combination of Advanced Oxidation Processes and biological treatments for wastewater decontamination – A review. Sci. Total Environ., 409 (20), 4141–4166.

47. Pala A., Erden G. 2005. Decolorization of a baker's yeast industry effluent by Fenton oxidation. J. Hazard. Mater. B, 127, 141–148.

48. Pera-Titus M., García-Molina M., Baños M.A., Giménez J., Esplugas S. 2004. Degradation of chlorophenols by means of advanced oxidation processes: a general review. Appl. Catal. B-Environ., 47 (4), 219–256.

49. Pérez M., Torrades F., Doménech X., Peral J. 2002. Fenton and photo-Fenton oxidation of textile effluents. Wat. Res., 36, 2703–2710.

50. Qin Z., Liang Y., Liu Z., Jiang W. 2011. Preparation of $InYO_3$ catalyst and its application in photodegradation of molasses fermentation wastewater. J. Environ. Sci., 23 (7), 1219–1224.

51. Rivas F.J., Beltrán F.J., Gimeno O., Alvarez P. 2003. Treatment of brines by combined Fenton's reagent-aerobic biodegradation II. Process modelling. J. Hazard. Mater., 96, 259–276.

52. Rizzo L. 2011. Bioassays as a tool for evaluating advanced oxidation processes in water and wastewater treatment. Wat. Res., 45, 4311–4340.

53. Roa-Morales G., Campos-Medina E., Aguilera-Cotero J., Bilyeu B., Barrera-Diaz C. 2007. Aluminium electrocoagulation with peroxide applied to wastewater from pasta and cookie processing. Sep. Purif. Technol., 54, 124–129.

54. Rodrigo M.A., Canizares P., Sanchez-Carretero A., Saez C. 2010. Use of conductive-diamond electrochemical oxidation for wastewater treatment. Catal. Today, 151, 173–177.

55. Sakthivel S., Neppolia B., Shankar M.V., Arabindoo B., Palanichamy M., Murugesan V. 2003. Solar photocatalytic degradation of azo dye: comparison of photocatalytic efficiency of ZnO and TiO_2. Sol. Energ. Mat. Sol. C., 77, 65–82.

56. Sangave P.C., Gogate P.R., Pandit A.B., Combination of ozonation with conventional aerobic oxidation for distillery wastewater treatment. Chemosphere, 2007, 68, 32–41.

57. Sanz J., Lombraña J.I., De Luis A.M., Ortueta M., Varona F. 2003. Microwave and Fenton's reagent oxidation of wastewater. Environ. Chem. Lett., 1, 45–50.

58. Sarayu K., Swaminathan K., Sandhya S. 2007. Assessment of degradation of eight commercial reactive azo dyes individually and in mixture in aqueous solution by ozonation. Dyes Pigments, 75, 362–368.

59. Satyawali Y., Balakrishnan M. 2008. Wastewater treatment in molasses-based alcohol distilleries for COD and colour removal: A review. Journal of Environmental Management 86, 481–497.

60. Selma M.V., Allende A., López-Gálvez F., Conesa M.A., Gil M.I. 2008. Heterogeneous photocatalytic disinfection of wash waters from the fresh-cut vegetable industry. J. Food Protection, 71 (2), 286–292.

61. Selma M.V., AllendeA., López-Gálvez F., Gil M.I. 2007. Different advanced oxidation processes for disinfection of wash waters from the fresh-cut industry. IOA Conference and Exhibition Valencia, Spain, 6.1, 1–8.

62. Shu H. 2006. Degradation of dyehouse effluent containing C.I. Direct Blue 199 by processes of ozonation, UV/H_2O_2 and in sequence of ozonation with UV/H_2O_2. J. Hazard. Mater., 133, 92–98.

63. Stasinakis A.S., Use of selected advanced oxidation processes (AOPs) for wastewater treatment – a mini review. Global NEST Journal, 2008, 10 (3), 376–385.

64. Tarek S. Jamil, Montaser Y. Ghaly, Ibrahim E. El-Seesy, Eglal R. Souaya, Rabab A. Nasr 2011. A comparative study among different photochemical oxidation processes to enhance the biodegradability of paper mill wastewater. J. Hazard. Mater., 185, 353–358.

65. Tokumura M., Ohta A., Znad H.T., Kawase Y. 2006. UV light assisted decolorization of dark brown colored coffee effluent by photo-Fenton reaction. Wat. Res., 40, 3775–3784.

66. Tokumura M., Znad H.T., Kawase Y. 2008. Decolorization of dark brown colored coffee effluent by solar photo-Fenton reaction: Effect of solar light dose on decolorization kinetics. Wat. Res., 42, 4665–4673.

67. Turhan K., Turgut Z. 2009. Decolorization of direct dye in textile wastewater by ozonization in a semi-batch bubble column reactor. Desalination, 242, 256–263.

68. Ugurlu M., Karaoglu M.H. 2011. TiO_2 supported on sepiolite: Preparation, structural and thermal characterization and catalytic behaviour in photocatalytic treatment of phenol and lignin from olive mill wastewater. Chem. Eng. J., 166, 859–867.

69. Ulson de Souza S.M., Santos Bonilla K.A., Ulson de Souza A.A. 2010. Removal of COD and colour from hydrolyzed textile azo dye by combined ozonation and biological treatment. J. Hazard. Mater., 179 (1-3), 35–42.

70. Ün Ü.T., Koparal A.S., Öğütveren Ü.B. 2009. Hybrid processes for the treatment of cattle-slaughterhouse wastewater using aluminum and iron electrodes. J. Hazard. Mater., 164, 580–586.

71. Wu J., Doan H. 2005. Disinfection of recycled red-meat-processing wastewater by ozone. J. Chem. Technol. Biotechnol., 80, 828–833.

72. Xu L.J., Sheldon B.W., Larick D.K., Carawan R.E. 2002. Recovery and utilization of useful by-products from egg processing wastewater by electrocoagulation. Poultry Sci., 81, 785–792.

73. Zaleska A., Grabowska E. 2008. Podstawy technologii chemicznej. Nowoczesne procesy utleniania – ozonowanie, utlenianie fotokatalityczne, reakcja Fentona. Politechnika Gdańska, Gdańsk.

74. Zayas P.T., Geissler G., Hernandez F. 2007. Chemical oxygen demand reduction in coffee wastewater through chemical flocculation and advanced oxidation processes. J. Environ. Sci., 19, 300–305.

75. Zhukova V., Sabliy L., Łagód G. 2010. Biotechnology of the food industry wastewater treatment from nitrogen compounds. Proceedings of ECOpole, 5 (1), 133–138.

76. Żmudziński W., Preliminary results of glucose oxidation by photocatalysis on titanium dioxide – primary intermediates. Physicochem. Probl. Miner. Process., 45141–45151.

77. Zorpas A.A., Costa C.N. 2010. Combination of Fenton oxidation and composting for the treatment of the olive solid residue and the olive mile wastewater from the olive oil industry in Cyprus. Bioresource Technol., 101, 7984–7987.

THE INFLUENCE OF RED WORMS (*E. FOETIDA*) ON COMPOST'S FERTILIZING PROPERTIES

Czesława Rosik-Dulewska[1], Tomasz Ciesielczuk[2], Urszula Karwaczyńska[2], Hanna Gabriel

[1] Institute of Environmental Engineering of the Polish Academy of Sciences, Skłodowskiej-Curie Str. 34, 41-819 Zabrze, Poland, e-mail: dulewska@ipis.zabrze.pl

[2] Opole University, Department of Land Protection, Oleska Str. 22, 45-052 Opole, Poland, e-mail: tciesielczuk@uni.opole.pl

ABSTRACT

Composting is becoming a more and more common way of biodegradable waste disposal. Composts should be characterized by high content of nutrients and low amount of pollutants. Vermicompost is a compost produced by overpopulated culture of earthworm *Eisenia foetida* (Savigny 1826). World scientific literature states that vermicompost has a high fertilizing value which often exceeds such value of conventional composts. The results showed that vermicompost has a much higher fertilizing value than the compost produced by the traditional pile method. However, prism vermicompost created with the participation of a less concentrated population of earthworms has an intermediate value as a fertilizer (nitrogen and heavy metals), it could be assessed as a lower value product due to the lower content of potassium and phosphorus than the material obtained without earthworms.

Keywords: vermicompost, earthworm compost, *Eisenia foetida*, fertilizing properties

INTRODUCTION

Vermicompost (earthworm compost) is the final product of organic matter after being broken down by earthworms in a mesophilic composting process called vermicomposting. The red worm Eisenia foetida is widely considered as the best species at converting organic matter into compost, what is more, it feeds intensively and breeds most quickly. Composting is a biochemical process that uses microorganisms to aerobic organic matter decomposition. In the vermicompost method microorganisms begin the process, but it is the red worm that plays the largest role in converting organic matter subjected to preliminary decomposition processes (e.g. hydrolysis or fermentation). Composting by those organisms are the most effective between 15 and 25 °C, and with humidity between 60 and 70%. If the conditions are beyond the optimum scope, the effectiveness declines [Edwards 1995, Suthar 2009]. During the typical composting process a high temperature (55–65 °C) guarantees composted hygienic organic waste but also significantly eliminates earthworms' activeness. Possible waste that can be utilized by vermicomposting is mainly kitchen waste, green waste, farm animals' excrement, and selectively gathered municipal waste's organic fraction and sewage residue.

Vermicompost method can be effectively used to utilize homogeneous waste that traditional composting could not handle, e.g. paper production waste or textiles [Kostecka 1999, Suthar 2009]. Nevertheless, it is important to add structural material that is rich in coal. The best conditions for breeding and generating active earthworms, are foods which contain 60% cellulose and 10 to 20% protein. Earthworms do not tolerate spicy or acidic food waste such as: onion, garlic, or citrus fruits. Furthermore, fats, bones, and animal waste have a negative effect to the composting process [Kostecka 1999, Munroe 2009, Songin 1994, Suthar 2009].

The main aim and advantage of the vermicompost method is a quick neutralization of organic matter along with generation of organic fertilizer which is a valuable source of organic coal for soil [Evanylo et al. 2008]. The two main advantages of vermicomposting are: creating a compost which is a valuable fertilizer and earthworm biomass, which can be used as fish bait or substrate to production of protein flour.

The compost, highly contained with macro- and microelements, not only enriches soil but can also be used as a bed for flowers or perfect matter for soil and soil-less beds, and finally as a sorbent of oil derivatives and heavy metals in low polluted soils [Olszewska 2001, Rosik-Dulewska and Ciesielczuk 2010, Wróbel and Nowak 2005]. Some researchers suggest to use vermicompost only as an additive to flower beds because of its salinity [Kostecka 1998]. However, its diluting reduces the salt to the levels that are harmless for plants. Other positive effects of vermicompost on plants include: increased sprouting, the limiting of fungi occurrence and limited stress while replanting [Edwards 1995, Olszewska 2001, Tripathi 2004].

LEGAL FRAMEWORK

Legal framework regulating waste management in Poland is the Act on waste of 21 April 2001 [Journal of Laws of the Republic of Poland 2010 No. 28 pos. 145 with further changes]. Among others, it regulates time horizons and sets the time when biodegradable waste mass should be reduced from 50 to 35% respectively in 2013 and 2010 in comparison to the base year: 1995. It means that more and more professional composting plants are founded, as well as composting in households is becoming a way to reduce the cost of mixed waste. The act from 10th July 2007 [Journal of Laws of the Republic of Poland No. 147 pos. 1033] on fertilizers and fertilizing regulates the conditions and the way of marketing fertilizers, additives encouraging plant growth and their usage in agriculture. The Act of Ministry of Agriculture and Rural Development of 18th June 2008 (Journal of Laws of the Republic of Poland No. 119 pos. 765], in respect to putting into effect several acts considering fertilizers and fertilizing, sets minimal quality requirements and permissible heavy metals concentrations in fertilizers. Using (mainly in reclamation process) low quality compost (not suitable for fertilizing) is al-

lowed by Act 10 from 22nd April 2011 [Journal of Laws of the Republic of Poland No. 86 476].

EXPERIMENTAL MATERIAL AND STUDY DESIGN

The studied area included composts generated in individual households. Compost samples were taken from kitchen and garden waste (KO) made by a prism method and two types of vermicompost: the first one was made of kitchen and garden waste (WK), and the second one from kitchen and garden waste (KM) that was created by vermicomposting using a condensed red worm E.foetida population. This sample was the only one from the tested samples that originated from Upper Silesia; the following two samples (WK and KM) were created in a rural area near Opole. All tested composts were made in sheltered composting plants that allowed rainwater to seep into the compost mass, while in the case of WK vermicompost, seeping rainwater was completely eliminated. For the sake of the experiment the KM composting plant used 500 adult red worms (E. foetida). In the examined composts the humidity, organic matter, nitrogen, phosphorus, acidity and the proper electric conduction (according to Polish Norms) were determined. Quantification of sodium, potassium, and calcium was determined by FES methods. Heavy metals were determined by an atomic absorption spectroscopy using the iCE-Thermo 3500 after previous wet microwave mineralization in aqua regia in MARS-X apparatus.

RESULTS AND DISCUSSION

Applying compost causes multiple, positive changes in the physicochemical properties of soil fertilized by the compost [van der Gaag et al. 2007]. A crucial direction of those changes shows a significant increase, stabilized reaction and, subsequently, limitation of soils susceptibility to acidification. Those changes are particularly crucial in cases of rehabilitation process of soil-less and saline areas [Lakhdar et al 2009]. Nevertheless, composts' reaction and organic matter content depend not only on the content of waste being composted but also on the effectiveness of waste selecting system [Lopez et al 2010 Montemurro et al 2010]. In Table 1 basic characteristics of the examined composts is presented.

Table 1. Characteristics of main investigated composts properties

Typ kompostu	WK	KO	KM	Dz.U. 2008 no 119, pos. 765
pH	6.83 (0.11)	6.84 (0.08)	7.00 (0.03)	nl
EC [mS/cm]	3.63 (1.07)	1.23 (0.15)	1.26 (0.07)	nl
Organic matter [%]	71.92 (5.36)	37.58 (1.13)	30.68 (8.46)	30
Nitrogen Kiejdahl $N_{(Kiejd)}$ [%]	2.97 (0.28)	1.10 (0.90)	1.49 (0.13)	0.3
Phosphorus P_2O_5 [%]	1.37 (0.04)	1.33 (0.16)	1.25 (0.09)	nl
Potassium K_2O [%]	2.34 (2.11)	0.80 (0.71)	0.67 (0.06)	0.2
Sodium Na_2O [%]	0.22 (0.01)	0.09 (0.00)	0.09 (0.00)	0.2
Calcium CaO [%]	1.29 (1.30)	0.90 (1.00)	0.25 (0.11)	nl
TOC [%]	34.48 (1.92)	16.07 (0.62)	13.01 (0.83)	nl

n = 3 (SD value in brackets).

nl – no limitated.

All tested samples are characterized by a neutral reaction and are nearly similar; nevertheless, in comparison to the scientific research data, the observed pH levels are slightly lower [Ciesielczuk et al. 2011].

The salinity of the tested composts (expressed with EC value) is low, only the WK samples are characterized by a slightly higher concentration of soluble salt which is 3.63 mS/cm, what is confirmed by other authors [Pączka and Kostecka 2013]. It is the effect of limited or no access of rainwater and, consequently, the leaching soil from compost which is proven by a nearly two times higher content of sodium and potassium ions in comparison to KO and KM composts. The conductivity values of KO and KM composts are typical for composts made by a prism method and not sheltered against precipitation, especially since the composting took place during heavy rain. However, the compost produced from municipal solid waste reached even higher EC values, exceeding even 7–8 mS/cm [Hargreaves et al. 2008, Saha et al. 2010]. A particularly important fertilizer's parameter is the content of organic matter. The vermicompost (WK) was characterized by the highest content of organic matter, while in composts KO and HP (produced in composts plants which allowed access of rain water) the value of this parameter was approximately two times lower. This may partly result from DOC losses resulting from the leaching by rainfall and a slightly different composition of the composted material.

WK compost is very rich in nitrogen, only vermicompost from municipal sewage sludge may contain more of it – up to 3.2% [Khwairakpam and Bhargava 2009]. Lower content of this element (0.8–1.5%) was observed in samples PS and KO made by the traditional prism method [Nagavallemma et al. 2004]. Despite significant nitrogen content (especially in samples WK) all tested composts were mature, characterized by the C / N ratio within 8.7–11.6.

The potassium content in the tested vermicompost WK (2.3%) far exceeds the amount of this element in other vermicomposts [Fernández-Gómez, 2009]. This is probably the result of waste composition being composted (kitchen waste) and the lack of leaching this element by rainfall. Potassium content in the samples of KM and KO (Table 1) are typical for the composts produced by the prism method [Evanylo et al. 2008].

The highest phosphorus content was determined in samples of WK compost, slightly less in the samples of composts KO and KM. These values are close but other organic fertilizers, such as manure is richer in this element [Bansal and Kapoor, 2000]. It is believed, however, that the enzymes in the digestive tract of earthworms (and remaining coprolites) convert phosphorus compounds to form what is available to plants – as a result, vermicomposts have a higher content of available phosphorus forms in comparison to the material formed without their participation [Suthar, 2009]. All the tested composts comply with the minimum nitrogen content (0.3%), potassium oxide and phosphorus oxide (0.2%) in organic fertilizer as defined in the Regulation of Ministry of Agriculture and Rural Development on the implementation of certain provisions of the Act on fertilizers and fertilization of 18 June 2008, including later amendments.

The highest calcium content was also determined in the compost WK (1.29%). This is, however, due to the addition of calcium carbonate to vermicompost, which was used to prevent pH lowering below the optimum value for earthworms and bacteria (including endosymbiotic). The recorded calcium content values in composts KO (0.90%) and KM (0.25%) are lower than in comparison to composts produced by the prism method. An over normative heavy metal content in compost is one of the reasons disqualifying compost from applying it as a fertilizer. Comparison of the results for the six heavy metals to the current standards is shown in Table 2. Segregation of municipal solid waste „at source", results in lowering the amount of heavy metals, even up to the level normally recorded in the soils [Rosik-Dulewska 2003, Smith 2009].

Among the tested samples, compost WK contains the lowest amounts of heavy metals, significantly below the maximum amounts allowed by the regulation of the Minister of Agriculture and Rural Development dated 18 June 2008 No 119 pos. 765. The content of chromium and zinc is the lowest of the compared data: two composts produced in households in Opole (GU) and Opole Voievodeship (KS), where chromium content was respectively (in mg / kg dm), 79.78 and 15.07, and zinc content 276.90 and 167.61 [Ciesielczuk and Kusza 2009]. It is expected that after the implementation of the act on fertilizers and fertilization, the major factor eliminating the use of compost as a fertilizer is going to be the zinc content. In the studied vermicompost only cadmium content is increased, but does not exceed the legally permitted limits. The reason for such a low content of heavy metals in vermicomposts seems to be the ability of cumulating metals in earthworms' tissues which argues for the popularization of this composting method [Edwards 1995].

This phenomenon is confirmed by the results obtained for composts KO and KM. The heavy metal content in compost KO is relatively high. Quantities of lead and cadmium content are high-er than allowed by Polish as well as Canadian legislation, and thus compost KO can not be used as a fertilizer, but it is possible to use it for rehabilitation purposes [Journal of Laws of the Republic of Poland No. 86 pos. 476]. The content of zinc and chromium are also the highest among the compared composts. Copper and nickel content is quite high, but does not exceed the permissible limits in Poland [Journal of Laws of the Republic of Poland No. 86 pos. 476 and Journal of Laws of the Republic of Poland from 2008 No 119 pos. 765]. The high content of heavy metals, especially cadmium and lead may appear due to the location where the composting process took place, which was in urban areas, near highly concentrated industrial zones (Upper Silesia), and a large part of the mass was biodegradable garden waste.

Compost KM which was made by red worms E. foetida contains small amounts of heavy metals, considerably below the maximum permitted by the law. Cadmium content is low and comparable for compost produced from biodegradable waste selected at the source [Ciesielczuk and Kusza 2009]. Nickel and chromium contents are very low – only compost WK contains fewer of them. The quantities of zinc and lead stays at an average level of the three compared to paper composts.

CONCLUSION

On the basis of the obtained results it was found that only two composts analyzed in this study (WK and KM) meet the requirements for organic fertilizers regarding the minimum content of nitrogen (0.3%), phosphorus and potassium in the oxide form (0.2%) and organic matter (30%) specified in the aforementioned regulation on the implementation of certain provisions of the act on fertilizers and fertilization. Vermicompost WK is a fertilizer particularly rich in nutrients, which is also confirmed by scientific data. This is due

Table 2. Heavy metals content in investigated composts and law concentrations limits (mg/kg d.w.)

Compost type	WK	KO	KM	Dz.U. 119, pos. 765	Dz.U. 86, pos. 476	CCME 2005 limits
Zinc (Zn)	154.35	1122.84	219.10	nl	2500	nl
Copper (Cu)	16.30	52.81	22.49	nl	800	< 400
Nickel (Ni)	6.98	15.70	7.51	< 60	< 200	< 62
Chromium (Cr)	5.16	39.72	6.30	< 100	< 800	< 210
Lead (Pb)	4.91	145.54	22.75	< 140	< 800	< 150
Cadmium (Cd)	1.52	5.46	0.53	< 5	< 25	< 3

nl – no limitated; CCME – Canadian Council of Ministers of the Environment, PIN 1340.

to the presence and activity of earthworms and, indirectly, their symbiotic bacteria. High salinity of vermicompost, which results not only from the composition of the composted waste, but also from a lack of salt leached by precipitation, which forces the use of this fertilizer in small doses, precisely targeted not only to the type of soil but also to its salinity. Among the tested materials, compost KO can not be classified as organic fertilizer, due to noted excess amounts of heavy metals such as cadmium and lead, but still can be used in accordance with the provisions of the regulation on the recovery of R10 22.4.2011 [Journal of Laws of the Republic of Poland No. 86 pos. 476]. Compost KM, despite the participation of a condensed earthworm population in the process of composting, is less rich in phosphorus, potassium and calcium. This may result from heavy rainfall and leaching of these elements from the compost. All in all, using E foetida earthworms favors rapid production of high quality compost, which can successfully be used as organic fertilizer.

REFERENCES

1. Bansal S., Kapoor K.K., 2000. Vermicomposting of crop residues and cattle dung with *Eisenia foetida*. Bioresource Technology 73: 95-98.

2. Ciesielczuk T., Kusza G. 2009. Zawartość metali ciężkich w kompostach z odpadów jako czynnik ograniczający ich wykorzystanie do celów nawozowych. Ochrona Środowiska i Zasobów Naturalnych, 41: 347-354.

3. Ciesielczuk T., Rosik-Dulewska Cz., Karwaczyńska U. 2011. Komposty z odpadów jako potencjalne źródło substancji organicznej i biogenów w produkcji roślinnej. [In:] Kompostowanie i mechaniczno-biologiczne przetwarzanie odpadów (red.) G. Siemiątkowski: 108-116.

4. Edwards C.A. 1995. Historical overview of vermicomposting. BioCycle, 36, 6: 56.

5. Evanylo G., Sherony C., Spargo J., Starner D., Brosius M., Haering K. 2008. Soil and water environmental effects of fertilizer-, manure-, and compost-based fertility practices in an organic vegetable cropping system. Agriculture, Ecosystems and Environment 127: 50-58.

6. Fernández- Gómez M.J., Romero E., Nogales R. 2010. Feasibility of vermicomposting for vegetable greenhouse waste recycling. Bioresource Technology 101: 9654-9660.

7. van der Gaag D.J., van Noort F.R., Stapel-Cuijpers L.H.M., de Kreij C., Termorshuizen A.J., van Rijn E., Zmora-Nahum S., Chen Y. 2007. The use of green waste compost in peat-based potting mixtures: Fertilization and suppressiveness against soilborne diseases. Scientia Horticulturae 114: 289-297.

8. Hargreaves J.C., Adl M.S., Warman P.R. 2008. A review of the use of composted municipal solid waste in agriculture. Agriculture, Ecosystems and Environment 123: 1–14.

9. Khwairakpam M., Bhargava R. 2009. Vermitechnology for sewage sludge recycling. Journal of Hazardous Materials 161: 948–954.

10. Kostecka J., 1999. Kompostowanie z udziałem dżdżownic – nowe możliwości. [In:] Kompostowanie i użytkowanie kompostu, I Konferencja Naukowo-Techniczna, (eds) Siuta J., Wasiak G., Puławy – Warszawa: 125-131.

11. Kostecka J. 1998. Dalsze obserwacje nad produkcją wermikompostu z domowych odpadów organicznych. Zeszyty Naukowe Akademii Rolniczej w Krakowie. Sesja Naukowa, 58: 25-34.

12. Lakhdar A., Rabhi M., Ghnaya T., Montemurro F., Jedidi N., Abdelly Ch. 2009. Efectiveness of compost use in salt-affected soil. Jounal of Hazardous Materials, 171: 29-37.

13. Lopez M., Soliva M., Martínez-Farre F.X., Fernández M., Huerta-Pujol O. 2010. Evaluation of MSW organic fraction for composting: Separate collection or mechanical storting. Resources, Conservation and Recycling, 54: 222–228.

14. Montemurro F., Charfeddine M., Maiorana M., Convertini G. 2010. Compost Use in Agriculture: The Fate of Heavy Metals in Soil and Fodder Crop Plants. Compost Science & Utilizatíon 18, 1: 47-54.

15. Munroe G., Manual of on-farm vermicomposting and vermiculture, strona internetowa http://www.agbio.ca/DOCs/Vermiculture_FarmersManual_gm.pdf, dostęp online: 6.06.2009.

16. Nagavallemma K.P., Wani S.P., Lacroix S., Padmaja V.V., Vineela C, Babu Rao M., Sahrawat K.L. 2004. Vermicomposting: Recycling wastes into valuable organic fertilizer. International Crops Research Institute for the Semi-Arid Tropics, SAT eJournal, 1, 2, 2006.

17. Olszewska B. 2001. Szkolna hodowla dżdżownic. Rolnik Mazowiecki, 6: 7-8.

18. Pączka G., Kostecka J. 2013. The influence of vermicompost from kitchen waste on the yield-enhancing characteristics of peas *pisum sativum* l. var. *saccharatum* ser. bajka variety. Journal of Ecological Engineering, 14(2): 49-53.

19. Rosik-Dulewska Cz. 2003. Forms of some heavy metals in composts of municipal wastes as an index of their impact on the environment. [In:] Municipal solid waste composts, production and influence on the environment.

20. Rosik-Dulewska Cz., Ciesielczuk T. 2010. The possibilities of using waste compost to remove

aromatic hydrocarbons from solution. [In] Environmental Engineering III, L. Pawłowski, M. Dudzińska & A. Pawłowski (eds) Taylor&Francis Group London: 237-242.

21. Saha J.K., N. Panwar N., Singh M.V. 2010. An assessment of municipal solid waste compost quality produced in different cities of India in the perspective of developing quality control indices. Waste Management 30: 192-201.

22. Songin W. 1994. Produkcja kompostu koprolitowego w ogrodzie przydomowym i działkowym, Postępy Nauk Rolniczych, 1, 41: 145-151.

23. Smith S.R. 2009. A critical review of the bioavailability and impacts of heavy metals in municipal solid waste composts compared to sewage sludge. Environment International 35: 142-156.

24. Suthar S. 2008. Bioconversion of post harvest crop residues and cattle shed manure into value-added products using earthworm *Eudrilus eugeniae* Kinberg. Ecol. Eng. 3, 2: 206-214.

25. Suthar S. 2009. Vermicomposting of vegetable-market solid waste using *Eisenia fetida*: Impact of bulking material on earthworm growth and decomposition rate. Ecological Engineering, 35: 914-920.

26. Tripathi G., Bhardwaj P. 2004. Decomposition of kitchen waste amended with cow manure using an epigeic species (*Eisenia fetida*) and an anecic species (*Lampito mauritii*), Biores. Technol. 92: 215-218.

27. Wróbel S., Nowak K. 2005. Ocena działania torfu i wermikompostu w łagodzeniu skutków fitotoksyczności cynku. Inżynieria Ekologiczna 11: 231-232.

3

INFLUENCE OF BIO-PREPARATION ON WASTEWATER PURIFICATION PROCESS IN CONSTRUCTED WETLANDS

Monika Puchlik[1], Katarzyna Ignatowicz[1], Wojciech Dąbrowski[1]

[1] Białystok University of Technology, 45A Wiejska Str., 15-351 Białystok, Poland; e-mail: m.puchlik@pb.edu.pl; k.ignatowicz@pb.edu.pl; dabrow@pbbialystok.pl

ABSTRACT

Technological system of analyzed wastewater treatment plant is in part a biological bed of soil-reed in parallel arrangement. Unusual application is the application of two independent purification lines where in the second line, a bio-preparation is additionally dosed. The constructed wetland provides high removal of organic compounds expressed as BOD_5 and COD, as well as reducing the concentration of ammonia nitrogen and phosphates. This indicates a high performance of such a sewage treatment plant.

Keywords: constructed wetland, bio-preparation, wastewater.

INTRODUCTION

In rural areas and areas with dispersed buildings, the issue of wastewater management is still unresolved. Many years of negligence in the construction of sewers and sewage treatment plants cause that many residents of rural areas are forced to solve the problem of sewage treatment individually. Only 63.6% of inhabitants used the wastewater treatment plant in the Podlaskie province in 2013 (CSO). The use of house sewage treatment plants in Podlaskie province is becoming an increasingly common method of wastewater management, wherever it is not possible to connect the sewerage network. The constructed wetland method of wastewater treatment is a process taking place with the participation of heterotrophic microorganisms as well as aquatic plants and hydrophytes existing in purposely designed objects, i.e. groundwater beds or ponds.

Constructed wetlands have the ability to accumulate and retain organic substances. The accumulation of these substances is the result of excess of the primary production in relation to the respiration of heterotrophic microorganisms active in the wetland system [www.ekologia24.biz25.05.2014, Ozimek et al. 1996]. Organic matter produced in the environment of wetland ecosystem is seasonally accumulated on the substrate surface, and then converted during microbiological processes into hard-degradable organic substances, such as humic compounds forming a new soil layer [Obarska-Pempkowiak et al. 2010, Obarska Pempkowiak 2002, Henze et al. 2000].

Bio-preparations are selected microorganisms, mainly bacteria, fungi, and enzymes, bound to the mineral medium, used for the biodegradation of organic pollutants. They are non-toxic to humans, animals and plants. Selection of bacterial cultures in bio-preparations is made in such a way to eliminate disease-forming pathogens. The purpose of the bio-preparations application is faster decomposition of organic pollutants in wastewater (manure, paper, plant debris, grease), better biodegradation of detergents and cleaners, as well as the elimination of unpleasant odors.

Along with the use of bio-preparations, part of impurities forms sediment or is permanently bound to the ground, while the rest leaves the bed with the outflow of purified wastewater. As a result, conditions that allow for development of heterotrophic microorganisms involved in the biochemical processes in the supplied sewage, are created. [Reddy, D'Angelo 1996, www.ekologia24.biz25.05.2014].

MATERIAL AND METHODS

The study was conducted in a household sewage treatment plant with constructed wetland bed in Dzierniakowo in Podlaskie province. The object is intended for use by 10 people. Thus, it is designed for the average wastewater flow of 1.2 m³/d. Exploitation of the household sewage treatment plant with constructed wetland started in January 2012. To determine the effect of the bio-preparation on the efficiency of wastewater treatment, the plant operates in parallel system and is made up of two lines operating independently of each other [Markowski 2011]. Wastewater is pre-purified in septic tanks. Wastewater from the main septic tank (1) of a volume of 2.3 m³ is separated by the well (2) evenly into two subsequent settlers of 1.5 m³ volume working in parallel, from which they are delivered to two pumping stations (P4, L4) through two independent lines. Wastewater from the pumping section flows to the wetland beds. Two identical beds (working in parallel) with a square base 4×4 m and the bed height of 90 cm each. The ground-plant beds are piled in three layers: bottom layer (gravel 2–16 mm, thickness 20 cm), middle layer (sand 0.5–2 mm, 55 cm), upper layer of drainage (gravel 8–16 mm), on which common reed (Phragmites communis), was planted. A block diagram of a sewage treatment plant is shown in Figure 1.

The research has been conducted since January 2012. In this paper, the average test results from 2013, when bed was exhausted and plants acclimatized, are presented. During the whole period of research, bio-preparation BIOSAN KZ 2000 was dosed into the septic tank of the II line every 14 days. The bio-preparation includes starting conditioner along with properly assembled set of non-pathogenic microorganisms of strictly directed action resulting in decomposition and degradation of harmful substances such as: ammonia, nitrites, hydrogen sulfide, indole, skatole, mercaptans, phosphorus compounds, and other single and multi-carbonaceous organic compounds. Microbial composite has the ability to bind heavy metals, directing the desired fermentation processes, changing pH of the sewage and liquid manure, and very strong inhibition of pathogenic microorganisms growth, as well as enzymatic destruction of eggs and spores of insects and endoparasites [www.ekologia24.biz25.05.2014]. Bio-preparation BIOSAN KZ 2000 is designed to microbial wastewater treatment of sewage in treat-

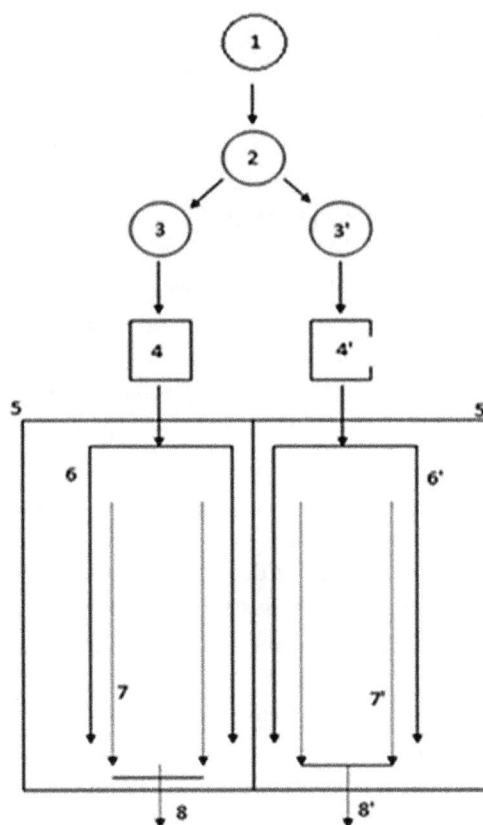

Figure 1. Scheme of constructed wetland plant: 1 – septic tank, 2 – separating well, P3 – septic settler – right line, L3 – septic settler – left line, P4 – sewage pumping section – right line, L4 – sewage pumping section – left line, P5 – ground-plant bed – right line, L5 – ground-plant bed – left line

ment plants, lagoons, septic tanks, liquid manure and slurry reservoirs, microbiological treatment of sewage and impurities in other facilities and devices of similar nuisances and loads, as well as elimination of unpleasant odors and microbial contamination in the environment of such objects and to improve the physicochemical and hygienic parameters of contaminated reservoirs and water intakes [www.ekologia24.biz25.05.2014]. The bio-preparation is used in an amount of 20 g per 1 m³ of wastewater or projected liquid waste every 14 days [www.ekologia24.biz25.05.2014].

The second line of sewage treatment plant worked independently without the addition of supporting preparations, as a control object. In order to assess the impact of bio-preparation on the efficiency of wastewater purification from both lines, samples were collected from: the septic tank before the sewage separation (1), septic tanks after the separator (2), and the outflow from constructed wetlands (3). Following items were determined in the wastewater samples in accordance with the applicable methodology: COD, BOD_5, phosphates (PO_4^{3-}), total phosphorus ($P_{tot.}$), ammonium nitrogen (NH_4^+), and Kjeldahl nitrogen (TKN).

RESULTS AND DISCUSSION

The COD to BOD_5 ratio for raw sewage is less than 2, which indicates that those wastewaters are readily biodegradable. The average purification effect on the bed for BOD_5 was 79.2% in the left line, while 96.2% in the right line with the addition of bio-preparation; the total effect in the left line amounted to 84.3% and 97.4% in the right line (Figure 2).

The high percentage of the purification effect for BOD_5 and COD provides the correct course of wastewater treatment from organic impurities, and achieved values at the outlet into the receiver comply with the requirements of 24 July 2006 (as amended in 2009), set by the Regulation of the Minister of Environment on the conditions to be met when discharging sewage into waters or soils, and on substances particularly harmful to the aquatic environment.

The tested sewage subjected to the treatment had pH = 6.5÷8, and therefore ammonia is present practically in 95 to 100% in an ionized form of NH_4^+. Municipal sewage contains biologically non-decomposable organic nitrogen in the amount of about 2 mg N/m^3. It remains in this form and concentration in treated wastewater [Kalinowska et al. 2005]. A similar amount of organic nitrogen was observed in the test samples, which was calculated on the basis of the difference between the concentration of TKN and $N-NH_4$. Organic nitrogen compounds undergo a process of ammonification already during the inflow to the sewage

treatment plant. As a result, the average amount of ammonia nitrogen in samples from the septic tank was 102 mg N/dm^3, and the Kjeldahl nitrogen 115 mg N/dm^3. Favorable conditions for nitrification (good oxygenation of sewage, pH close to 7) during the flow of wastewater through the constructed wetland contributed to the removal of ammonia nitrogen in the left line in 81.6% and in the right line with use of the bio-preparation – in 97.1% 9 (Figure 3). Due to the microorganisms that can assimilate nitrogen only in the inorganic form, and for which ammonia is easily absorbed form, the concentration of NH_4^+ in the effluent flowing out to the receiver was in the left line 18 mg N/dm^3 and in the right line 2.4 mg N/dm^3 (Table 1).

Mean values of the Kjeldahl nitrogen concentration in the treated wastewater is for the left line 19 mg N/dm^3, while for the right line with applied bio-preparation, a decrease to 5.3 mg N/dm^3 was recorded. This indicates the presence of a small amount of nitrogen in organic form and ammonia nitrogen that was not subject to nitrification. The activity of Nitrosomonas and Nitrobacter bacteria is provided not only by reduction in the amount of ammonia nitrogen in the outflowing sewage, but also an increase in nitrate concentration at the measurement point (after beds). Comparing the achieved values and taking into account the fact that heterotrophic bacteria in the struggle for substrates displace nitrifying bacteria, it can be assumed that a significant portion of organic load was removed on the hydrophilic bed.

A small amount of wastewater flowing into the sewage treatment plant daily makes a considerable concentration of certain pollutants. This is evident in the case of phosphates in the studied wastewater. Phosphates in wastewater have their origin mainly from synthetic detergents, wash-

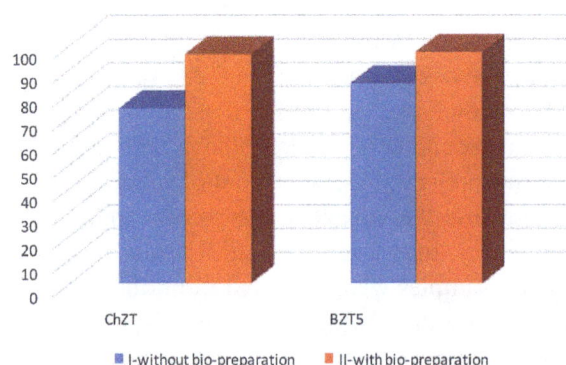

I-without bio-preparation   II-with bio-preparation

Figure 2. Efficiency of total COD and BOD [%] in Dzierniakowo in 2013

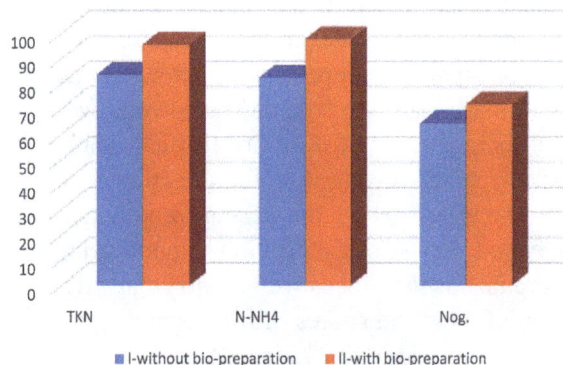

I-without bio-preparation   II-with bio-preparation

Figure 3. The effectiveness of the total forms of nitrogen [%] in Dzierniakowo in 2013

ing agents, and their quantity affects the total phosphorus level. In the studied raw sewage, the average phosphate concentration was 39 mg P/dm³. During the flow of wastewater through the constructed wetland, the phosphate concentration decreased to 26 mg P/dm³ in the left line, whereas up to 10 mg P/dm³ in the line where bio-preparation was applied. Complete effect of phosphate removing from wastewater during the study period reached 33.3% in the left, while 74.3% in the right line. Effect of the removal of total phosphorus in left line was 35.7%, and 71.4% when bio-preparation was used (Figure 4). In none of tested samples, there was any increase in the phosphate concentration. No secondary release of phosphorus compounds by bacteria testifies maintaining the aerobic conditions during the wastewater treatment on the bed. Some of the phosphorus compounds were assimilated by bacteria and plants, expanding root system of which provides an environment rich in aerobic and anaerobic bacteria.

According to Sadecka and Myszograj [Myszograj 2011, Sadecka 2008], in the case of sewage from rural areas, one should be prepared to determine pollutants concentrations, other than

the literature, particularly in relation to the content of biogenic compounds. For the analysis of the wastewater composition, it should be remembered that they are constantly changing, what is caused by a high non-uniformity resulting from the operation cycles or habits of the people using a given treatment plant.

CONCLUSIONS

The analysis of performed measurements and calculation allows for drawing the following conclusions:

1. The use of bio-preparation Biosan KZ 2000 contributes to the increase in the efficiency of treated wastewater in household sewage treatment plants with constructed wetland.

2. Improved total efficiency of organic compounds removal expressed as COD by 22.65% and BOD$_5$ by 13.12% with the application of bio-preparation Biosan KZ 2000, was observed.

3. The bio-preparation slightly affects the enhanced removal of total nitrogen (7.40%) and ammonia nitrogen (15.30%).

4. Efficiency of organic compounds removal in the constructed wetland for the line supported by the bio-preparation is higher by 23.72% for COD and by 16.99% for BOD$_5$.

5. Removal of nitrogen forms for the bio-preparation supported line in the constructed wetland is higher by 15.43% for ammonia nitrogen and by 6.39% for total nitrogen.

6. Twice as effective phosphorus removal was observed on the bed supported by bio-preparation Biosan KZ 2000 during the whole study period.

7. Applied bio-preparation Biosan KZ 2000 clearly improves the quality of treated wastewater.

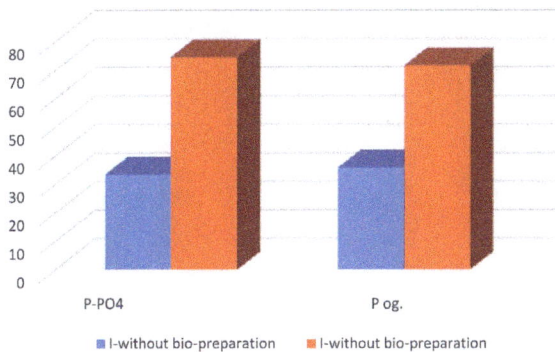

Figure 4. The effectiveness of the total phosphorus forms [%] in Dzierniakowo in 2013

Table 1. Parameters of wastewater during the treatment process in household sewage ground-plant bed

Factor / Sample	BOD mg O$_2$/dm³		COD mg O$_2$/dm³		TKN mg N/dm³		N–NH$_4^+$ mg N/dm³		N–NO$_3^-$ mg N/dm³		Nog. mg N/dm³		Pog. mg P/dm³		P–PO$_4^{3-}$ mg P/dm³	
Separating well (1)	435–520 427		642–851 746		74–165 115		65–139 102		0.1–4.4 2.7		74.1–169.4 117.7		22–65 42		20–60 39	
Series	I	II	I	II	I	II	I	II	I	II	I	II	I	II	I	II
Septic settler (2)	190–450 323	130–400 499	557–765 674	251–770 499	76–156 102	68–117 96	66–136 98	58–106 83	0.1–7.0 4.8	0.2–5.8 5.1	76.1–163 106.8	68.2–122.8 101.1	3.1–67 43	23–58 33	21–58 39	19–47 29
Ground–plant bed (3)	27–107 67	5.0–20 11.0	107–235 195	5.0–36 26	5.0–38 19	3.0–11.0 5.3	3.0–37 18	1.4–7.0 2.4	0.5–45 23	3.0–45 28	5.5–83 42	7.0–56 33.3	6.0–37 27	4.0–26 12	5.0–36 26	2.0–23 10

min–max
medium

Acknowledgments

Financial support for this research was provided by Ministry of Science and Higher Education within the project N N304 274840 and W/WBiIŚ/8/2013.

REFERENCES

1. Obarska-Pempkowiak H., Gajewska M., Wojciechowska E. 2010. Hydrofitowe oczyszczanie wód i ścieków. Wydawnictwo Naukowe PWN 2010.

2. Ozimek T., Renman G. 1996. Rola heliofitów w oczyszczalniach hydrobotanicznych. [In:] II Międzynarodowa Konferencja Naukowo-Techniczna, Akademia Rolnicza w Poznaniu, Poznań, 109–118.

3. http://www.ekologia24.biz/oczyszczalnie/biopreparat-biosan-kz-2000-25.05.2014r.

4. Obarska-Pempkowiak H. 2002. Oczyszczalnie hydrofitowe. Wydawnictwo Politechniki Gdańskiej, Gdańsk.

5. Reddy K.R., D'Angelo E.M. 1996. Biochemical indicator to evaluate pollutant removal efficiency in constructed wetland. [In:] 5th International Conference on Wetland Systems for Water Pollution Control. Universität für Bodenkultur Wien and International Association on Water Quality. Vienna.

6. Markowski W. 2011. Wpływ biopreparatów na efektywność usuwania związków węgla w przydomowej oczyszczalni ścieków ze złożem hydrofitowym. Praca magisterska. Politechnika Białostocka, 76–86.

7. Kalinowska E., Bonar G., Duma J. 2005. Zasady i praktyka oczyszczania ścieków. Wyd. LEMTECH Konsulting, Kraków.

8. Henze M., Harremoes P., Jansen J., Arvin E. 2000. Oczyszczanie ścieków. Procesy biologiczne i chemiczne. Wyd. Politechniki Śląskiej, Kielce.

9. Myszograj S. 2008. Zmiany ilościowe i jakościowe ścieków dopływających do małych oczyszczalni. [In:] Oczyszczanie ścieków i przeróbka osadów ściekowych, Tom 2, Zielona Góra.

10. Sądecka Z. 2008. Oczyszczanie ścieków z małych miejscowości. [In:] Oczyszczanie ścieków i przeróbka osadów ściekowych, Tom 2, Zielona Góra.

4

THE EFFECT OF WINTER CATCH CROPS ON WEED INFESTATION IN SWEET CORN DEPENDING ON THE WEED CONTROL METHODS

Robert Rosa[1]

[1] Department of Vegetable Crop, Siedlce University of Natural Sciences and Humanities, Bolesława Prusa 14, 08-110 Siedlce. Poland, e-mail: robert.rosa@uph.edu.pl

ABSTRACT

An experiment was carried out in east-central Poland (52°06' N, 22°55' E) over 2008–2011 to study the effect of winter catch crops on the weed infestation, number, and fresh matter of weeds in sweet corn (*Zea mays* L. var. *saccharata*). The following winter catch crops were grown: hairy vetch (*Vicia villosa* Roth.), white clover (*Trifolium repens* L.), winter rye (*Secale cereale* L.), Italian ryegrass (*Lolium multiflorum* L.) and winter turnip rape (*Brassica rapa* var. *typica* Posp.). The catch crops were sown in early September and incorporated in early May. The effect of the catch crops was compared to the effect of FYM (30 t·ha⁻¹) and control without organic manuring (NOM). Three methods of weed control were used: HW – hand weeding, twice during the growing period, GCM – the herbicide Guardian Complete Mix 664 SE, immediately after sowing of corn seeds, Z+T – a mixture of the herbicides Zeagran 340 SE and Titus 25 WG applied at the 3–4-leaf stage of sweet corn growth. Rye and turnip rape catch crops had least weeds in their fresh matter. Sweet corn following winter catch crops was less infested by weeds than corn following farmyard manure and non-manured corn. Least weeds and their lowest weight were found after SC, BRT and VV. LM and BRT reduced weed species numbers compared with FYM and NOM. The greatest weed species diversity, determined at the corn flowering stage, was determined after SC and FYM. The number and weight of weeds were significantly lower when chemically controlled compared with hand weeding. The best results were observed after a post-emergent application of the mixture Z+T. The weed species diversity on Z+T-treated plots was clearly lower compared with GCM and HW.

Keywords: weed infestation, green manure, organic fertilisation, weed control method, *Zea mays* L. var *saccharata*.

INTRODUCTION

In the integrated and ecological agriculture systems, more attention is being paid to the cultivation of catch crops, which should be a regular part of crop rotation. The cultivation of catch crops for ploughing in is a precursor of the long-lasting organic matter, which is an energy source for micro-organisms, and influences physical and chemical properties of the soil [Vos and van der Putten 2001, Marshall et al.2003, Clark et al. 2007]. When catch crop organic matter is regularly supplied to the soil, its biological activity can be preserved. Catch crops incorporated as green manures prior to sweet corn cultivation favourably affect the quantity and quality of ear and kernel yields [Zhang et al. 2010, Zaniewicz-Bajkowska et al. 2011, Rosa 2014].

Sweet corn is a poor competitor with weeds in the initial period of growth. Weed infestation can contribute to decreased yields of ears by up to 85% [Williams 2010]. It is therefore crucial to control weeds from the very beginning of sweet corn cultivation. The most effective method of weed control is to apply herbicides which, however, have an adverse effect on the natural environment and may negatively affect yield quality.

Because of environmental and human health concerns, worldwide efforts are being made to reduce the heavy reliance on synthetic herbicides that are used to control weeds. The modern approach to the issue of weed control in integrated agriculture involves a rational (that is taking into account the economic calculus) combination of effective methods, which are safe for the environment and consumers, applied to reduce the number of unwanted segetal plants to the level known as the economic threshold for weeds occurring in a crop plant [Singh et al. 2003, Armengot et al. 2013].

Many authors stress that well established catch crops efficiently compete with weeds; some plants species cultivated for incorporation are even able to hamper weed seed sprouting and initial growth [Teasdale et al. 1991, Akemo et al. 2000, Caporali et al. 2004, O'Reilly et al. 2011]. What is more, Liebman and Davis [2000] as well as Barberi [2002] believe that catch crops can be an alternative to chemical weed control.

The objective of this work was to assess the effect of winter catch crop ploughed in and an application of different weed control methods on weed infestation of sweet corn.

MATERIAL AND METHODS

The experiment was carried out over 2008–2011 at the Experimental Station of the Siedlce University of Natural Sciences and Humanities, which is located in east-central Poland (52°03'N, 22°33'E). According to the international system of FAO classification, the soil was classified as a Luvisol (LV) [World Reference... 2006]. The experiment was established in a split-block design with three replicates.

The effect of winter catch crops and weed control methods on the weed infestation, number, and fresh matter in 'Sweet Nugget F_1' sweet corn was investigated. The following winter catch crops were grown: VV – hairy vetch (*Vicia villosa* Roth.), TR – white clover (*Trifolium repens* L.), SC – winter rye (*Secale cereale* L.), LM – Italian ryegrass (*Lolium multiflorum* L.) and BRT – winter turnip rape (*Brassica rapa* var. *typica* Posp.). Their seeds were sown in early September in the years 2008–2010, at the following rates: VV – 70 kg, TR – 20 kg, SC – 180 kg, LM – 35 kg, BRT – 12 kg per 1 ha. Different nitrogen rates were applied: 30 kg N·ha^{-1} for VV and TR, and 60 kg N·ha^{-1} for SC, LM and BRT. The rates of phos-

phorus and potassium applied to all the catch crop plants were 40 kg P_2O_5·ha^{-1} and 80 kg K_2O·ha^{-1}, respectively. The effect of winter catch crops was compared to a control plot without organic manure (NOM) and farmyard manure (FYM) at a rate of 30 t·ha^{-1}. Green matter of the catch crops (roots + above ground parts) and FYM were incorporated in early May.

Directly before catch crop incorporation, the samples of plant material (roots + above ground parts) were taken to assess fresh (FM) and dry matter (DM) yields. Samples were taken from an area of 1 m^2 at three randomly selected places in each experimental combination. The dry matter contents in the catch crops were determined using the oven-drying gravimetric method. Moreover, the percentage share of weeds in catch crop biomass and dominating weeds were determined.

The field for cultivation of the catch crop plants and sweet corn was prepared in accordance with the principles of proper agricultural technology.

Seeds of sweet corn were sown in the years 2009–2011 between 11 and 24 May, at a spacing of 60 × 25 cm. Before sowing, mineral fertilisers were applied at the following rates: 60 kg N (urea), 50 kg P_2O_5 (superphosphate), 180 kg K_2O (60% potassium chloride) per 1 ha. When plants of sweet corn were 20 cm high, top dressing with 60 kg N·ha^{-1} (ammonium nitrate) was applied. Three weed control methods were applied: HW – hand weeding (twice during the growing season), GCM – herbicide Guardian CompleteMix 664 SE (acetochlor + terbuthylazine) immediately after sowing of sweet corn seeds, at the rate of 3.5 l·ha^{-1} per 250 dm^3 water, Z+T – a mixture of the herbicides Zeagran 340 SE (bromoxynil + terbuthylazine) (1.6 l·ha^{-1}) + Titus 25 WG (rimsulfuron) (40 g·ha^{-1}) + adjuvant Trend 90 EC (0.1%) per 250 dm^3 water, applied at the 3–4 leaf stage of sweet corn growth. Herbicide treatments were performed by knapsack sprayer. Other cultivation practices followed the generally established rules of sweet corn agrotechnology.

The effect of the examined factors on weed infestation was assessed twice each year. The primary infestation was studied in the initial period of sweet corn growth, 21 days after seed sowing. After that, manual (HW) and chemical (Z+T) weeding was performed. Another hand weeding followed in HW plots after 42-49 days from sweet corn sowing, depending on the study year. The assessment of secondary infestation was performed 72 days after sweet corn sowing when tassels appeared on corn plants. Weed infestation was determined by the quantitative-weighing method. This method entailed determining the number of individual weeds species and their fresh mass in each plot. Samples were taken from an area of a selected 0.5 m² at three randomly selected places in each plot. The number and weight of the weeds were expressed per 1 m².

The following indices of weed species diversity were calculated:

- Shannon-Wiener index of species diversity (*H'*), which takes into account species evenness and richness. The index reflects the probability that two individuals chosen randomly from a sample will represent different species. Its value depends on the species number and proportions between species numbers [Zanin et al. 1992]:

$$H' = -\sum (p_i \ln p_i)$$

- Simpson's index of domination (*D*), reflects the probability of choosing two individuals representing the same species. It takes into account species number and relative abundance of each species [Zanin et al. 1992]:

$$D = \sum p_i^2$$

where p_i is the share of ith species in the sample.

The results were statistically analysed by ANOVA following the model for the split-block design. The significance of differences was determined by the Tukey test at the significance level of $P \leq 0.05$. All the calculations were performed in Statistica 10.0.

RESULTS AND DISCUSION

Biomass yields of the catch crops are presented in Figure 1. The greatest fresh (FM) and dry matter (DM) yields were produced by winter rye (35.5 t·ha⁻¹ FM and 7.3 t·ha⁻¹ DM) and winter turnip rape catch crops (29.1 t·ha⁻¹ FM and 4.9 t·ha⁻¹ DM). Rye produced over twice as much DM as hairy vetch and Italian ryegrass and over three times more DM than white clover. Farmyard manure at a rate of 30 t·ha⁻¹ supplied the soil with 7.6 t·ha⁻¹ dry matter.

Analysis of the percentage share of weeds in the fresh matter of winter catch crops demonstrated substantial differences in their resistance to infestation by weeds (Figure 2). The infestation ranged from 2.8 to 25.3%, being the lowest in winter rye (2.8%) and winter turnip rape (4.1%) catch crops, and the greatest in Italian ryegrass catch crop (25.3%). In general, the dominant weedy species in the catch crops were *Viola arvensis* L., *Amaranthus retroflexus* L., *Anthemis arvensis* L., *Matricaria perforata* Mérat., *Vicia sativa* L. and *Elymus repens* (L.) Gould. However, not all the aforementioned species occurred in each catch crop (Table 1). The most weedy species were noted in white clover, winter rye and Italian ryegrass (four in each catch crop) and the least in hairy vetch (two). What is more, self-seeded *Secale cereale* plants were found in hairy vetch. Only *Viola arvensis* L. accompanied all the catch crops.

The species composition of weed communities depends largely on soil and climate conditions [Zarzecka and Gąsiorowska 2001] and agrotechnical practices [Pszczółkowski 2003]. In this study, 21 days after sweet corn sowing, 14

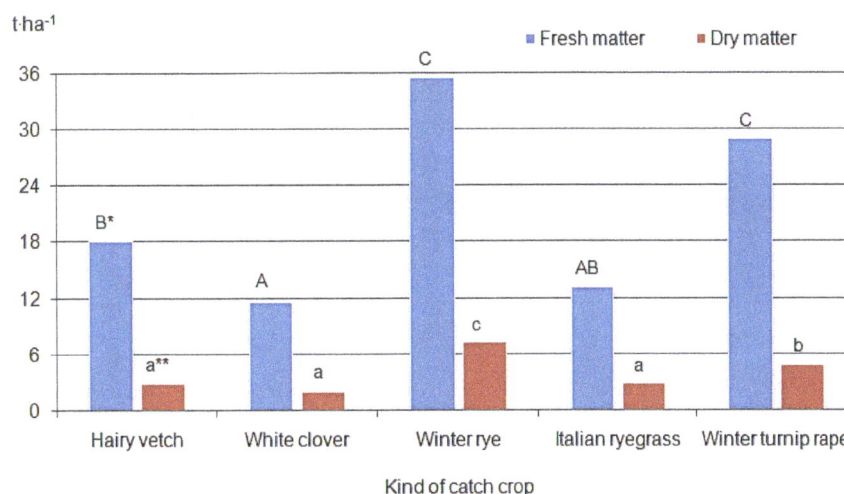

* Values followed by the same uppercase letters are not significantly different at $P \leq 0.05$
** Values followed by the same lowercase letters are not significantly different at $P \leq 0.05$

Figure 1. The amount of biomass (t•ha⁻¹) produced by the winter catch crops (mean for 2009–2011)

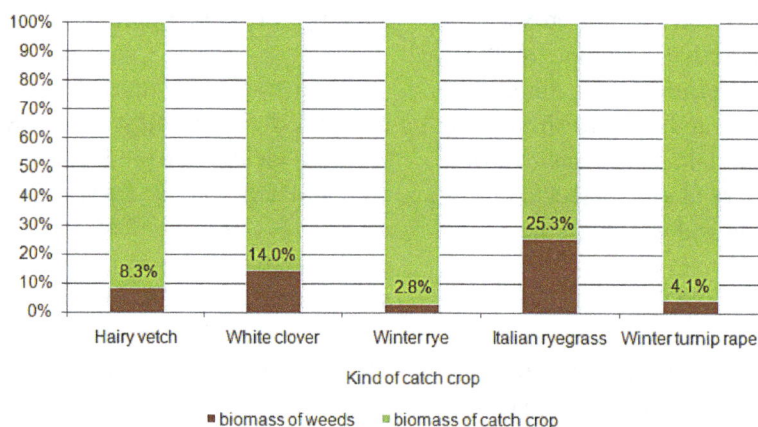

Figure 2. Share of weeds (%) in the catch crops biomass (mean for 2009–2011)

Table 1. The dominant weed species in the catch crops biomass (mean for 2009–2011)

Weed species	Kind of catch crop				
	Hairy vetch (VV)	White clover (TR)	Winter rye (SC)	Italian ryegrass (LM)	Winter turnip rape (BRT)
Viola arvensis L.	+	+	+	+	+
Amaranthus retroflexus L.	+	+	+	+	–
Anthemis arvensis L.	–	–	+	+	+
Matricaria perforata Mérat.	–	+	–	–	–
Vicia sativa L.	–	–	+	–	–
Secale cereale (self-seeded)	+	–	–	–	–
Elymus repens (L.) Gould.	–	+	–	+	+

+ – species presence, – – species absence.

Table 2. The species composition of weeds (plant·m^{-2}) 21 days after sweet corn sowing (mean for 2009–2011)

Weed species	Kind of organic manure							Weed control method		Mean
	NOM[1]	FYM	Catch crops					no weeding[2]	GCM	
			VV	TR	SC	LM	BRT			
Annual species										
Echinochloa crus-galli (L.) P. Beauv.	38.5	52.1	32.6	40.9	33.8	41.5	51.0	76.6	6.3	41.5
Chenopodium album L.	33.2	26.7	27.9	11.3	48.0	39.7	37.3	35.3	28.7	32.0
Veronica arvensis L.	22.5	31.4	24.9	30.8	32.0	23.7	13.6	45.1	6.8	26.0
Fallopia convolvulus (L.) Á. Löve	53.9	8.9	3.0	17.2	14.2	11.3	12.4	24.6	9.9	17.3
Amaranthus retroflexus L.	5.9	17.8	7.1	13.0	5.9	9.5	11.3	20.2	-	10.1
Geranium pusillum L.	1.2	3.6	9.5	14.2	3.6	10.7	24.9	19.4	-	9.7
Matricaria perforata Mérat.	-	2.4	-	3.0	3.6	7.1	4.7	4.8	1.3	3.0
Viola arvensis L.	2.4	1.2	1.2	4.7	1.2	2.4	-	2.7	1.0	1.9
Galium aparine L.	-	-	1.2	-	-	-	-	0.4	-	0.2
Persicaria maculosa Gray	-	-	-	1.2	-	-	-	0.4	-	0.2
Stellaria media (L.) Vill.	-	-	-	1.2	-	-	-	0.4	-	0.2
Number of annual species	7	8	8	10	8	8	7	11	6	11
Perennial species										
Elymus repens (L.) Gould.	7.1	1.2	8.3	8.3	3.6	4.1	-	5.6	3.8	4.7
Sonchus arvensis L.	4.7	2.4	3.6	2.4	-	-	-	3.2	0.5	1.9
Cirsium arvense (L.) Scop.	1.2	-	-	-	-	-	-	0.4	-	0.2
Number of perennial species	3	2	2	2	1	1	0	3	2	3
Total number of species	10	10	10	12	9	9	7	14	8	14

[1] NOM – control without organic manure, FYM – farmyard manure, VV – hairy vetch, TR – white clover,
  SC – winter rye, LM – Italian ryegrass, BRT – winter turnip rape.

[2] no weeding – no weeding to 21 days from sweet corn sowing, GCM – Guardian CompleteMix 664 SE.

weed species including 11 annuals and 3 perennials, were noted (Table 2). They belonged to the species typically establishing in sweet corn crop. Irrespective of the examined factors, the dominant species among the annuals were *Echinochloa crus-galli* (L.) P. Beauv., *Chenopodium album* L., *Veronica arvensis* L., *Fallopia convolvulus* (L.) Á. Löve, *Amaranthus retroflexus* L. and *Geranium pusillum* L. Of the perennials, the most common weed was *Elymus repens* (L.) Gould. The composition of the weed species was similar to that found in the study carried out in east-central Poland (Zarzecka and Gugała 2005, Kosterna 2014). On day 21 after sowing, most weedy species (12) were recorded in TR plots and the least (7) in BRT plots.

Pre-emergent GCM resulted in by 7 times less weedy species compared with the treatment where weeds were not controlled for 20 days of sweet corn cultivation. The most noxious weeds establishing in sweet corn include: *Ch. album*, *F. convolvulus, A. retroflexus, E. crus-galli* oraz *Polygonum aviculare* [Waligóra et al. 2008]. No *A. retroflexus, G. pusillum, G. aparine, P. maculosa, S. media* or *C. arvense* species were found in GCM-treated plots. Additionally, GCM reduced

the number of *E. crus-galli, V. arvensis, F. convolvulus* and *Ch. album* by 92, 85, 60 and 24%, respectively.

Number of weeds recorded 72 days after sweet corn planting was the same as during the first assessment. However, the species composition was different (Table 3). During the second assessment, annual *G. aparine*, was not found but *S. arvensis* and *M. arvensis* were present. *C. arvense* was not found either. *E. crus-galli, F. convulvulus, V. arvensis* were the most abundant species. Of the dominating species, least *E. crus-galli* plants were noted for NOM control, and *F. convulvulus* and *V. arvensis* after LM catch crop. Least weedy species were recorded after LM and BRT catch crops (8), and most (11) after FYM. For all the organic manuring treatments, the average number of weeds representing each species declined compared with the first assessment. It was due to the weed control treatments applied to HW and Z+T plots.

An application of Z+T on day 21 of sweet corn growing killed most weedy species observed at the beginning of sweet corn growing season. Also, hand weeding performed twice contributed to a reduced number of weedy species as well as

Table 3. The species composition of weeds (plant·m⁻²) 72 days after sweet corn sowing (mean for 2009–2011)

Weed species	Kind of organic manure							Weed control methods			Mean
	NOM[1)	FYM	Catch crops					HW[2)	GCM	Z+T	
			VV	TR	SC	LM	BRT				
Annual species											
Echinochloa crus-galli (L.) P. Beauv.	10.7	27.3	20.1	16.6	23.7	20.1	20.1	25.1	21.8	12.4	19.8
Fallopia convolvulus (L.) Á. Löve	35.0	11.3	13.6	7.1	8.9	5.9	10.7	12.7	25.4	1.5	13.2
Viola arvensis L.	18.4	15.4	11.3	16.0	7.7	5.9	7.7	9.7	23.9	1.8	11.8
Chenopodium album L.	-	8.3	2.4	1.2	9.5	11.3	2.4	10.9	4.1	-	5.0
Sinapis arvensis L.	3.6	4.7	2.4	3.6	1.2	2.4	3.6	9.1	-	-	3.0
Matricaria perforata Mérat.	4.7	4.7	1.2	-	1.2	-	7.7	6.1	2.3	-	2.8
Amaranthus retroflexus L.	-	2.4	2.4	2.4	1.2	5.9	-	3.0	3.0	-	2.0
Geranium pusillum L.	1.2	-	-	2.4	2.4	-	3.6	2.5	1.5	-	1.4
Stellaria media (L.) Vill.	1.2	2.4	-	-	-	-	-	1.5	-	-	0.5
Myosotis arvensis (L.) Hill.	-	1.2	1.2	-	-	-	-	-	1.0	-	0.3
Persicaria maculosa Gray	-	-	-	1.2	-	-	-	-	-	0.5	0.2
Veronica arvensis L.	-	-	-	1.2	-	-	-	-	0.5	-	0.2
Number of annual species	7	9	8	9	8	6	7	9	9	4	12
Perennial species											
Elymus repens (L.) Gould.	4.7	17.2	5.9	19.0	8.9	9.5	3.6	13.0	2.0	14.5	9.8
Sonchus arvensis L.	1.2	1.2	1.2	-	-	1.2	-	2.0	-	-	0.7
Number of perennial species	2	2	2	1	1	2	1	2	1	1	2
Total number of species	9	11	10	10	9	8	8	11	10	5	14

[1) NOM – control without organic manure, FYM – farmyard manure, VV – hairy vetch, TR – white clover,
    SC – winter rye, LM – Italian ryegrass, BRT – winter turnip rape
[2) HW – hand weeding, GCM – Guardian CompleteMix 664 SE, Z+T – Zeagran 340 SE + Titus 25 WG + Trend 90 EC

number of plants of a given species. In GCM-treated plots, the number of weedy species increased during the second compared with the first counting. What is more, some species increased in abundance (in particular *E. crus-galli*, *Viola arvensis*, *F. convolvulus*) whereas the abundance of *Ch. album*, *Veronica arvensis*, *E. repens* declined, *S. media* and *S. arvensis* being totally eliminated.

Number of weeds and amount of their dry matter per unit of area, reflecting weed infestation in sweet corn cultivated in the experimental plots, was significantly affected by organic manuring and weed control methods (Tables 4–5).

Regardless of weed control method, on day 21 after sweet corn sowing the least weeds (119.2 no.·m^{-2}) were found following an incorporation of VV catch crop (Table 4). In contrast, most weeds (170.7 no.·m^{-2}) were noted for NOM and a similar number after LM and BRT catch crops (149.9 and 155.0 no.·m^{-2}, respectively). No significant changes in the number of weeds and amount of their fresh matter were observed in GCM-treated plots due to an application of different organic manuring. In plots which were not treated for 20 days after corn sowing, least weeds were recorded after VV and most after BRT and in NOM. All the catch crops which were followed by sweet corn reduced the weight of fresh matter compared with FYM and NOM, the greatest decrease being observed after BRT and SC. It amounted to, respectively, 61% and 55% compared with FYM, and 67% and 62% compared with NOM.

Pre-emergent application of GCM reduced the number of weeds and amount of their fresh matter compared with non-treated plots by 78 and 89%, respectively.

The average number of weeds, determined 72 days after corn planting, was by 77.4 no.·m^{-2} lower and the average fresh matter weight by 355.6 g·m^{-2} greater compared with the first assessment (Table 5). The decrease in weed abundance was due to weed control practices applied in HW and Z+T plots. Also, developing corn plants and weeds that remained between rows competed with newly established weeds. A lower number of weeds per unit area resulted in them freely increasing in size and weight. According to Jodaugienė et al. [2006], weed sprouting is poorer in the period starting in mid-summer compared with spring and early summer so the effect of weed control methods is the most pronounced during the first part of the growing season. In turn Armengot et al. [2013] stress that total eradication of weeds in cultivated fields is not necessary but weeds should be controlled to the level when they do not affect negatively the crop plant.

Regardless of the weed control method, significantly most weeds were counted after farmyard manure (on average 93.6 no.·m^{-2}). FYM is widely believed to be a source of weed infestation in arable fields [Wichrowska and Jaskólski 2014]. The number of diaspores spread with FYM may in extreme cases exceed 420 th no.·mg^{-1}. Bedding and faeces are the major sources of weeds in farmyard manure [Pleasant and Schlather 1994].

Just like during the first assessment, the number of weeds after all the catch crops was significantly lower compared with FYM and NOM.

Table 4. Number and fresh mass of weeds (g·m^{-2}) 21 days after sweet corn sowing (mean for 2009–2011)

Kind of organic manures	Number of weeds per m^2			Fresh matter of weeds (g·m^{-2})		
	Weed control methods		Mean	Weed control methods		Mean
	no weeding [2]	GCM		no weeding	GCM	
NOM [1]	272.0 b*	69.3 a	170.7 c	670.2 b	47.1 a	358.7 d
FYM	250.1 ab	45.3 a	147.7 b	560.0 b	41.8 a	300.9 c
VV	197.5 a	40.9 a	119.2 a	289.8 a	13.3 a	151.6 ab
TR	232.6 ab	64.0 a	148.3 b	275.6 a	69.3 a	172.4 b
SC	243.3 ab	48.0 a	145.7 b	247.1 a	23.1 a	135.1 a
LM	235.8 ab	64.0 a	149.9 bc	263.1 a	42.7 a	152.9 ab
BRT	270.0 b	40.0 a	155.0 bc	208.0 a	28.4 a	118.2 a
Mean	243.0 B**	53.1 A	148.1	359.1 B	38.0 A	198.5

[1] NOM – control without organic manure, FYM – farmyard manure, VV – hairy vetch catch crop, TR – white clover catch crop, SC – winter rye catch crop, LM – Italian ryegrass catch crop, BRT – winter turnip rape catch crop
[2] no weeding – no weeding to 21 days from sweet corn sowing, GCM – Guardian CompleteMix 664 SE,
* Values within columns followed by the same lowercase letters are not significantly different at $P \leq 0.05$
** Values within rows followed by the same uppercase letters are not significantly different at $P \leq 0.05$

Table 5. Number and fresh mass of weeds (g·m⁻²) 72 days after sweet corn sowing (mean for 2009–2011)

Kind of organic manures	Number of weeds per m²				Fresh matter of weeds (g·m⁻²)			
	Weed control methods			Mean	Weed control methods			Mean
	HW [2)	GCM	Z+T		Hw	GCM	Z+T	
NOM [1)	108.4 b**	99.6 b	30.2 ab	79.4 c	1060.9 ab	695.7 c	52.6 a	603.1 cd
FYM	152.9 c	94.2 ab	33.8 ab	93.6 d	2023.8 d	451.3 b	182.9 a	886.0 e
VV	69.3 a	80.0 ab	39.1 ab	62.8 ab	849.2 a	361.8 ab	82.5 a	431.2 ab
TR	71.1 a	83.6 ab	53.3 b	69.3 b	1151.3 b	358.5 ab	86.4 a	532.1 bc
SC	87.1 ab	85.3 ab	32.0 ab	68.1 b	865.5 a	196.0 a	53.0 a	371.5 a
LM	92.4 ab	64.0 a	30.2 ab	62.2 a	1670.1 c	295.8 ab	77.0 a	681.0 d
BRT	71.1 a	92.4 ab	14.2 a	59.3 a	878.8 a	211.1 a	30.6 a	373.5 a
Mean	93.2 B**	85.6 B	33.3 A	70.7	1214.2 C	367.2 B	80.7 A	554.1

[1) NOM – control without organic manure, FYM – farmyard manure, VV – hairy vetch catch crop, TR – white clover catch crop, SC – winter rye catch crop, LM – Italian ryegrass catch crop, BRT – winter turnip rape catch crop
[2) HW – hand weeding, GCM – Guardian CompleteMix 664 SE, Z+T – Zeagran 340 SE + Titus 25 WG + Trend 90 EC
* Values within columns followed by the same lowercase letters are not significantly different at $P \leq 0.05$
** Values within rows followed by the same uppercase letters are not significantly different at $P \leq 0.05$

Weed abundance was most effectively reduced by BRT and LM catch crops. In GCM-treated plots the most weeds were observed in NOM and in Z+T-treated plots after TR. The greatest differences in the number of weeds and their fresh matter weight between the experimental organic manures were observed in hand weeded (HW) plots. Weed weight was significantly lower after catch crops than FYM. Compared with FYM, the greatest reduction in weed weight was observed after SC and BRT which had produced most biomass. Singh et al. [2003] have reported that catch crops reduce weed infestation because they overgrow weeds, and due to allelopathy. The more green matter is produced by catch crops, the greater the reduction. Bogužas et al. [2010] found that, of the catch crops they examined, winter rape and white mustard were the best, and red clover was the worst (it produced least biomass) at reducing the number of weeds and weight of their fresh matter. Malik et al. [2008], Jędrszczyk and Poniedziałek [2009] and O'Reilly et al. [2011] pointed to rye as a good catch crop which was very efficient at reducing weed load in sweet corn. In the present study, the weight of weeds after VV, SC and BRT was significantly lower compared with NOM. Abdin et al. [2000] reported lower numbers of weeds and their weight after fall rye, hairy vetch, white clover + ryegrass catch crops compared with corn cultivated without catch crop. Also Caporali et al. [2004] claimed that the number and weight of weeds establishing between rows were lower after ryegrass, subclover and hairy vetch catch crops compared with control where no catch crops were incorporated, ryegrass being the best weed competitor.

Olorunmaiye [2010] has suggested that leguminous plants can potentially be effective at controlling weeds but the amount of their biomass is crucial here. Under the growing conditions of Poland, Jędrszyk and Poniedziałek [2009] found that weed infestation of sweet corn was lower after white clover compared with the control without catch crop but much higher than after a rye catch crop. Special weed control properties of winter turnip rape and other *Brassica* plants have been highlighted by e.g.: Al-Khatib et al. [1997], Petersen et al. [2001] and O'Reilly et al. [2011]. They are connected with the fact that these plants secrete isothiocyanates which are toxic to some weedy species. It has been proven that winter turnip rape secretes substances which hinder seed sprouting of *Sonchus asper* (L.) Hill, *Matricaria inodora* L., *Amaranthus hybridus* L., *Echinochloa crus-galli* (L.) Beauv. and *Alopecurus myosuroides* Huds. [Petersen et al. 2001]. Also Caamal-Maldonado et al. [2001] and Weih et al. [2008] have pointed to a possibility of using allelopathic properties of plants representing other botanical families to reduce an occurrence of weeds.

In the present study, the herbicides Z+T were the best suppressors of weed number and weight. The superiority of chemical weed control over hand weeding repeated several times has been demonstrated by Abdin et al. [2000] and Malik et al. [2008]. Brandsæter et al. [2012] have suggested that post-emergent herbicides are more effective at reducing weed biomass compared with pre-emergent chemicals.

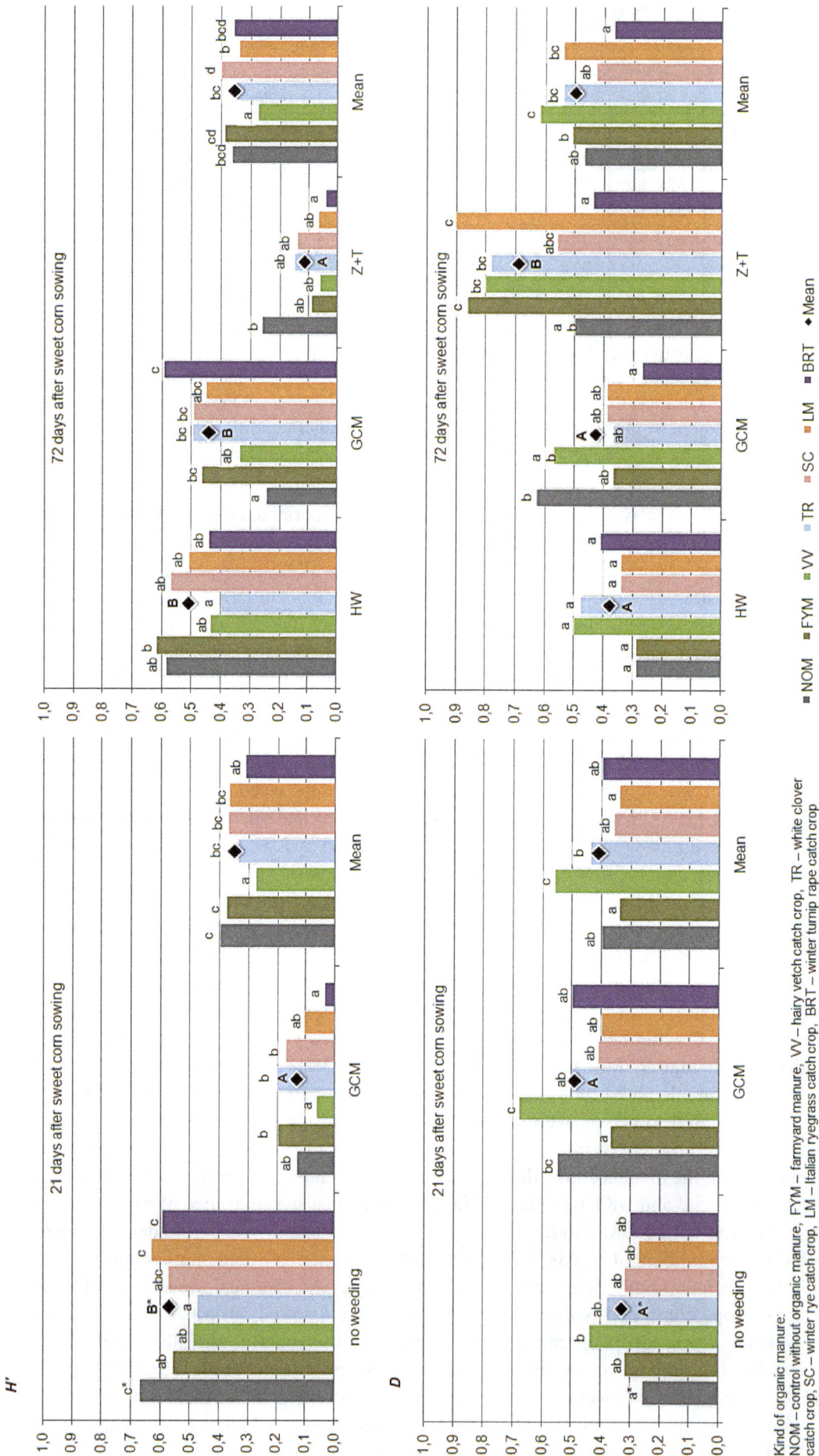

Kind of organic manure:
NOM – control without organic manure,   FYM – farmyard manure,   VV – hairy vetch catch crop,   TR – white clover
catch crop,   SC – winter rye catch crop,   LM – Italian ryegrass catch crop,   BRT – winter turnip rape catch crop

Weed control method:
HW – hand weeding,   GCM –Guardian CompleteMix 664 SE,   Z+T – Zeagran 340 SE + Titus 25 WG + Trend 90 EC

*Values followed by the same letters are not significantly different at $P \leq 0.05$

Figure 3. Shannon-Wiener index of species diversity (H') and Simpson's index of domination (D) in sweet corn depending on the factors examined (mean for 2009–2011)

Figure 3 demonstrates segetal flora diversity reflected by the Shannon-Wiener index of diversity (*H'*) and Simpson's index of domination (*D*). Values of index *D* range from 0 to 1; values approximating 1 indicate that one or several species are clearly dominant and diversity of the community is low. By contrast, the higher index of diversity (*H'*), the greater diversity of phytocenose [Zanin et al. 1992]. In the experiment discussed here, the average value of index *H'* at the beginning of the growing season and sweet corn flowering did not change and amounted to 0.35. By contrast, the value *D* index increased from 0.41 to 0.50. According to Stupnicka-Rodzyniewicz et al. [2004] and Kostrzewska et al. [2011], the species biodiversity of weeds changes during the growing season and is affected by agrotechnological factors. The authors observed increased biodiversity during the mid-growing season of cereals. Regardless of the weed control method, on day 21 after sweet corn sowing the greatest diversity of weeds was found after FYM (*H'*=0.38) and for NOM (*H'*=0.40), and the lowest after VV (*H'*=0.27). During the second assessment, SC plots had the most diverse species composition of weeds (*H'*=0.40) whereas the least diversity was found for VV plots (*H'*=0.27) (as was the case during the first assessment). Majchrzak and Skrzypczak [2007] found no significant differences between index *H'* values for corn cultivated after spring vetch and control without catch crop. Wanic et al. [2004] pointed out that floristic composition of weed communities got impoverished due to the effect of Italian ryegrass whereas Kuraszkiewicz and Pałys [2003] reported a similar influence of red clover and annual ryegrass. In the study discussed here, weed diversity increased in GCM-treated plots on day 72 from corn planting, compared with the first assessment, the greatest diversity being observed after BRT catch crop.

The highest values of the index of domination (*D*) on day 21 and 72 from sweet corn planting were obtained after VV. Of all the GCM-treated plots, the greatest index *D* was calculated for NOM. In the case of Z+T plots, it was the greatest after FYM and LM. O'Reilly et al. [2011] cultivated sweet corn after rye and oat catch crops and found a significantly lower value of index *D* compared with cultivation without catch crop.

A post-emergent application of the herbicides Z+T definitely reduced species diversity of weeds compared with GCM and hand weeding repeated twice, as revealed by a decline in the value of index of biodiversity (*H'*) and an increase in the index of domination (*D*). Also Yao et al. [2010] observed declining values of index *H'* and increasing values of index *D* after chemical weed control compared with hand weeding.

CONCLUSIONS

1. Winter catch crops incorporated prior to sweet corn planting reduced weed infestation of the crop compared with cultivation including an application of farmyard manure and without catch crop. The most effective weed suppressors were rye, turnip rape and hairy vetch catch crops.

2. Italian ryegrass and turnip rape catch crops reduced weed richness compared with cultivation including an application of farmyard manure and without catch crop.

3. The greatest species diversity of weeds, determined at the flowering stage of sweet corn, was observed after rye catch crop and farmyard manure.

4. Chemical weed control more effectively reduced the number and weight of weeds than hand weeding. The best effects were obtained after post-emergent application of the mixture of herbicides Zeagran 340 SE + Titus 25 WG.

5. Species diversity of weeds following an application of Zeagran 340 SE + Titus 25 WG was definitely lower compared with a pre-emergent application of Guardian CompleteMix 664 SE and hand weeding performed twice.

Acknowledgements

The research was supported by the Polish Ministry of Science and Higher Education as part of the statutory activities of the Department of Vegetable Crops, Siedlce University of Natural Sciences and Humanities.

REFERENCES

1. Abdin O.A., Zhou X.M., Cloutier D., Coulman D.C., Faris M.A., Smith D.L. 2000. Cover crops and interrow tillage for weed control in short season maize (*Zea mays* L.). Eur. J. Agron., 12, 93–102.

2. Akemo M., Regnier E., Bennet M. 2000. Weed supresion in spring-sown rye (*Secale cereale*) – Pea (*Pisum sativum*) cover crop mixes. Weed Technol., 14, 545–549.

3. Al-Khatib K., Libbey C., Boydston R. 1997. Weed suppression with *Brassica* green manure crops in green pea. Weed Sci., 45 (3), 439–445.

4. Armengot L., José-María L., Chamorro L., Sans F.X. 2013. Weed harrowing in organically grown cereal crops avoids yield losses without reducing weed diversity. Agron. Sustain. Dev., 33, 405–411.

5. Barberi P. 2002. Weed management in organic agriculture: are we addressing the right issues. Weed Res., 42, (3), 177–193.

6. Bogužas V., Marcinkevičienė A., Pupalienė R. 2010. Weed response to soil tillage, catch crops and farmyard manure in sustainable and organic agriculture. Žemdirbystė – Agriculture, 97 (3), 43–50.

7. Brandsæter L.O., Mangerud K., Rasmussen J. 2012. Interactions between pre- and post-emergence weed harrowing in spring cereals. Weed Res., 52, 338–347.

8. Caamal-Maldonado J.A., Jiménez-Osornio J.J., Torres-Barragán A., Anaya A.L. 2001. The use of allelopathic legume cover and mulch species for weed control in cropping systems. Agron. J., 93 (1), 27–36.

9. Caporali, F., Campiglia, E., Mancinelli, R., Paolini, R. 2004. Maize performances as influenced by winter cover crop green manuring. Ital. J. Agron., 8 (1), 37–45.

10. Clark A.J., Meisinger J.J., Decker A.M., Mulford F.R. 2007. Effects of a grass-selective herbicide in a vetch-rye cover crop system on corn grain yield and soil moisture. Agron. J., 99, 43–48.

11. Jędrszczyk E., Poniedziałek M. 2009. Wpływ żywych ściółek na wybrane właściwości gleby i zachwaszczenie w uprawie kukurydzy cukrowej. Zesz. Problem. Post. Nauk Rol. 539, 265–271.

12. Jodaugienė D., Pupalienė R., Urbonienė M., Pranckietis V., Pranckietienė I. 2006. The impact of different types of organic mulches on weed emergence. Agron. Res., 4, Special issue, 197–201.

13. Kosterna E. 2014. The effect of soil mulching with organic mulches, on weed infestation in broccoli and tomato cultivated under polypropylene fibre, and without a cover. J. Plant Prot. Res., 54 (2), 188–198.

14. Kostrzewska M.K., Wanic M., Jastrzębska M., Nowicki J. 2011. Wpływ życicy wielokwiatowej jako wsiewki międzyplonowej na różnorodność zbiorowisk chwastów w jęczmieniu jarym. Fragm. Agron., 28 (3), 42–52.

15. Kuraszkiewicz R., Pałys E. 2003. Wpływ wsiewek międzyplonowych na zachwaszczenie łanu roślin ochronnych na glebie lekkiej. Ann. UMCS, Sec. E, 58, 53–67.

16. Liebman M., Davis A. S. 2000. Integration of soil, crop and weed management in low-external-input farming systems. Weed Res., 40 (1), 27–47

17. Majchrzak L., Skrzypczak G. 2007. Wpływ sposobu przygotowania gleby do siewu i rodzaju pozostawionej biomasy na zachwaszczenie kukurydzy. Prog. Plant Prot./Post. Ochr. Roślin, 47 (3), 191–198.

18. Malik M.S., Norsworthy J.K., Culpepper A.S., Riley M.B., Bridges W. Jr. 2008. Use of wild radish (*Raphanus raphanistrum*) and rye cover crops for weed suppression in sweet corn. Weed Sci., 56 (4), 588–595.

19. Marshall E.J.P., Brown V.K., Boatman N.D., Lutman P.J.W., Squire G.R., Ward L.K. 2003. The role of weeds in supporting biological diversity within crop fields. Weed Res., 43, 77–89.

20. Olorunmaiye P.M. 2010. Weed control potential of five legume cover crops in maize/cassava intercrop in a Southern Guinea savanna ecosystem of Nigeria. Aust. J. Crop Sci., 4 (5), 324–329.

21. O'Reilly K.A., Robinson D.E., Vyn R.J., Van Eerd L.L. 2011. Weed populations, sweet corn yield, and economics following fall cover crops. Weed Technol., 25, 374–384.

22. Petersen J., Belz R., Walker F., Hurle K. 2001. Weed suppression by release of isothiocyanates from turnip-rape mulch. Agron. J., 93 (1), 37–43.

23. Pleasant J., Schlather K. J. 1994. Incidence of weed seed in cow (Bos sp.) manure and its importance as a weed source for cropland. Weed Technol., 8, 304–310.

24. Pszczółkowski P. 2003. The attempts to control weed infestation in potatoes cultivated under shields. Part I. Plant response to herbicides. Biul. IHAR, 228, 249–260.

25. Rosa R. 2014. The structure and yield level of sweet corn depending on the type of winter catch crops and weed control method. J. Ecol. Eng., 15 (4), 118-130.

26. Singh H.P., Batish D.R., Kohli R.K. 2003. Allelopathic interactions and allelochemicals: new possibilities for sustainable weed management. Critic. Rev. Plant Sci., 22 (3-4), 239–311.

27. Stupnicka-Rodzynkiewicz E., Stępnik K., Lepiarczyk A. 2004. Wpływ zmianowania, sposobu uprawy roli i herbicydów na bioróżnorodność zbiorowisk chwastów. Acta Sci. Pol., Agricultura, 3 (2), 235–245.

28. Teasdale J., Beste E., Potts W. 1991. Response of weeds tillage and cover crop residue. Weed Sci., 39, 195–199.

29. Wanic M., Kostrzewska M.K., Jastrzębska M., Brzezin G. 2004. Rola wsiewek międzyplonowych w regulacji zachwaszczenia jęczmienia jarego w płodozmianach zbożowych. Fragm. Agron., 21 (1), 85–100.

30. Vos J., van der Putten P.E.L. 2001. Field observations on nitrogen catch crops. III. Transfer of nitro-

gen to the succeeding main crop. Plant and Soil, 236, 263–273.

31. Yao H.J., Jin Z.L., Yang W.B., Zhao J.H., Zhang F. 2010. Effects of weeding methods on weed community and its diversity in a citrus orchard in southwest Zhejiang. Ying yong sheng tai xue bao = The journal of applied, 21 (1), 23–28.

32. Waligóra H., Skrzypczak W., Szulc P. 2008. Zachwaszczenie i plonowanie kukurydzy cukrowej po zastosowaniu pielęgnacji mechanicznej. J. Res. Appl. Agric. Engng., 53 (4).

33. Weih M., Didon U.M.E., Ronnberg-Wastljung A.C., Bjorkman C. 2008. Integrated agricultural research and crop breeding: Allelopathic weed control in cereals and long-term productivity in perennial biomass crops. Agr. Syst., 97, 99–107.

34. Wichrowska D., Jaskólski D. 2014. Effect of organic and mineral fertilization and soil fertilizer on the weed infestation of potato plantation. Acta Sci. Pol., Agricultura, 13 (1), 61–71.

35. Williams M.M. 2010. Biological significance of low weed population densities on sweet corn. Agron. J., 102, 464–468.

36. World Reference Base for Soil Resources. 2006. A Framework for International Classification, Correlation and Communication. World Soil Resources Report FAO, 103. Food and Agriculture Organization of the United Nations, Rome, 116.

37. Zanin G., Mosca G., Catizone P. 1992. A profile of the potential flora in maize fields of the Po Valley. Weed Res., 32 (5), 407–418.

38. Zarzecka K., Gąsiorowska B. 2001. Wpływ metod pielęgnacji na zachwaszczenie i plonowanie ziemniaka. [The effect of weed control methods on the infestation and potato yields]. Zesz. Nauk. Akademii Podlaskiej w Siedlcach, Ser. Rolnictwo, 59, 15–25.

39. Zarzecka K., Gugała M. 2005. Population and species composition of weed under differentiated conditions of weed control method. Acta Agrobot., 58 (1), 291–302.

40. Zaniewicz-Bajkowska A., Rosa R., Kosterna E., Franczuk J. 2011. Serradella and faba bean catch crops as a kind of organic manuring in sweet corn cultvaton. (in:) Nowoczesne metody analizy surowców rolniczych. Red. Puchalski Cz. i Bartosz G., Wyd. Uniwersytetu Rzeszowskiego, Monografia Nauk., 227–240.

41. Zhang Y., GaoL., Zhou W., Li Z. 2010. Effects of intercropping clover on yield, quality of sweet corn and soil mineral N in field. Acta Agric. Boreali-Sinica, S1, 236–238.

5

THE OCCURRENCE OF ENTOMOPATHOGENIC FUNGI IN SOILS FROM CULTIVATED PERENNIAL RYEGRASS (*LOLIUM PERENNE* L.)

Roman Kolczarek[1]

[1] Institute of Agronomy, University of Natural Sciences and Humanities, B. Prusa 14, 08-110 Siedlce, Poland, e-mail: rk@uph.edu.pl

ABSTRACT

An important role in the agricultural agrocenosis is attributed to entomopathogenic fungi. They limit the occurrence of certain populations of soil pests and insects overwintering in the soil environment, or held in the pupation. Fungi are the only pathogens of arthropods witch have the ability to infect plant pests directly by body. However, bacteria are the largest group of microorganisms inhabiting the soil. The aim of this study was to compare the species composition and the severity of the occurrence of entomopathogenic fungi in soils of monoculture crops perennial ryegrass (*Lolium perenne* L.). The material consisted of soil samples taken from the experimental research conducted in two experimental stations of the Central Research Centre for Cultivar. The fungi isolated from soil insecticides using a method developed by the selective medium.

Keywords: soil, entomopathogenic fungi, perennial ryegrass (Lolium perenne L.), density of infective units.

INTRODUCTION

An important role is played by fungi agrocenosis agricultural insecticides that are capable of reducing most of the soil-dwelling pests and pests of crop plants. The main site of entomopathogenic fungi is soil, which has a huge impact on both the incidence and spread of mikoz insects. Currently, there are over 1000 known species of fungi parasitic on insects [Vanninen 1999; Bałazy 2006]. According to Gaugler [1988], it is estimated that about 90% of harmful plant arthropods spent at least part of their life cycle in the soil. Both the occurrence and the development and pathogenicity of entomopathogenic fungi in the soil are conditioned by a number of biotic and abiotic factors in the environment, as well as by agricultural and non-agricultural human activity. This is due to, among others, temperature, humidity, soil type, structure, and the manner of its use and the time of year as well as, crop species [Tkaczuk, Miętkiewski 1996]. Fungi insecticides as one of the first pathogens used in the biological control of pests on plants [Lipa 1967]. The aim of this study was to compare the species composition and the severity of the occurrence of entomopathogenic fungi in soils of monoculture crops of *Lolium perenne* L.

MATERIAL AND METHODS

The material consisted of soil samples taken from the experiments on close-growing Perennial ryegrass (*Lolium perenne* L.) conducted in two experimental stations of the Research Centre for Cultivar Testing (COBORU) in Krzyżewo (Podlaskie Province) and Uhnin (Lublin Province). The tests were taken in two periods: spring and autumn (25.04.2012 and 28.09.2012). The experiment was carried out in two sites and concerned the assessment of yielding varieties of perennial ryegras. Detailed information on the experiment design and soil and hydrothermal conditions are shown in Table 1. The samples were taken at random from 10 points of a plot. The soil was collected using canes to a depth of 15 cm. From the material collected from the area of an untreated mixed and stored in plastic bags at 0–4 °C. The Fungi isolated from the soil insecticides using a method developed by the selective medium [Strasser et al. 1996]. The medium is commonly used for the isolation of entomopathogenic fungi in the soil [Keller et al. 2003, Tkaczuk, 2008]. From each sample originating from a given mixed plot 2 g was weighed and added to 18 ml of distilled water with addition of 0.05 ml a solution of preparation Trithon X-100, and vigorously shaken for about 35 seconds. Then a solution of 0.1 ml of the soil was poured and spread using a glass spatula on a surface of the substrate selectively in three petri dishes, which were repeated. The dishes were placed in incubators at 22 °C and after 8–10 days colonies were counted for each species of fungi. The results are expressed as the number of infectious units (CFU) of entomopathogenic fungi in 1 g of soil.

Meteorological data for the research were obtained from the Hydrological and Meteorological Station in Krzyżewo and Uhnin. However, in order to determine the temporal variability of meteorological elements and their influence on plant growth, the coefficient of hydrothermal Sielianinov [Bac et al. 1993] and the classification Skowera and Pula [2004] were used.

The data in Table 1 show that the distribution of rainfall underwent temporal and spatial variation. Drought was observed only in Uhninie in the May (K = 0.84). While in June, both in Krzyżewo and Uhninie the weather conditions should be considered quite humid and it was an optimum period for Krzyżewo and slightly less moisture of the Uhninie. Howewer the August was more moistly in Krzyżewo. In September, there were dry periods for both locations of Selianinov coefficient with value K = 0.40 for Krzyżewo.

RESULTS

The reaserch experiment showed that using a selective medium with meadow soils three species of entomopathogenic fungi *Beauveria bassiana*, *Metarhizium anisopliae* and *Isaria fumosorosea* were isolated. Species composition and density of the colony-forming units of these fungi varied depending on the variety and the place from which soil samples were derived. Spring in soils from five varieties of perennial ryegrass grown in the station COBORU in Krzyżewo recorded dominance of the fungus *Metarhizium anisopliae*. This species occurred in all soil samples analyzed from all varieties of perennial ryegrass (*Lolium perenne* L.), and worked an average of 1.5×10^3 CFU g^{-1} in 1 g of soil. The presence of the fungus *Isaria fumosorosea* reported in soils from 2 of 5 perennial ryegrass (*Lolium perenne* L.) grown varieties - Lacerta and Argona. *Beauveria bassiana* occurred only in the soil sampled from the varieties of Amarant and Argona. These fungi formed an average of 1.6×10^3 g^{-1} and 0.3×10^3 g^{-1} of colony forming units in the soils (Figure 1).

The total insecticides, most infectious units (CFU) fungiformedin soils collected from within the perennial ryegrass varieties (*Lolium perenne* L.) on spring: Lacerta, Amarant, Argona a minimum of Tivoli and Bah 209 (Table 2). Within the autumn in soil samples collected varieties of perennial ryegrass (*Lolium perenne* L.) grown in Krzyżewo included fungus *Metarhizium an-*

Table 1. Conditions of experimentation - Perennial ryegrass (*Lolium perenne* L.)

Lp.	Name of place	Height above sea level (m)	Agriculture Value in 100 point scale IUNG	Complex of agricultural suitability	Type	Species	pH 1 n KCl	Forecrop
			Soil					
1	Krzyżewo	130	70	4	A	gl	6.4	spring barley
2	Uhnin	155	80	1z	M	N	6.6	meadow

	Hydrothermal coefficient of Selianinov (K) / Month						
	IV	V	VI	VII	VIII	IX	X
1 Krzyżewo	1.63 (fm)	1.09 (qd)	1.83 (fm)	1.55 (o)	3.18 (em)	0.40 (ed)	2.27 (m)
2 Uhnin	1.06 (qd)	0.84 (d)	1.92 (fm)	0.81 (d)	1.25 (qd)	0.79 (d)	4.90 (em)

Symbols: (ed) – extremely dry, (vd) – very dry, (d) – dry, (qd) – quite dry, (o) – optimal, (fm) – fairly moist, (m) – moist, (vm) – very moist, (em) – extremely moist.
Col.4; 4 – rye very good, 1z – grassland very good and good.
Col.5; A – podzolic; M – muck.
Col.6; gl – loam, n – deep peat with thickness of organic matter >30 cm.

isopliae, which made an average of 0.9×10^3 g^{-1} CFU. Its presence was found in soil from all varieties of perennial ryegrass (*Lolium perenne* L.). Fungi *Isaria fumosorosea* also occurred in all tested samples of soil and constituted an average of 0.6×10^3 g^{-1} CFU in 1 g of soil (Figure 1). The presence of the fungus *Beauveria bassiana* was observed only in soils from one cultivar perennial ryegrass (*Lolium perenne* L.)-Tivoli. Overall fungi insecticides formed more CFU than in the spring for fungi *Isaria fumosorosea* and Metarhizium anisopliae from all analyzed varieties. (Table 3). In the studied soil the samples collected from plots in the spring at the experimental station COBORU-Uhnin where cultivated varieties of perennial ryegrass 5 (*Lolium perenne* L.), the most infectious units fungus *Beauveria bassiana* made an average of 0.8×10^3 g^{-1}, *Metarhizium anisopliae* (0.2×10^3 g^{-1}). The fungus *Isaria fumosorosea* did not occur in the soil sampled within spring from any of the five varieties analyzed (Figure 2). In conclusion, the general concluded

that fungi insecticides formed most infectious units in the soil sampled within the spring from growing varieties of perennial ryegrass (*Lolium perenne* L.) as Tivoli (1.7×10^3 g^{-1}) and Argona (1.2×10^3 g^{-1}) (Table 4).

The analysis of soil samples taken at the Uhnin under cultivation of different varieties perennial ryegrass (*Lolium perenne* L.) by autumn showed that the majority of tested samples had fungus *Isaria fumosorosea*. The presence of this species was observed in soils from all cultivated varieties, where created average of 1.8×10^3 g^{-1} CFU per 1 gram of soil (Figure 2). Fungus *Beauveria bassiana* and *Metarhizium anisopliae* starred in the soil sampled from four varieties, and these fungi formed an average of 1.9×10^3 g^{-1} and 0.7×10^3 g^{-1} colony-forming units in the soils. (Figure 2). Comparing the occurrence of units of infectious fungi entomopatogennych depending on the time of soil sampling it was found that the most infectious units formed in the spring in soils under cultivation of such varieties perennial rye-

Table 2. The density of infectious units (CFU$\times10^3$ g^{-1}) of entomopathogenic fungi in soil under cultivation of different varieties perennial ryegrass (*Lolium perenne* L.) (Krzyżewo, spring)

Variety	Species of fungi			
	Beauveria bassiana	*Isaria fumosorosea*	*Metarhizium anisopliae*	Total
Lacerta	–	2.7e	2.2d	4.9
Tivoli	–	–	3.2f	3.2
Bah 209	–	–	0.5bc	0.5
Amarant	0.4ab	–	0.7c	1.1
Argona	0.2a	0.4ab	0.7c	1.3

* The value marked with the same letters do not differ significantly.

Table 3. The density of infectious units (CFU$\times10^3$ g^{-1}) of entomopathogenic fungi in soil under cultivation of different varieties perennial ryegrass (*Lolium perenne* L.) (Krzyżewo,autumn)

Variety	Species of fungi			
	Beauveria bassiana	*Isaria fumosorosea*	*Metarhizium anisopliae*	Total
Lacerta	–	0.2a	1.2d	1.4
Tivoli	0.2a	0.4ab	1.0d	1.6
Argona	–	0.4ab	1.2d	1.6
Amarant	–	1.7e	0.4ab	2.1
Bah 209	–	0.4ab	0.7c	1.1

* The value marked with the same letters do not differ significantly.

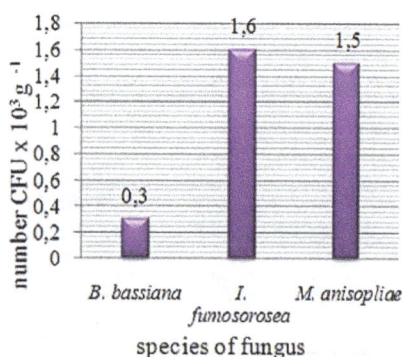

(KS)

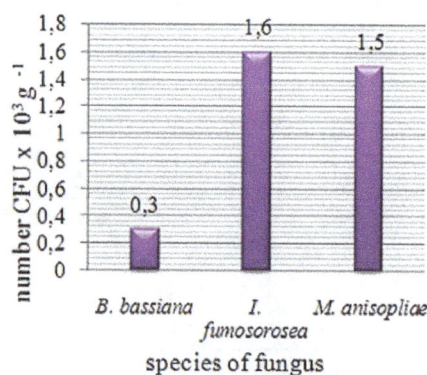

(KS)

Figure 1. Average concentration of infectious units (CFU$\times10^3$ g^{-1}) of individual entomopathogenic fungi in soil from cultivation of perennial ryegrass (*Lolium perenne* L.): Krzyżewo – (K), spring – (S), autumn – (A)

Table 4. The density of infectious units (CFU×10^3g⁻¹) of entomopathogenic fungi in soil under cultivation of different varieties perennial ryegrass (*Lolium perenne* L.) (Uhnin, spring)

Variety	Species of fungi			Total
	Beauveria bassiana	*Isaria fumosorosea*	*Metarhizium anisopliae*	
Argona	1.0d	–	0.2ab	1.2
Tivoli	1.5e	–	0.2ab	1.7
Bah 209	0.7bc	–	–	0.7
Lacerta	0.3a	–	–	0.3
Amarant	0.5b	–	–	0.5

* The value marked with the same letters do not differ significantly

Table 5. The density of infectious units (CFU×10^3g⁻¹) of entomopathogenic fungi in soil under cultivation of different varieties perennial ryegrass (*Lolium perenne* L.) (Uhnin, autumn)

Variety	Species of fungi			Total
	Beauveria bassiana	*Isaria fumosorosea*	*Metarhizium anisopliae*	
Argona	2.4f	0.9cd	–	3.3
Tivoli	0.7c	4.7h	0.7c	6.1
Bah 209	3.4g	1.9e	0.9cd	6.2
Lacerta	0.9cd	0.7c	0.9cd	2.5
Amarant	–	0.7c	0.4b	1.1

* The value marked with the same letters do not differ significantly

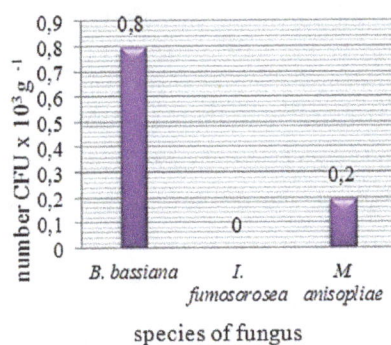

(US)

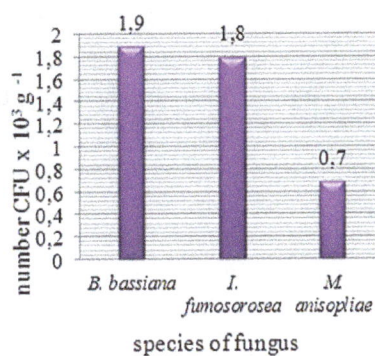

(UA)

Figure 2. Average concentration of infectious units (CFU×10^3g⁻¹) of individual entomopathogenic fungi in soil from cultivation perennial ryegrass (*Lolium perenne* L.): Uhnin – (U), spring – (S), autumn – (A)

grass (*Lolium perenne* L.) as Tivoli and Bah 209, (6.1×10^3g⁻¹ and 6.2×10^3g⁻¹ CFU) (Table 5). At the same time it was noted that the soil samples taken from the growing perennial ryegrass (*Lolium perenne*L.) by spring there was no presence of the fungus *Isaria fumosorosea*, which is found in soils collected within the autumn time.

DISCUSSION

The soils on which cultivated varieties of grasses perennial ryegrass (*Lolium perenne* L.) was isolated three species of entomopathogenic fungi *Beauveria* bassiana, *Metarhizium anisopliae* and *Isaria fumosorosea* by using a selective medium. They belonged to the most common entomopathogenic fungi in the soil environment. This is confirmed by many studies carried out by other authors [Miętkiewski et al, 1991; Bajan et al. 1997]. Species composition and density of the colony-forming units of these fungi varied depending on the variety, the seasons and the place

from which soil samples were derived. The soil samples collected in Krzyżewo in both terms, from all varieties of perennial ryegrass (*Lolium perenne* L.), *Metarhizium anisopliae* fungus occurred while *Isaria fumosorosea* also starred in the soil sampled from all varieties of perennial ryegrass (*Lolium perenne* L.), but only within autumn. *Beauveria bassiana* fungus and *Isaria fumosorosea* occurred in the collected soil sampled in the spring time only two of the five cultivated varieties of perennial ryegrass (*Lolium perenne* L.). In soils collected within *Beauveria bassiana* occurred only in soil sampled from the growing variety of Tivoli. The research by Tkaczuk [2008] has shown, that in Polish soils meadow and pasture fungus *Metarhizium anisopliae*, *Beauveria bassiana* and then *Isaria* fumosorosea are most common. This species seems to be characteristic of meadow and pasture environments in the UK [Chandler et al 1997], Switzerland [Keller et al 2003], and was previously listed as dominant in these soils. The soil samples collected in the spring at Station Variety Assessment in Uhnin from all varieties of perennial ryegrass most colony-forming species of fungi was *Beauveria bassiana*was. However *Isaria fumosorosea* did not occurr in anyof the analyzed soil samples collected in the spring. In turn, the density of the fungus Metarhizium infectious units anisopliae was found only in soils from cultivated varieties of Lacerta and Tivioli. concentration of infectious units in soils collected within the fall was much higher for all three species of fungi. In the case of the fungus *Isaria fumosorosea*, it was found that the fungus occurred in the soil sampled in the period from the fall in all cultivated varieties of perennial ryegrass (*Lolium perenne* L.). B*eauveria bassiana* and *Metarhizium anisopliae* fall within not only occurred in the case of two varieties of perennial ryegrass (*Lolium perenne* L.) – Lacerta and Argona. In a similar study to Kolczarek and Jankowski [2014] the same soil habitat for various varieties of meadow fescue (*Festuca pratensis*) it was found that fungus *Beauveria bassiana* did not occur in mineral soil collected within the autumn. Other fungi were present in amounts similar to the habitats of perennial ryegrass (*Lolium perenne* L.). It is observed that the cause of infectious units at greater concentration of the soil which was characterized by a high content of organic matter, in contrast and light mineral soil from station COBORU Variety marks in Krzyżewo. Just as in research Ignoffo et al. [1978] who argue that clay soils are richer in entomopathogenic fungi because clay particles have an increased capacity ion, so that they have the ability to easily adsorb spores with opposite electric charge. The soil rich in organic matter are mainly rich in species that have the ability to grow in saprophytic phase. These include fungi of the genus *Beauveria* [Müller-Kögler, Zimmermann 1986]. Research Miętkiewski et al. [1992] indicate a higher abundance of grassland soils in fungi insecticides than adjacent soils of farmland. They compared the incidence of fungi in the soil insecticides both arable and meadow. Were studied with the cultivation of wheat fields and adjacent meadows. The authors have isolated four species of fungi *Beauveria bassiana, Metarhizium anisopliae, Isaria farinosa* and *Isaria fumosorosea*. The incidence of entomopathogenic fungi in the soils was quite varied. In the meadows fungus *Metarhizium anisopliae* was present, other species, besides *Isaria fumosorosea,* were more numerous in grassland soils. The studies on mineral soil under cultivation of perennial rye-

grass (*Lolium perenne* L.) were also represented, including fungus *Metarhizium anisopliae.*

CONCLUSIONS

1. Soils under perennial ryegrass (*Lolium perenne* L.) contained three species of entomopathogenic fungi, that is *Beauveria bassiana, Isaria fumosorosea* and *Metarhizium anisopliae*, which were isolated using a selective medium.

2. Species composition and density of the colony-forming units varied depending on the variety, sampling date and site.

3. The fungi *Beaveria bassiana* and *Isaria fumosorosea* formed the most infectious units in organic soils and *Metarhizium anisopliae* in mineral soils, the number of infectious units being influenced by the soil type.

4. The fungus *Beaveria bassiana* formed the most infectious units (CFU) in organic soil under the variety Bah 209, *Isaria fumosorosea* in the autumn-sampled soil under the variety Tivoli and *Metarhizium anisopliae* in the spring-sampled mineral soil under Tivoli.

REFERENCES

1. Bac S., Koźmiński C., Rojek M., 1993. Agrometeorology. PWN, Warsaw, 32–33.

2. Bałazy S. 2006. Diagnosis and attempt to estimate the role of entomopathogenic fungi in the stands. Studies and Research Centre for Education Natural Sciences and Forest. 4 (14), 154–165.

3. Bajan C., Kmitowa K. 1997. Thirty years studies on entomopathogenic fungi in the Institute of Ecology, PAS. Pol. Ecol. Stud., 23(3-4), 133–154.

4. Chandler D., Hay D., Reid A.P., 1997. Sampling and occurrence of entomopathogenic fungi and nematodes in UK soils. Appl Soil Ecol. 5, 133–141.

5. Gaugler R. 1988. Ecological consideration In the Biol.Contr. of soilinhabiting insects with entomopathogenic nematodes. Agric., Ecosystem. Environ. 24, 351–361.

6. Ignoffo C.M., Garcia C., Hostetter D.L., Pinnel R.E. 1978. Stability of conidia of entomopathogenic fungus Nomuraea rileyi in a soil. J. Invertebr. Pathol. 28, 259–262.

7. Keller S., Kessler P., Schweizer C. 2003. Distribution of insect pathogenic soil fungi in Switzerland with special reference is Beauveria brongniartii and Matarhizium anisopliae. Biocontrol, 48, 307–319.

8. Keller S., G. Zimmermann 1989. Macopathogen of Siol insects. [In:] Insect-Fungus Interactions. Wilding N., Collins N.M., Hammon P.H., Webber J.F. (Eds.) Acadamic Press, London, 204–270.

9. Kolczarek R., Jankowski K. 2014. Occurrence of entomopathogenic fungi in soils from Festuca pratensis Huds. Crop. J. Ecol. Eng., 15 (2), 73–77.

10. Linden J. 1967. Outline of insect pathology. PWRiL. Warsaw.

11. Miętkiewski R. Tkaczuk C., Badowska-Czubik T. 1992. Entomogenous fungi isolated from strawberry plantation soil infested by Otiorhynchus ovatus L. Annals of Agricultural Sciences Ser. E, 22 (1/2) 39–46.

12. Miętkiewski R. Tkaczuk C., Zurek M., Bałazy S. 1991. Occurrence entomopatogenic fungi in soil arable and forest litter. Rocz. Agricultural Sciences. Ser. E, 21 (1/2), 61–68.

13. Müller-Kögler E., Zimmermann G. 1986. Go to Beauveria bassiana von Lebensdauer in kontaminiertem Boden unter Freiland-Und Laboratoriumsbedungungen. Entomophaga, 31, 285–292

14. Skowera B., Puła J. 2004. Extreme pluvio-thermal conditions in the spring on Polish territory in the 1971–2000. Acta Agrophysica, 3(1), 171–177.

15. Strasser H., Forrer A., Schinner F. 1996. Development of media for the selective isolation and maintenance of virulence of Beauveria brongniartii. In: Microbial control of soil dwelling pests. (Eds. TA Jackson and TR Glare), AgResearch, Lincoln, New Zealand, 125–130.

16. Tkaczuk C., Miętkiewski R., 1996, Occurrence of entomopathogenic fungi in different kinds of soil. Rocz. Agricultural Sciences., Series E, 25 (1/2), 44–48.

17. Tkaczuk C. 2008. Occurrence and potential infectious fungal insecticidesin soils and environments agrocenoz seminaturalnych in the agricultural landscape. Treatise No. 94, Ed. University of Podlasie in Siedlce.

18. Vanninen I. 1999. The distribution, ecological fitness and virulence of entomopathogenic fungideuteromycetous In Finland. Ph. D. Thesis, University of Helsinki.

6

COMPARISON OF DETEMINISTIC INTERPOLATION METHODS FOR THE ESTIMATION OF GROUNDWATER LEVEL

Agnieszka Kamińska[1], Antoni Grzywna[2]

[1] Department of Applied Mathematics and Computer Science, University of Life Sciences in Lublin, Poland

[2] Department of Environmental Engineering and Geodesy, University of Life Sciences in Lublin, Leszczyńskiego 7, 20-950 Lublin, Poland, e-mail: agrzywna@wp.pl

ABSTRACT

This paper compares two spatial interpolation techniques – Radial Basis Functions (RBF) and Inverse Distance Weighting (IDW) – with the goal of determining which method creates the best representation of reality for measured groundwater levels in catchment area. The study used the results of research and field observations from the year 2011, in Sosnowica (West Polesie). The data set consists of groundwater levels measured at 15 points in three series of tests. Surface generation was obtained for each method. The water prediction maps showed spatial variation in the groundwater level in the study area and they are quite different. RBF method resulted in a smoother map. The analysis of the methods of interpolation of analyzed data with the help of cross validation statistics and plots showed that Radial Basis Functions creates better representation of reality for measured groundwater levels.

Keywords: Radial Basis Functions, Inverse Distance Weighting, groundwater levels, prediction maps.

INTRODUCTION

Interpolation is a method or mathematical function that estimates the values at locations where no measured values are available. Spatial interpolation assumes the attribute data are continuous over space. This allows the estimation of the attribute at any location within the data boundary. Another assumption is that the attribute is spatially dependent, indicating the closer values are more likely to be similar than the values farther apart. The goal of spatial interpolations is to create a surface that is intended to best represent empirical reality thus the method selected must be assessed for accuracy [1].

There is no single preferred method for data interpolation. Aspects of the algorithm selection criteria need to be based on the actual data, the level of accuracy required, and the time and computer resources available. Selecting an appropriate spatial interpolation method is fundamental to surface analysis since different methods of interpolation can result in different surfaces and ultimately different results.

This paper compares two spatial interpolation techniques – Radial Basis Functions (RBF) and Inverse Distance Weighting (IDW) – with the goal of determining which method creates the best representation of reality for measured groundwater levels.

MATERIALS AND METHODS

The drainage area of the ditch K-2 discharging surface waters to the Peony river was selected to study the variability of water. It is located in the Sosnowica village in West Polesie [6]. Ditch drainage area is 0.46 km² and is 86% used as a once-semi-natural grasslands, the remaining 14% are birch and pine woodlands. The basin are 75% moorshed and moist habitats characterized by high levels of ground water and the position of small variations in retention. The catchment area of the trench has a very small decrease of 1.1‰

and includes a flat bottom valley. In 2011, at 60 days in the vegetation grassland (30.03–31.10) at 15 points was measured depth of the groundwater table. The measurements were carried out in the middle distances drainage ditches in piezometric wells [10]. In that paper the data set consists of groundwater levels measured at 15 points in three series of tests: spring, summer and autumn.

On the run in the 2006–2009 the research shows that the depth of the water table depends on the size of evaporation and modified the effects of the hydrographic network. We recorded the smallest water depths position at the beginning of the growing season, while the largest depth of the water table was recorded in the height of summer. The most stable water levels occurred in autumn [3].

There are two main grouping of interpolation techniques: deterministic and geostatistical. Deterministic interpolation techniques create surfaces from measured points, based on either the extend of similarity (Inverse Distance Weighted) or the degree of smoothing (Radial Basis Functions). IDW and RBF are exact interpolators, predict a value identical to the measured value at a sampled location [4].

Radial basis function methods are considered as exact interpolation techniques. The exact interpolators predict values identical with those measured at the same point and the generated surface

requires passing through each measured points. The predicted values can vary above the maximum or below the minimum of the measured values [9]. There are five different basic functions: thin-plate spline, spline with tension, completely regularized spline, multiquadric function, and inverse multiquadric spline. Each function has a different shape and results in a different interpolation surface. While there are more entry points specified, the greater the influence of distant points and the smoother the surface [5]. The estimated values of the methods are based on a mathematical function that minimizes the total surface curvature, generating quite a smooth surface. The smoothness of the resulting surface is controlled by a smoothing parameter. Radial basis function are described in Bishop [2].

Inverse Distance Weighting is based on the assumption that the nearby values contribute more to the interpolated values than distant observations. In other words, for this method the influence of a known data point is inversely related to the distance from the unknown location that is being estimated.

The general formula is:

$$\hat{Z}(s_0) = \sum_{i=1}^{N} \lambda_i Z(s_i)$$

where: $\hat{Z}(s_0)$ – predicted value for location s_0,
N – number of measured sample points surrounding the prediction location that will be used in the prediction
λ_i – weights assigned to each measured point
$Z(s_i)$ – observed value at location s_i,

The formula to determine the weights is the following:

$$\lambda_i = \frac{d_{i0}^{-p}}{\sum_{i=1}^{N} d_{i0}^{-p}}$$

where: p – arbitrary positive real number called the power parameter (typically $p = 2$),
d_{i0} – distance between the prediction location s_0 and each of the measured locations.
d_{i0} is given by:

$$d_{i0} = \sqrt{(x_0 - x_i)^2 + (y_0 - y_i)^2}$$

where: (x_0, y_0) are the coordinates of the interpolation point s_0 and (x_i, y_i) are the coordinates of each dispersion point s_i.

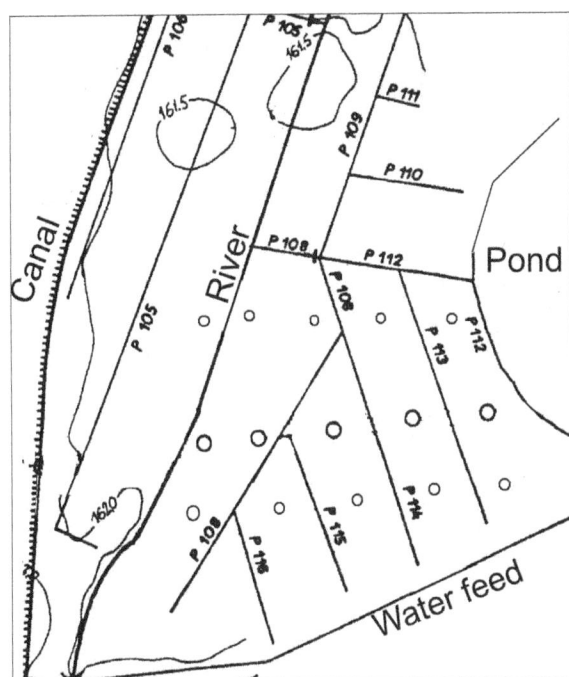

Figure 1. Location of measurement points groundwater level. o – points

The power parameter p influences the weighting of the measured location's value on the prediction of the location's value, that is, as the distance between the measured sample locations and the prediction location increase, the weight (or influence) that the measured point has on the prediction will decrease exponentially [4].

The adequacy of the fitted models was checked on the basis of validation tests. In this method, known as jackknifing procedure, interpolation is performed at all the data points, ignoring, in turn, each one of them one by one. The differences between estimated and observed values are summarized using cross-validation statistics [8].

For all points, cross-validation sequentially omits a point, predicts its value using the rest of the data, and then compares the measured and predicted values. The calculated statistics serve as diagnostics that indicate whether the model is reasonable for map production. In addition to visualizing the scatter of points around this 1:1 line (cross-validation scatter plot), a number of statistical measures can be used to assess the model's performance.

The differences between estimated and observed values are summarized using the cross-validation statistics: mean prediction error (ME), root-mean-square prediction error (RMSE). The summary statistics should meet the following criteria [7, 11]:

$$ME = \frac{1}{N}\sum_{i=1}^{N}\left(\hat{z}(x_i) - z(x_i)\right) \cong 0;$$

$$RMSE = \sqrt{\frac{1}{N}\sum_{i=1}^{N}\left(\hat{z}(x_i) - z(x_i)\right)^2} \text{ minimum;}$$

where: $\hat{z}(x_i)$ – predicted value;

$z(x_i)$ – observed value;

A GIS software package ArcGIS 10 and ArcGIS Geostatistical Analyst extension were used for the interpolation methods in this study. The maps were produced with the ArcMap module of the ArcGIS.

RESULTS

Figure 2 and 3 shows the spatial distribution of groundwater level for analyzed three time series in the study area obtained by Radial Basis Functions (with the multiquadric function) and

Inverse Distance Weighting (with power parameter $p = 2$). The prediction map provided by two interpolation methods (Figure 2 and Figure 3) are different. The comparison of IDW and RBF maps indicated that RBF method has resulted in a smoother map.

More quantitative comparison of these two techniques was obtained by comparing the cross-validation statistics (Table 1). The best model was selected based on two criteria: the mean prediction error (ME) nearest zero, the smallest root-mean-square prediction error (RMSE). IDW resulted in ME of -1.18 m to -2.18 m whereas RBF gave ME of -0.84 m to 0.46 m. Similarly, IDW gave RMSE of 11.90 m² to 17.44 m² and RBF 2.36 m² to 11.93 m². The ME values are closer to 0 and RMSE are smaller for RBF.

Table 1. Cross validation results for interpolation methods

Series	RBF		IDW	
	ME	RMSE	ME	RMSE
I	0.46	5.92	-2.18	16.37
II	-0.84	11.93	-2.21	17.44
III	-0.32	2.36	-1.18	11.90

The relationship between the interpolated values and the true observed data was also evaluated. Figure 3 presents the scatter plots of predicted versus measurement values obtained for used interpolation methods. It is expected that these should scatter around the 1:1 line. The cross-validation scatter plots provided by two interpolation methods (Figure 4) are different. The deviation from the 1:1 line are greater for the IDW method. It shows that within interpolation methods used, the RBF method is the one that best estimated the measurements results of the groundwater level.

CONCLUSIONS

In this study, two spatial interpolation techniques – Radial Basis Functions (RBF) and Inverse Distance Weighting (IDW) were applied to the groundwater level data in three series of tests. Surface generation was obtained for each method. The water maps showed the spatial variation in the groundwater level in the study area and they are quite different. RBF method

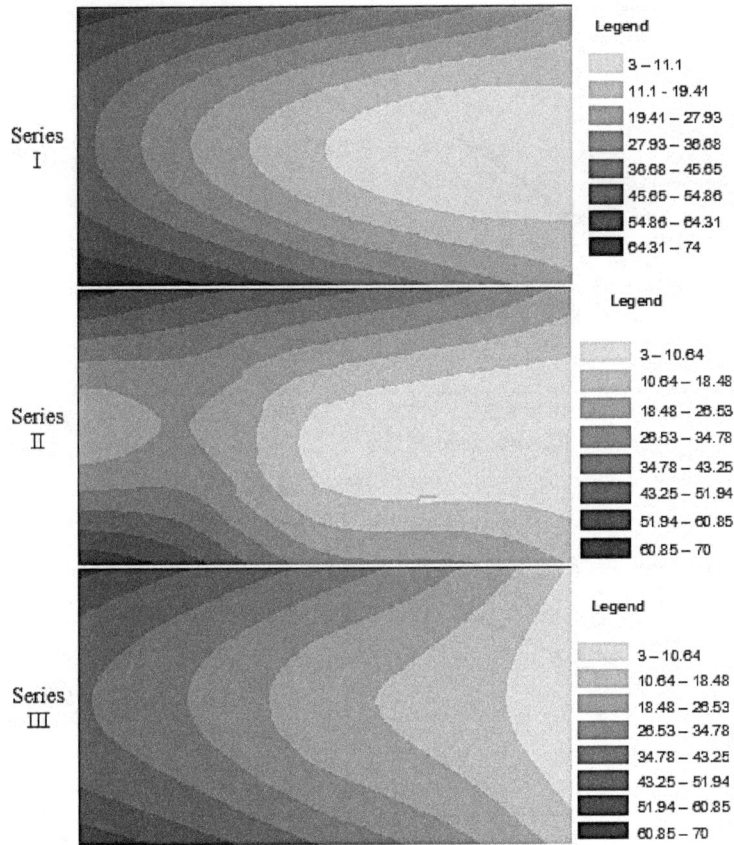

Figure 2. Prediction maps for RBF

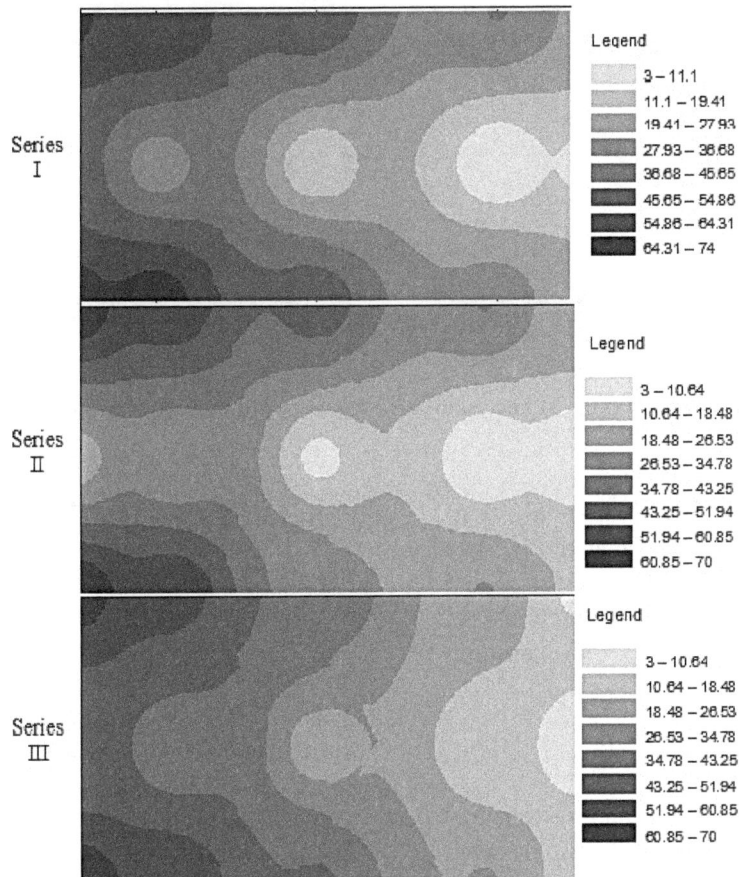

Figure 3. Prediction maps for IDW

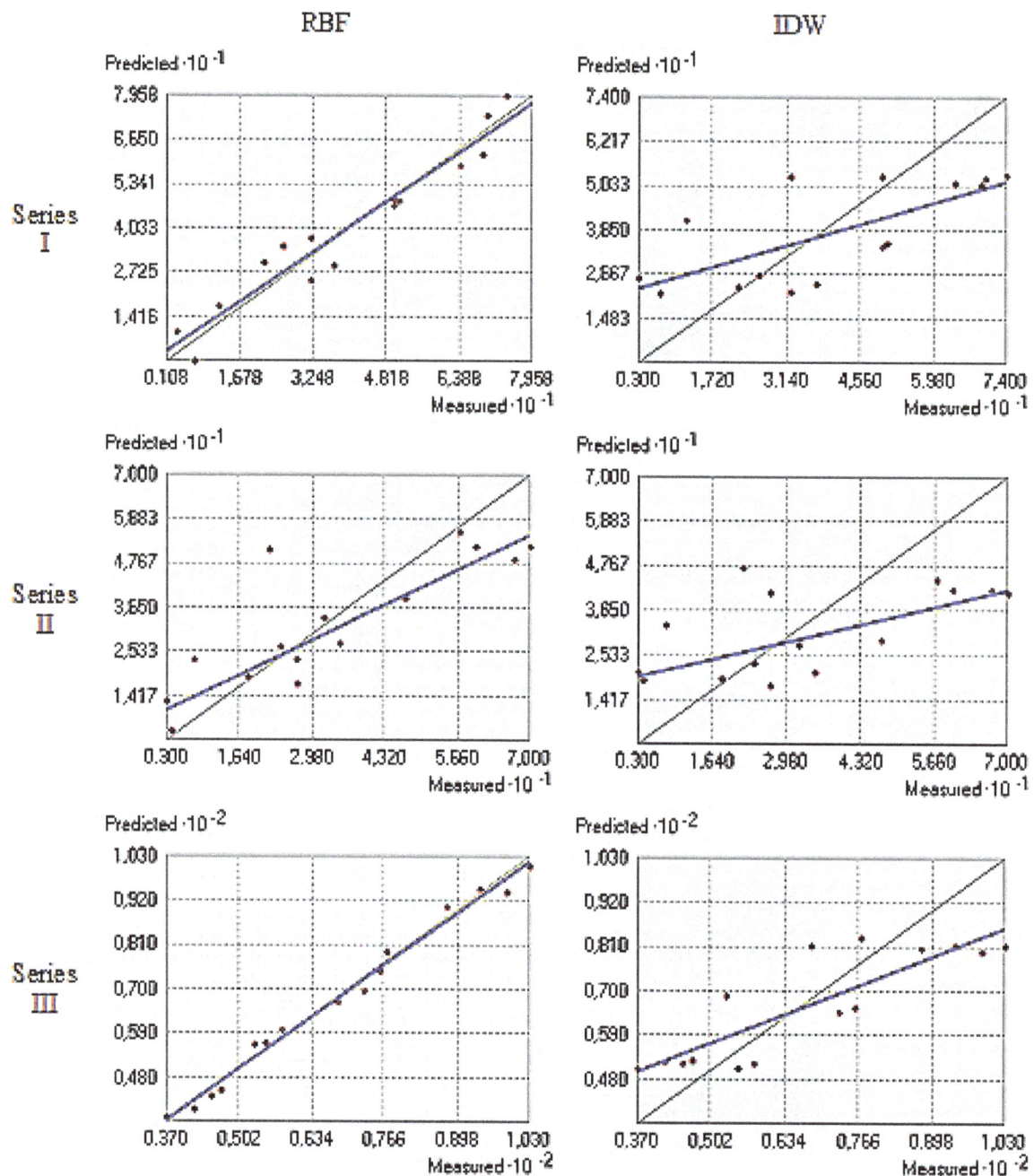

Figure 4. Scatter plots of interpolation methods of groundwater level data

has resulted in smoother map. The analysis of the methods of interpolation of analyzed data with the help of cross validation statistics and plots showed that Radial Basis Function creates better representation of reality for measured groundwater levels.

Acknowledgements

Research work from budget funds for science, as a research project N N313 439239 in the years 2010–2013.

REFERENCES

1. Azpurua M., Dos Ramos K., 2010. A comparison of spatial interpolation methods for estimation of average electromagnetic field magnitude. Progress In Electromagnetic Research M, 14, 135–145.

2. Bishop C.M., 1995. Neural Networks for Pattern Recognition. Oxford Press, Oxford, 164.

3. Grzywna A., 2011. Zmiany położenia zwierciadła wody gruntowej w latach 2006–2009 na zmeliorowanym obiekcie Sosnowica [Changes the level of the groundwater table in the years 2006–09 on

reclaimed object Sosnowica]. Gaz, Woda i Technika Sanitarna, 10, 359–360 [In Polish].

4. Johnston K., Ver Hoef J.M., Krivoruchko K., 2003. ArcGIS 9. Using ArcGIS Geostatistical Analyst ESRI.

5. Karydas C.G., Gitas I.Z.., Koutsogiannaki E., Lydakis-Simantiris N., Silleos G. N., 2009. Evaluation of spatial interpolation techniques for mapping agricultural topsoil properties in Crete. EARSeLe-Proc. 8, 26–39.

6. Kondracki J., 2002. Geografia regionalna Polski [Regional Geography in Poland]. PWN Warszawa [In Polish].

7. Krivoruchko K., 2006. Introduction to Spatial Data Analysis in GIS. ESRI Press.

8. Kumar V., Remadei., 2006. Kriging of groundwater levels – a case study. Journal of Spatial Hydrology, 6, 1, 81–92.

9. Nikolova N., Vassilev S., 2006. Mapping precipitation variability using different interpolation methods. Proceedings of the Conference on Water Observation and Information System for decision Support (BALWOIS), Ohrid, Macedonia, 25–29.

10. Nyc K., Pokładek R., 2004. Rola małej retencji w kształtowaniu ilości i jakości wód [The role of small retention in shaping the quality and quantity of water]. Zeszyty Naukowe AR we Wrocławiu 502, Seria Inżynieria Środowiska XIII, 72–79 [In Polish].

11. Zawadzki J., 2011. Metody geostatystyczne [Geostatistical methods]. Oficyna Wydawnicza Politechniki Warszawskiej, Warszawa [In Polish].

7

THE OCCURRENCE OF ENTOMOPATHOGENIC FUNGI IN SOILS FROM FIELDS CULTIVATED IN A CONVENTIONAL AND ORGANIC SYSTEM

Cezary Tkaczuk[1], Anna Król[1], Anna Majchrowska-Safaryan[1], Łukasz Nicewicz[1]

[1] Department of Plant Protection, Siedlce University of Natural Sciences and Humanities, B. Prusa 14, 08-110 Siedlce, Poland, e-mail: tkaczuk@uph.edu.pl

ABSTRACT

The occurrence of entomopathogenic fungi, involved in regulating the population of insects preying on arable crops, depends on numerous factors, such as using crop protection chemicals, which are prohibited in the organic tillage system. Using two methods: selective medium and the *Galleria mellonella* bait method, the species composition and the occurrence of entomopathogenic fungi in soils deriving from organic and conventional fields sown with winter cereals were compared. Four entomopathogenic fungal species were identified in the samples: *Beauveria bassiana*, *Isaria fumosorosea*, *Metarhizium anisopliae* and *Lecanicillium* sp.. *M. anisopliae* was the most frequently detected one. *M. anisopliae* and *B. bassiana* formed more colony forming units in soils from organic fields, whereas *I. fumosorosea* in soils from the conventional ones.

Keywords: entomopathogenic fungi, soil, arable field, conventional system of cultivation, organic system of cultivation.

INTRODUCTION

Enthomopathogenic fungi participate in the regulation of insects populations, including agricultural pests. Soil, where they find the best growing conditions, is their major dwelling place. Moreover, insects infections take place there.

The enthomopathogenic fungi occurrence in the soil can depend, inter alia, on soil type [Tkaczuk and Miętkiewski 1996], on the cultivated plant species [Krysa et al. 2012, Tkaczuk 2008], or on the agricultural practices [Oliveira et al. 2013, Jabbour and Barbercheck 2009, Tkaczuk 2008, Quesada-Moraga et al. 2007, Hummel et al. 2002]. Because of their low resistance to environment changes, the degree of soil colonization by these fungi can be the indicator of its condition [Meiling and Eilenberg 2006]. Among the factors limiting the occurrence of these insect pathogens, there are plant protection chemicals, particularly fungicides and herbicides [Tkaczuk et al. 2013, Poprawski and Majchrowicz 1995]. Taking care of good soil conditions and the prohibition of

chemical pesticides usage is an organic farming characteristic.

The transition from conventional to organic farming method predominantly favours the increase of the soil microorganisms diversity [Mäder et al 2002]. However, in the case of entomopathogenic fungi the impact of such changes has not been sufficiently elucidated yet. Organic cultivation has been shown to increase the abundance and diversity of entomopathogens in, soils in comparison to the conventional tillage [Klingen et al.2002; Mader et al. 2002], but other authors report no significant differences in assemblages of fungal entomopathogens in soils from conventionally cultivated fields after their transition to organic tillage [Jabbour and Barbercheck 2009].

Lubelskie Voivodeship, which was chosen as a research area, is characterized by a large number of organic farms, in comparison to other regions of the country [Kuś and Jończyk 2009]. Furthermore, the strong farm fragmentation is observed in this region and, therefore, due to the small agricultural area it is difficult for the farms to live off the agricultural production. Limited financial resources is the reason why mineral fertilization is here at one of the lowest levels in the country [Dziaduch 2010].

Studies on the occurrence of entomopathogenic fungi populations in soils from fields differing in the level of farming intensity can help to monitor the functioning of the agri-environmental packages. Knowledge of the species composition of entomopathogenic fungi in different regions of the country can be also useful in assessing the potential of individual fungal species to regulate the populations of crop pests [Meyling i Eilenberg 2006].

The aim of the study was to compare the species composition and the intensity of entomopathogenic fungi occurrence in soils from selected agricultural fields cultivated in the organic and conventional system in the Lublin voivodeship.

MATERIAL AND METHODS

Soil samples from 28 arable fields were collected in September 2012. Research plots were located in the north-estern region of Lubelskie Voivodeship, near the following locations:

Kuzawka, Łomnica, Liszna, Dołhobrody, Szuminka, Kołacze, Pożarów, Krynica, Skrychiczyn, Kępa, Połoski, Zahorów, Wereszczyn, Wola Wereszczyńska, Sławatycze, Różnaka, Suchawa, Iżyce, Brzozowiec, Janostrów, Kępa, Piszczac, Trojanów and Tarnów; 14 in organic field crops of winter cereals (rye, triticale, wheat and spelt) falling within agri-environmental program package (2.0) and 14 in conventional field crops (rye, wheat and triticale). Fields from the organic farms had been organically cultivated for at least four years. The objective was to obtain pairs of samples (conventional and organic) from farms at neighboring sites. These sites differed in agricultural practices and fertilizers and pesticide usage, but environmental factors, such as climate and soil conditions were as comparable as possible. At the conventional farms, cereals seeds were treated with fungicides, and insecticides and herbicides were sprayed as needed during vegetation period. Mineral fertilization of soil also was used on the conventional farms. On the organic farms, by contrast, no pesticides or mineral fertilizers were used.

The samples were taken using a shovel, to a depth of up to 15 cm, from 10–15 random points on the tested field. The mixed samples were prepared from the collected material and stored in plastic bags at the temperature of 3–4 °C. Immediately before starting the experiment in the laboratory, the soil was sieved to separate the larger particles of impurities, and dried up to a moisture content of approximately 25–30% (which is optimal for fungal growth and limits the growth of entomopathogenic nematodes).

Fungi were isolated from the soil using two methods: the insect bait method and the selective medium. The Greater Wax Moth (*Galleria mellonella* L.) in the penultimate instar larvae, coming from a laboratory culture in the Plant Protection Department from Siedlce University of Natural Sciences and Humanities, was used as a bait insect. Five plastic boxes of 200ml capacity were filled with the soil taken from each field. Ten *G. mellonella* larvae were put into each box, then the boxes were placed in an incubator in the temperature of 20–22 °C. The first mortality control was conducted after 7 days, and then at the three-day intervals until the death of all larvae. Dead larvae with symptoms of fungal infection were transferred directly into moist Petri dishes. Dead larvae, without symptoms of fungal infection, were surface-sterilized in 1% sodium hypochlorite solution, and then rinsed three times in distilled water. After that, the larvae were put in Petri dishes with moistened filter paper. Microscopic preparations from fungal mycelium and spores which grew on the insects surface provided the basis for the determination of fungal species.

To examine the concentration of colony-forming units (CFU) of entomopathogenic fungi in tested soil samples, a selective medium developed by Strasser et al. was applied (1996). This is a commonly used method for the isolation of entomopathogenic fungi from soil environment [Tkaczuk 2008, Meyling and Eilenberg 2006, Keller et al.2003]. Two grams of soil were weighed out of each sample, then 18 ml of distilled water with addition of 0.05 Triton X-100, which reduces the surface tension, were added. The resulting solution was vigorously shaken for 30-40 seconds. Then, 0.1 ml of the soil solution was poured out on a selective medium and spread using a glass spatula. The selective medium consisted of: 1 litre of water, 20 g of glucose, 18 g of agar and 10g of peptone. After sterilization and cooling, the following selective components were added to the medium: 0.6 g of streptomycin sulfate, 0.005g of chlortetracycline, 0.05 g of cycloheximide and 0.1 g of dodine. These components were limiting the growth of saprophytic bacteria and fungi and foster the growth of entomopathogenic fungi. Experiment has been performed in three Petri dishes per sample. Dishes were transferred into incubators at temperature of 22 °C, and after 10–12 days colonies of individual fungal species were counted. The results were expressed as a number of colony-forming units (CFU) of entomophatogenic fungi in 1 g of soil. The results concerning density of CFU in soil were analyzed statistically by performing a 2-factorial analysis of variance and detailed comparison of average values was made using the Tukey's test at significance level α = 0.05.

RESULTS AND DISCUSSION

A total of four species of entomopathogenic fungi were isolated from the tested soils: *Beauveria bassiana*, *Isaria fumosorosea*, *Metarhizium anisopliae* and species of the genus *Lecanicillium*. All of them commonly occur and infect insects in soils from cultivated fields [Tkaczuk 2008, Tkaczuk and Mietkiewski1996, Bajan et al. 1995, Steenberg 1995, Miętkiewski et al. 1991, Vänninen et al. 1989, Keller and Zimmerman 1989].

The use of Galleria bait method enabled the isolation all four species from soil samples collected from organic fields, and only three from conventional cultivations: *I. fumosorosea*, *M. anisopliae*, and *Lecanicillium* sp. (Table 1 and Table 2). Frequency of individual species occurrence, isolated using insect bait method, is presented in Figure 1. In both cases, *M. anisopliae* was the most frequently isolated fungus. It was detected in 92% of the soil either from organic or conventional fields. *I. fumosorosea* occurred in 50% of soil from conventional fields, and in 57,1% of soil samples from organic cultivations. Using this method no *B. bassiana* occurrence was detected in soil from conventional cultivations, whereas it was isolated from 37% of soils from organic fields. Fungal species from *Lecanicillium* genus was detected in 7,1% of the tested soils, both from conventional and organic fields.

M. anisopliae caused most fungal infestations of *G. mellonella* as well (Table 1 and Table 2). Percentage of insect larvae infected by individual species of fungi in soils from conventional fields was not significantly different from that observed in soils from the organic fields (Figure 2). *M. anisopliae* is often considered as a dominant in soil from cultivated fields, however, it is rarely isolated from natural habitats [Bidochka et al. 1998, Chandler et al. 1997]. According to Vänninen [1996] it is the most tolerant species to agricultural treatments, such as plowing or using pesticides. It tolerates a periodic absence of potential hosts in the environ-

Table 1. *G. mellonella* larvae mortality (%) in soils from fields cultivated in the organic system

Mortality factor	Study plot													
	1a	2a	3a	4a	5a	6a	7a	8a	9a	10a	11a	12a	13a	14a
Entomopathogenic fungi														
Beauveria bassiana	–	–	4	2	–	2	2	–	–	–	2	–	–	–
Isaria fumosorosea	86	80	26	–	–	4	–	–	4	12	2	–	–	–
Metarhizium anisopliae	–	10	8	80	98	30	92	76	42	66	46	82	98	92
Lecanicillium sp.	–	–	–	–	–	–	–	–	–	–	–	2	–
Total	**86**	**90**	**38**	**82**	**98**	**36**	**94**	**76**	**46**	**78**	**50**	**82**	**100**	**92**
Fungi of unproved entomopathogenic abilities														
Aspergillus sp.	–	–	4	–	–	2	–	–	–	4	–	–	–	2
Gliocladium sp.	–	–	–	–	2	6	–	–	–	–	–	–	–	–
Fusarium sp.	–	2	–	4	–	6	2	–	2	–	–	–	–	–
Mucor sp.	–	–	–	–	–	4	–	2	–	–	2	–	–	4
Unfrutiful mycelium	12	6	30	4	–	30	2	2	14	6	6	4	–	–
Total	**12**	**8**	**34**	**8**	**2**	**48**	**6**	**4**	**16**	**10**	**8**	**4**	**–**	**6**
Other causes														
Nematodes	–	–	14	2	–	2	–	18	16	2	20	2	–	2
Unspecified causes	2	2	14	8	–	14	–	2	22	10	22	12	–	–
Total	**2**	**2**	**28**	**10**	**–**	**16**	**–**	**20**	**38**	**12**	**42**	**14**	**–**	**2**

Table 2. *G. mellonella* larvae mortality (%) in soils from fields cultivated in the conventional system

Mortality factor	Study plot													
	1b	2b	3b	4b	5b	6b	7b	8b	9b	10b	11b	12b	13b	14b
Entomopathogenic fungi														
Beauveria bassiana	–	–	–	–	–	–	–	–	–	–	–	–	–	–
Isaria fumosorosea	2	80	–	6	10	–	–	–	–	–	14	96	16	2
Metarhizium anisopliae	96	14	78	82	64	94	92	98	82	80	44	–	64	4
Lecanicillium sp.	–	–	2	–	–	–	–	–	–	–	–	–	–	–
Total	**98**	**94**	**80**	**88**	**74**	**94**	**92**	**98**	**82**	**80**	**58**	**96**	**80**	**6**
Fungi of unproved entomopathogenic abilities														
Aspergillus sp.	–	–	–	–	–	–	–	–	–	–	–	2	–	8
Gliocladium sp.	–	–	–	–	–	–	–	–	–	–	–	–	6	2
Fusarium sp.	2	–	4	–	–	2	2	–	6	4	–	–	4	2
Mucor sp.	–	–	6	2	2	–	2	–	2	12	2	–	–	2
Unfrutiful mycelium	–	6	4	10	4	2	2	–	8	4	22	–	10	20
Total	**2**	**6**	**14**	**12**	**6**	**4**	**6**	**–**	**16**	**20**	**24**	**2**	**20**	**34**
Other causes														
Nematodes	–	–	–	–	8	–	–	–	–	–	18	–	–	36
Unspecified causes	–	–	6	–	12	2	2	2	2	–	–	2	–	24
Total	**–**	**–**	**6**	**–**	**20**	**2**	**2**	**2**	**2**	**–**	**18**	**2**	**–**	**60**

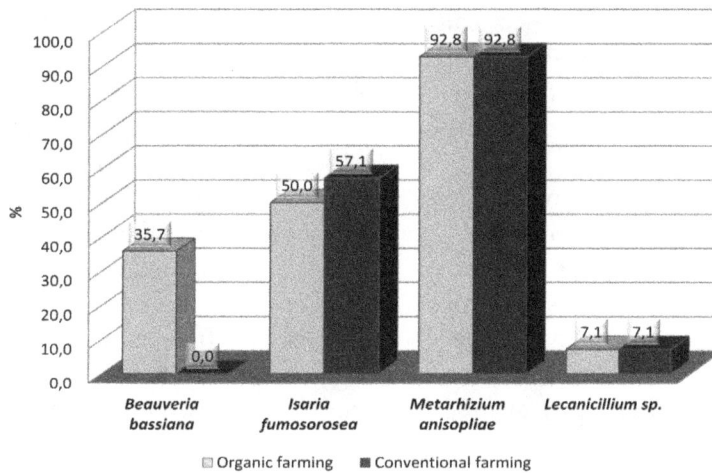

Figure 1. The frequency of isolation of entomopathogenic fungi in soils from fields cultivated in the organic and conventional system (Galleria bait method)

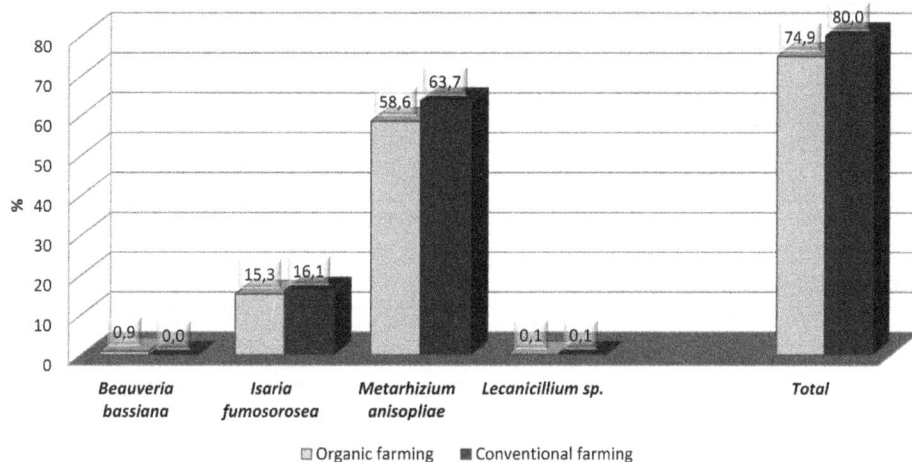

Figure 2. Mortality *G. mellonella* larvae in the soil of the fields cultivated in the organic and conventional system (the average of all samples)

ment too. Comparing occurrence of *M. anisopliae* in soils from fields of varied cultivation systems, Meyling et al. [2011] found no significant differences in occurrence of this fungus in soil from organic and conventional cultivations.

Using the insect bait method for isolation of entomopathogenic fungi from soil, Tkaczuk [2008] and Tkaczuk et al. [2012] indicate two species: *M. anisopliae* and *I. fumosorosea* as dominants in agricultural soils of Poland. Contrary to these results in soils from numerous countries such as Spain, Denmark, Italy or Albania dominant species is *B. bassiana* [Qesada-Moraga et al. 2007, Tarasco and Poliseno 2005, Tarasco et al. 1997].

The use of selective medium method provided slightly different results in comparison to the insect bait method. Using selective medium method both in the case of soils from organic and conventional fields three species were isolated: *B. bassiana*, *I. fumosorosea*, and *M. anisopliae* (Table 3 and Table 4). Concerning the frequency of individual species isolation this method was more effective than the insect bait method (Figure 3). In contrast to Galleria bait method, this method allowed to detect the colony-forming units of *B. bassiana* fungus in soils from conventional cultivations.

Medo i Cagan [2011], who examined occurrence of entomopathogenic fungi in soils of Slovakia, have also found differences in the obtained results using different methods. The authors reported that *M. anisopliae* was isolated approximately four times more frequently using selective medium than insect bait method.

Bruck [2004], however, reported the greater sensibility of insect bait method for *B. bassiana*. Keller et al. [2003] observed no significant differences using these two methods to examine the occurrence of entomopathogenic fungi in soils from farmlands in Switzerland. According to Miętkiewski et al. [1991] in order to determine the abundance of entomopathogenic fungi in the environment in a more full way, it is advised to use simultaneously several species of insects as a trap, because susceptibility to infections is different in various insects.

In soils collected from organic fields entomopathogenic fungi formed, on average, $5,8 \times 10^3$ CFU in 1 gram of soil (from 2,7 to $11,9 \times 10^3$ on individual fields), and slightly less in conventionally cultivated soils: from 1,3 to $8,6 \times 10^3$, on average, $5,3 \times 10^3$ g^{-1} (Table 3 and Table 4, Figure 4). *M. anisopliae* and *B. bassiana* occurred in significant higher density in soils from organic fields, forming $2,7 \times 10^3$ g^{-1} and $1,0 \times 10^3$ g^{-1} CFU respectively. In soils from conventional cultivations *M. anisopliae* formed $2,1 \times 10^3$ g^{-1} CFU, whereas *B. bassiana* formed 0.6×10^3 CFU in 1 gram of soil. *I. fumosorosea* was the only species forming more CFU in soils from conventional cultivation (on average $2,6 \times 10^3$ g^{-1}) than in soils from organic fields (on average 2.1×10^3) (Figure 4).

I. fumosorosea fungus is a common species in soils from different environments in Poland [Miętkiewski et al. 1992, 1998]. However, it is rarely isolated from soils in other European countries [Tkaczuk 2008, Keller et al. 2003, Chandler et al. 1997, Vänninen 1996, Steenberg 1995].

Table 3. The density of colony forming units of entomopathogenic fungi (CFU×10^3 g^{-1}) in soils from fields cultivated in the organic system

Fungal species	Study plot													
	1a	2a	3a	4a	5a	6a	7a	8a	9a	10a	11a	12a	13a	14a
Beauveria bassiana	0,7	–	1,3	–	1,5	1,8	1,5	–	0,3	–	2,0	0,5	2,0	2,0
Isaria fumosorosea	5,0	4,5	2,3	0,2	–	0,3	0,2	3,3	3,0	4,2	3,2	0,7	–	3,2
Metarhizium anisopliae	6,2	1,2	3,8	2,5	2,3	1,2	5,0	2,2	1,5	1,8	1,0	5,5	2,2	1,0
Total	**11,9**	**5,7**	**7,4**	**2,7**	**3,8**	**3,3**	**6,7**	**5,5**	**4,8**	**6,0**	**6,2**	**6.7**	**4,2**	**6,2**

Table 4. The density of colony-forming units of entomopathogenic fungi (CFU×10^3 g^{-1}) in soils from fields cultivated in the the conventional system

Fungal species	Study plot													
	1b	2b	3b	4b	5b	6b	7b	8b	9b	10b	11b	12b	13b	14b
Beauveria bassiana	0.5	–	0.8	1.2	0.7	1.5	2.3	–	–	–	0.2	–	0.5	0.3
Isaria fumosorosea	4.1	4.1	0.2	3.3	1.2	–	–	1.0	4.7	4.3	1.8	4.6	3.3	4.0
Metarhizium anisopliae	4.0	1.0	0.3	–	2.2	3.8	4.0	5.5	1.0	1.2	4.6	–	1.2	1.0
Total	**8.6**	**5.1**	**1.3**	**4.5**	**4.1**	**5.3**	**6.3**	**6.5**	**5.7**	**5.5**	**6.6**	**4.6**	**5.0**	**5.3**

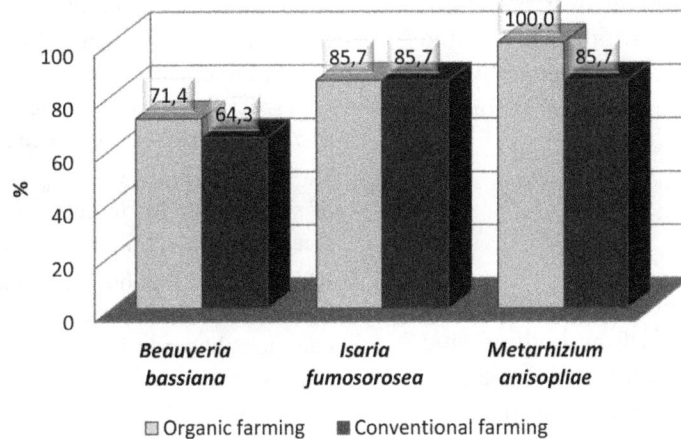

Figure 3. The frequency of isolation of entomopathogenic fungi in soils from fields cultivated in the organic and the conventional system (isolation on selective medium)

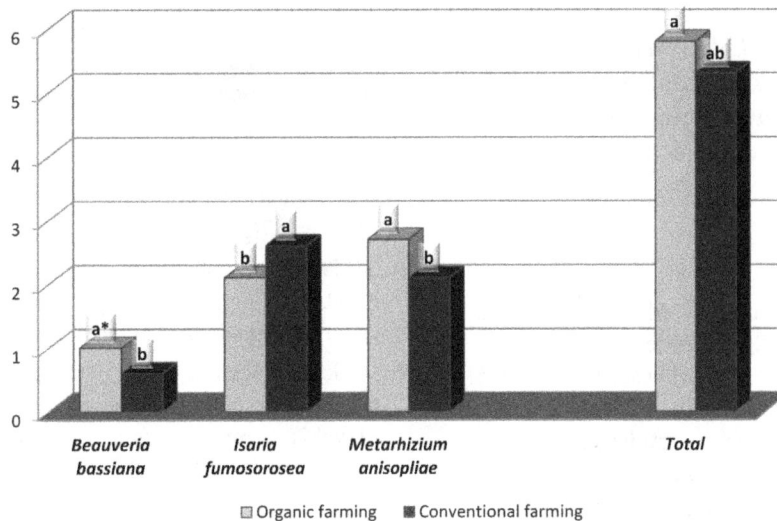

Figure 4. The density of colony forming units of entomopathogenic fungi (CFU×10^3 g^{-1}) in soils from fields cultivated in the organic and conventional system

The occurrence of entomopathogenic fungi in soils from cultivated fields may depend on soil type [Tkaczuk 2008]. In sandy soils, which are common in Poland, *I. fumosorosea* and *M. anisopliae* are the dominant species, in clay soils *B. bassiana*, *I. fumosorosea*, and *M. anisopliae* dominate. The dominance of the *B. bassiana* species is more typical to organic soils. The dominance of *I. fumosorosea* and *M. anisopliae* and lower prevalence of *B. bassiana* in sandy soils was also reported by Tkaczuk and Miętkiewski [1996] and Kleespies et al. [1989].

Hummel et al. [2002] using insect bait method in a long-term field study found that the application of certain pesticides significantly reduces the occurrence of entomopathogenic fungi in the soil. According to some authors, the discontinuance of the use of chemical plant protection products in organic cultivations can have a positive effect on

the occurrence of these fungi [Klingen et al. 2002, Vänninen and Hokkanen 1988, Miętkiewski et al. 1997]. Tkaczuk [2008] found that fungus *I. fumosorosea* was the most resistant to pesticides of the studied species under *in vitro* conditions. It can be assumed that *I. fumosorosea* is the best species to be concomitantly used with pesticides in integrated crop protection systems.

A more frequent occurrence of *B. bassiana* species in soil from organic fields may be the result of using organic fertilizers, such as manure or green manure, which enrich the soil with organic matter. Using higher doses of manure may favorably affect the efficiency of *B. bassiana* as a biological control agent of soil pests [Rosin et al. 1996].

The presence of individual entomopathogenic fungi in different cultivations is an indicator of their capability to survive in such environments. Research on the species composition of soil microorganisms in fields cultivated in different tillage may be useful for selecting the suitable species for biological pest control. Generally, indigenous dominants are the most suitable for this purpose [Meyling and et al. 2011].

CONCLUSIONS

1. *M. anisopliae* was the most frequent entomopathogenic fungus in soils from both organic and conventional fields. It infected the largest number of *G. mellonella* larvae as well.

2. Entomopathogenic fungi formed on average more colony forming units in 1 g of soil in soil from organic cultivations than in soil from conventional fields.

3. *M. anisopliae* and *B. bassiana* occurred in higher density in soils from organic fields, whereas *I. fumosorosea* formed more CFU in soils conventionally cultivated.

REFERENCES

1. Bajan C., Kmitowa K., Mierzejewska E., Popowska-Nowak E., Miętkiewski R., Górski R., Miętkiewska Z., Głowacka B., 1995. Występowanie grzybów owadobójczych w ściółce i glebie borów sosnowych w gradiencie skażenia środowiska leśnego. Prace IBL, ser. B. 24, 87–97.

2. Bidochka M. J., Kasperski J. E., Wild G. A. M. 1998. Occurrence of the entomopathogenic fungi *Metarhizium anisopliae* and *Beauveria bassiana* in soils from temporate and near-notherns habitats. Can. J. Bot. 76, 1198–1204.

3. Bruck D.J. 2004. Natural occurrence of entomopathogens in Pacific Northwest nursery soils and their virulence to the black vine weevil, *Otiorhynchus sulcatus* (F.) (Coleoptera: Curculionidae). Environ. Entomol. 33, 1335–1343.

4. Chandler D., Hay D., Reid A.P. 1997. Sampling and occurrence of entomopathogenic fungi and nematodes in UK soils. Appl. Soil Ecol., 5, 133–141.

5. Dziaduch S. 2010. Rolnictwo w województwie lubelskim w 2009 r. Urząd Statystyczny w Lublinie, 82s.

6. Hummel R.L., Walegenbach J. F., Barbercheck M.E., Kennedy G.G., Hoyt G.D., Arellano C. 2002. Effects of production practices on soil- borne entomopathogens in western North Karolina vegetable systems. Environ. Entomol., 31, 84–91.

7. Jabbour R., Barbercheck M.E. 2009. Soil management effects on entomopathogenic fungi during the transition to organic agriculture in a feed grain rotation. Biological Control. 51, 435–443.

8. Keller S., Kessler P., Schweizer C., 2003. Distribution of insect pathogenic soil fungi in Switzerland with special reference to *Beauveria brongniartii* and *Metarhizium anisopliae*. Biocontrol, 48, 307–319.

9. Keller S., Zimmermann G., 1989. Mycopathogen of soil insects. In: Insects – Fungus Interactions, Academic Press, London, 240–270.

10. Kleespies R., Bathon H., Zimmermann G. 1989. Untersuchen zum naturlichen Vorkommen von entomopathogenen Pilzen und Nematoden in verschiedenen Boden in der Umgebung von Darmstad. Gesunde Pflanzen, 41, 350–355.

11. Klingen I., Eilenberg J., Meadow R. 2002. Effects of farming system, field margins and bait insect on the occurence of insect pathogenic fungi in soils. Agric. Ecosyst. Environ. 91, 191–198.

12. Kuś J., Jończyk K. 2009. Rozwój rolnictwa ekologicznego w Polsce. Journal of Research and Application in Agricultural Engineering. 54(3), 178–182.

13. Krysa A., Ropek D., Kuźniar T. 2012. The occurence of entomopathogenic fungi depending on season in selected organic farm. J. Res. Appl. Agric. Eng. 57(3), 226–230.

14. Mäder P., Fliessbach, A., Dubois, D., Gunst, L., Fried, P., Niggli, U., 2002. Soil fertility and biodiversity in organic farming. Science 296, 1694–1697.

15. Medo J., Cagan L. 2011. Factors affecting the occurrence of entomopathogenic fungi in soils of Slovakia as revealed using two methods. Biological Control 59, 200–208.

16. Meyling N.V., Eilenberg J. 2006. Occurrence and distribution of soil borne entomopathogenic fungi within a single organic agroecosystem. Agric. Ecosyst. Environ. 113, 336–341.

17. Meyling N.V., Thorup-Kristensen K., Eilenberg J. 2011. Below- and aboveground abundance and distribution of fungal entomopathogens in experimental conventional and organic cropping systems. Biological Control. 59, 180–186.

18. Miętkiewski R., Dzięgielewska M., Janowicz K. 1998. Entomopathogenic fungi isolated in the vicinity of Szczecin. Acta Mycol. 33(1), 123–130.

19. Miętkiewski R.T, Pell J.K, Clark S.J., 1997. Influence of pesticides use on the natural occurence of etomopathogenic fungi in arable soils in the UK. Field and laboratory comparisons. Biocontr. Sci. Technol. 7, 565–575.

20. Miętkiewski R., Tkaczuk C., Zasada L., 1992. Występowanie grzybów entomopatogennych w glebie ornej i łąkowej. Acta Mycol. 27, 197–203.

21. Miętkiewski R., Tkaczuk C., Żurek M., Bałazy S., 1991. Występowanie entomoppatogennych grzybów w glebie ornej, leśnej oraz ściółce. Rocz. Nauk Rol. 21(1/2), 61–68.

22. Oliveira I., Pereira J.A., Quesada-Moraga E., Lino-Neto T., Bento A., Baptista P. 2013. Effect of soil tillage on natural occurence of fungal entomopathogens associated to *Prays oleae* Bern. Sci. Hort. 159, 190–196.

23. Poprawski T.J., Majchrowicz I. 1995. Effects of herbicides on *in vitro* vegetative growth and sporulation of entomopathogenic fungi. Crop Prot. Vol 14. 1, 81–87.

24. Quesada-Moraga E., Navas- Cortes J.A., Maranhao A.A., Ortiz- Urquiza A., Santiago-Alvarez C. 2007. Factors affecting the occurence and distributio of entomopathogenic fungi in natural and cultivated soils. Mycol. Res. 111, 947–966.

25. Rosin F., Shapiro D.I., Lewis L.C, 1996. Effects of fertilizers on the survival of *Beauveria bassiana*. J. Invertebr. Pathol. 68, 194–195.

26. Steenberg T. 1995. Natural occurence of *Beauveria bassiana* (Bals,) Vuill. with focus on infectivity to Sitona species and other insects in Lucerne. Ph.D. Thesis, The Royal Veterinary and Agricultural University. Copenhagen.

27. Strasser H., Forrer A., Schinner F. 1996. Development of media fort the selective isolation and maintenance of viruence of *Beauveria brongniartii*. W: Microbial control of soil dwelling pests. (ed. T.A Jackson and T.R. Glare), AgResearch, Lincoln, New Zeland, 125–130.

28. Tarasco E., de Bievre C., Papierok B., Poliseno M., Triggiani O. 1997. Occurence of entomopathogenic fungi in soils in Southern Italy. Entomol. Bari. 31, 157–166.

29. Tarasco E., Poliseno M. 2005. Preliminary survey on the occurrence of entomopathogenic nematodes and fungi in Albanian soils. IOBC/WPRS Bull. 28(3), 165–168

30. Tkaczuk C. 2008. Występowanie i potencjał infekcyjny grzybów owadobójczych w glebach agrocenoz i środowisk seminaturalnych w krajobrazie rolniczym. Rozprawa naukowa nr 94, 160 s.

31. Tkaczuk C., Krzyczkowski T, Wegensteiner R. 2012. The occurrence of entomopathogenic fungi In soils from mid-field woodlots and adjacent small-scale arable fields. Acta Mycol. 47 (2), 191–202.

32. Tkaczuk C., Majchrowska-Safaryan A., Miętkiewski R. 2013 Wpływ wybranych fungicydów oraz wyciągów glebowych na wzrost owadobójczego grzyba *Metarhizium anisopliae*. Prog. Plant Prot./ Post. Ochr. Roślin, 53(4), 751–756.

33. Tkaczuk C., Miętkiewski R. 1996. Occurrence of entomopathogenic fungi in different kinds of soil. Rocz. Nauk Rol. Seria E 25(1/2), 44–48.

34. Vänninen I., Hokkanen H. 1988. Effect of pesticides on four species of entomopathogenic fungi *in vitro*. Ann. Agric. Fen. 27, 345–353.

35. Vänninen I. 1996. Distribution and occurrence of four entomopathogenic fungi in soil. J. App. Entomol. 2, 213–215.

36. Vänninen I., Husberg G.B., Hokkanen H.M.T. 1989. Occurrence of entomopathogenic fungi and entomopathogenic nematodes in cultivated soils in Finland. Effect of geographical location, habitat type and soil type. Mycol. Res. 100, 93–101.

8

THE PILOT STUDY OF CHARACTERISTICS OF HOUSEHOLD WASTE GENERATED IN SUBURBAN PARTS OF RURAL AREAS

Aleksandra Steinhoff-Wrześniewska[1]

[1] Institute of Technology and Life Sciences, Lower Silesian Research Centre in Wrocław, Zygmunta Berlinga 7, 51-209 Wrocław, Poland, e-mail: aleksandra.sw@gmail.com

ABSTRACT

The subject of the studies were waste generated in suburban households, in 3-bag system. The sum of wastes generated during the four analyzed seasons (spring, summer, autumn, winter – 1 year), in the households under study, per 1 person, amounted to 170,3 kg (in wet mass basis). For 1 person, most domestic waste was generated in autumn – 45,5 kg per capita and the least in winter – 39,0 kg per capita. The analysis performed of sieved composition (size fraction) showed that fractions: >100 mm, 40–100 mm, 20–40 mm constituted totally 80% of the mass of wastes (average in a year). The lowest fraction (<10 mm), whose significant part constitutes ashes, varied depending on the season of year: from 3.5% to 12.8%. In the morphological composition of the households analyzed (on average in 4 seasons), biowastes totally formed over 53% of the whole mass of wastes. A significant part of waste generated were also glass waste (10,7% average per year) and disposable nappies (8,3% average per year). The analysis of basic chemical components of biowastes showed that in case of utilizing them for production of compost, it would be necessary to modify (correct) the ratios C/N and C/P. Analysis of the chemical composition showed that the biowastes were characterized by very high moisture content and neutral pH.

Keywords: household waste, biodegradably waste, rural areas, suburban areas.

INTRODUCTION

In rural areas, constituting 93% of the country's total area and inhabited by 39.8% of the population, about 20% of the total mass of municipal wastes was collected. In the prevailing legal system, the districts are the owners of wastes generated on their terrain. It is therefore advisable to study the current state (quantity and kinds) of municipal wastes so as to plan correctly short- and long-term actions concerning waste management. In the case of rural communities located in the vicinity of the city "subzones" are created. These areas are significantly different in terms of population density and building, land fragmentation, pollution from a typical rural municipality. This diversity undoubtedly creates difficulties in standardizing waste management system in such a rural community. There is only one waste management system in the municipality under current law. Organizing one common system for different areas and one rate for all residents of the municipality is difficult.

The share of municipal wastes collected from households in rural areas, in the total mass of municipal wastes collected in rural localities (villages) amounted to 75.5%. However, there is lack of data regarding the quantity of municipal wastes generated on administrative areas considered as rural. The growth observed in the number of people living in rural areas, presented by Central Bureau for Statistics GUS [The results of National Census Population and Housing 2011] is to a large extent, associated with migrations from large urban centers to outskirts of towns already belonging to administrative areas distinguished as rural. Intensive housing (mostly family), the influx of urban inhabitant related to the city, very large differences in the structure of land use, co-existence of rural and urban settlement forms are features highly distinctive suburban areas. Formed as a result of such migration are informal (not included within administrative borders) districts, and such terrains lose their agricultural character and can significantly influence the characteristics of the rural district.

Due to large differences in households in rural districts directly bordering with large agglomerations, studies were undertaken which were aimed at analyzing the quantities, kinds and possibilities of utilizing the wastes generated in households in such areas.

RESEARCH AND METHODS

The studies were performed in 21 households administratively located in areas of rural districts directly neighbouring Wrocław. None of the analyzed households was engaged in agricultural production. All properties had areas where garden waste were formed during the growing season. Only four households were equipped with backyard composters. Neither one of the analyzed farms could develop biowaste on their property. The subject of the studies constituted all wastes generated in suburban households (with the exception of large-sized and buildings renovated) according to methodology of Jędrczak and Szpadt [Szpadt, Jędrczak, 2006]. The total number of residents in the households amounted to 83. The number of households covered by the studies was as follows: two-person – 3, three person – 3, four-person – 10, five-person – 2, six-person – 3. Selected farms reflect the demographic composition of population in Poland (Population. Status and demographic and social structure). In the study group of people the children under 4 years were 6.0%, older children and adolescents – 30.1%, of working age – 56.6% and seniors – 7,2%.

The analyses were conducted during four consecutive seasons (spring, summer, autumn and winter) and the wastes were collected through 7 consecutive days, accumulated in 3-bag system:
- biowastes – kitchen and garden wastes – viz. wet,
- hygienic and utilized health-protection wastes,
- all other wastes.

The total amount of analyzed wastes was 1148 kg. For the purpose of conducting sieved composition (i.e. granulometric, sized) analysis, the wastes collected were sieved through sieve of mesh size: > 100 mm, 40 mm, 20 mm,

10 mm (sieved fraction < 10 mm was obtained in this way). Next, the wastes were segregated by hand and sorted to obtain 34 material fractions and subfractions (morphological). Division into individual groups and their denotation was performed on the basis of Jędrczak and Szpadt [Szpadt, Jędrczak, 2006] (description in Table 1). In the biowastes accumulated moisture content, pH, Corg, Nog, Pog and ions: Na, K, Ca, Mg, Zn, Cu were determined. The value of pH was measured in water extracts of analyzed samples, by potentiometric method. Organic carbon content was determined by the Tiurin method. Total nitrogen was determined colorimetrically by means of indophenol reaction using spectrophotometer UV/VIS 916 of the firm GBC. Quantity of total phosphorus was established applying the colorimetric method using molybdenum blue. Micro and macro content of the components in organic wastes was determined by using ICP Integra spectrometer of the firm GBC. Quantity of heavy metals in the ashes analyzed was established by ASA method using spectrometer Solaar 6M of the firm Thermo.

RESULTS AND DISCUSSION

The sum of wastes generated during the four analyzed seasons – 1 year, in the households under study, per 1 person, amounted to 170,3 kg (in wet mass basis). The result is very similar to the data of the statistical yearbook of waste collected from household presented by GUS [Environment, 2013]. Research at households at rural areas in Poland conducted by Strzelczyk (2013) showed that, average amount of waste generated per capita is 180 kg per year. In own study for 1 person, most domestic waste was generated in autumn – 45,5 kg per capita, and the least in winter – 39,0 kg per capita (Table 2). According to the data [Burnley, 2007; Hoornweg, Bhada-Tata, 2012] the average production of wastes in countries of OECD amounts to 2.1 kg/person/day, although in extreme cases even up to 14 kg/person/day. In the light of such data, the results obtained from our studies indicate that in such households significantly less wastes is generated than the average in OECD countries [Hoornweg, Bhada-Tata, 2012].

Not without significance is the location of the households analyzed. Statistical data obtained for the whole country may differ considerably from

Table 1. Material structure of size fractions from household wastes – on average for the 4 seasons (spring, summer, autumn and winter)

Material structure of size fractions (%)	Symbol	Fraction				
		>100 mm	40–100 mm	20–40 mm	10–20 mm	<10 mm
Kitchen waste	OR1 01	9.13	44.18	45.25	28.28	13.23
Garden waste	OR1 02	4.82	2.86	47.99	57.90	18.75
Other organic wastes	OR1 03	1.66	0.22	0.84	1.72	5.97
Non-treated wood	W2 01	–	0.03	0.01	0.02	0.01
Treated wood	W2 02	0.03	0.04	0.01	0.03	–
Glittering paper. wallpaper	PC3 01	0.21	0.12	–	–	–
Packaging paper and cardboard	PC3 02	3.50	1.69	0.07	0.15	–
Newspapers and magazines	PC3 03	5.70	0.13			–
Other paper and cardboard non-packaging	PC3 04	1.37	4.34	0.87	1.79	–
Packaging film	PL4 01	1.96	0.86	0.08	0.17	–
Non-packaging film	PI4 02	1.73	1.60	0.05	0.11	–
Other packaging plastic	PL4 04	9.53	5.79	0.58	1.20	–
Other non-packaging plastic	PL4 05	1.08	1.24	0.21	0.44	–
Packaging glass – white	G5 01	5.56	20.43	0.57	1.17	–
Packaging glass – brown	G5 02	0.79	2.43	–	–	–
Other packaging glass	G5 03	3.89	2.86	0.15	0.31	–
Other non-packaging glass	G5 04	1.43	0.44	0.03	0.06	–
Textiles – clothes	T6 01	2.15	0.62	0.02	0.05	–
Textiles – other	T6 02	1.24	0.45	0.05	0.10	
Ferrous metal packaging	M7 01	0.62	2.99	0.19	0.37	–
Non–ferrous packaging	M7 02	0.87	0.82	0.06	0.13	0.01
Other ferrous metal	M7 03	0.31	0.21	0.10	0.22	0.03
Other non-ferrous metal	M7 04	0.20	0.20	0.04	0.08	0.06
Batteries	H8 01	–	–	0.17	0.34	–
Other potentially hazardous	H8 02	–	0.20	0.02	0.01	–
Multi component packaging	C9 01	4.98	3.92	0.36	0.73	–
Non-multicomponent packaging	C9 02	1.59	0.76	0.19	0.40	–
WEEE	C9 03	0.30	0.16	0.21	0.43	–
Ashes		–	0.05	0.97	1.97	59.24
Disposable nappies	U11 01	29.56	–	–	–	–
Identifiable clinical wastes	U11 02	2.30	–	0.03	0.06	–
Other categories	U11 03	3.49	0.60	0.85	1.74	0.58
Fine fraction	F12	–	–	0.01	0.01	1.57
Inert waste	IN1002	–	–	0.01	0.02	0.56
Total	%	100.00	100.00	100.00	100.00	100.00

the results obtained in studies on outskirts or of specific character. According to statistical data, the coefficient of wastes collected in Denmark amounts to 650 kg/person/year, however, according to the studies conducted in Sisimut, the mass of wastes amounts only to 133 kg/person/year [Eisted, Christensen, 2011; Environment, 2013]. In Poland total mass of municipal waste collected in 2012 year was 248,6 kg per capita/year [Environment, 2013]. Estimating the quantity of wastes generated (which is most often performed on the basis of statistical indicators) has basic significance for planning all investments in waste management. It must therefore be based on most reliable data. It should be noted that the waste stream is affected by several elements (e.g. location, financial capability of the residents, cultural factors and demography, season of year, dietary habits, having a pet, waste disposal in their own estate, type of heating fuel used). Seasons obvi-

Table 2. Material composition of wastes generated in suburban households in the seasons

Material composition of wastes in 4 seasons (%)	Spring	Summer	Autumn	Winter
Kitchen waste	27.7	30.9	34.1	39.3
Garden waste	28.6	27.1	13.9	11.3
Wood (total)	0.03	0.04	0.02	0.04
Paper and Cardboard (total)	3.18	6.24	6.56	4.88
Packaging plastic (total)	3.98	5.80	6.30	5.01
Non-packaging plastic (total)	1.68	1.73	2.21	1.06
Galss (total)	11.21	10.07	12.06	9.56
Textiles (total)	1.40	1.12	1.25	1.15
Ferrous waste (total)	0.70	1.15	2.27	1.40
Non-ferrous waste (total)	0.50	0.51	0.96	0.63
Hazardous waste (total)	0.03	0.02	0.42	0.10
Multi component packaging	2.01	2.71	3.24	2.49
Non-multicomponent packaging	0.71	0.62	0.88	0.65
WEEE	0.56	0.04	0.01	0.02
Ashes	6.46	1.51	1.15	10.15
Disposable nappies	7.23	6.13	10.45	9.26
Identifiable clinical wastes	0.38	0.81	0.64	0.72
Other categories	2.76	1.17	0.30	0.09
Fine fraction	0.14	0.33	0.19	0.20
Inert waste	–	0.24	–	0.01
The weight (kg) of waste packaging produced per capita per season	10.9	10.8	8.2	7.3
The weight of waste produced per capita (kg). Avg±SD	44.2 ± 27.6	41.6 ± 17.5	45.5 ± 33.2	39 ± 17.3
Average yearly total weight of waste produced per capita (kg). Avg±SD	170.3 ± 58.8			

* Analyzed wastes made from one kind of material are listed summarily.
Avg – average of analyzed households, SD – standard deviation.

ously influenced the size of furnace and garden waste stream in analyzed households (Table 2).

The studies of wastes from the angle of size composition belong to the basic type. They are significant for planning the technological line for sorting wastes. The largest fraction (>100 mm) contains mainly secondary raw materials (specially packaging), as well as garden wastes (branches, trees), easy for mechanical separation. They mainly constitute combustible wastes of high calorific value and low moisture content. The next important fraction, from the point of view of mechanical sorting, is the group of wastes of size up to 20 mm, since the mesh of sieves of this size forms a practical limit of mechanical segregation. Sieves of lower mesh are blocked by organic material in the wastes. The analysis performed of sieved composition (size fraction) showed that fractions: >100 mm, 40–100 mm, 20–40 mm constituted totally 80% of the mass of wastes (average in a year) (Figure 1). The lowest fraction (<10 mm) whose significant part constitutes ashes, varied depending on the season of year: from 3.5% to 12.8%.

One of the largest problem of mechanical segregation is separating kitchen and garden wastes (viz. biowastes) and polluting other morphological fractions with them. Sorting biowastes "at source" enables obtaining secondary raw materials of better quality and more attractive for processing units. This concerns not only packaging waste paper, etc. It is also very important for further procedures with biowastes towards their biological utilization. Separation of "bio" fractions by residents was practiced mainly in rural or other areas, if the owner of the property had composting facility by the house and wanted to utilize biowastes himself for composting.

In the morphological composition of the households analyzed (on average in 4 seasons), biowastes (kitchen and garden waste) totally formed over 53% of the whole mass of wastes and this value is higher than data from literature. Depending on the applied mode of collecting wastes, the share of this group of wastes in other countries of Europe amounted to approx. 40% [Burnley, Flowerdew, Poll, Prosser, 2007; Gallardo,

Bovea, Colomer, Prades, 2012]. Application of the 3-bag method in the study conducted was significantly more effective for the segregated wet fraction. Seasonal composition of biodegradable part of wastes disposed in the investigated households was variable. The share of biodegradable waste in the whole mass of waste generated in analyzed households was 61% (annual average). The households analyzed did not conduct any cultivation by the house for own requirements. The garden wastes formed came from nurturing procedures of lawns and flower beds. Their share in the mass of biowastes amounted to about 11% to 23%, depending on the season of the year (Figure 2). The main part of biowaste are kitchen waste but seasonal variability, size and way of property organization changes the proportion of biowaste. In our study, the share of garden waste (average per year) was 20% and was similar to the Czech Republic results in Prague [Hanc, Novak, Dvorak, Habart, Svehla, 2011]. It also showed that the weight of kitchen waste per capita per year was 50 kg.

Additionally, detached houses formed approximately 1 kg of garden waste / 1 m² property. Studies conducted by Kotvicova (2010, 17.04.2014) have also shown that biowastes dominated in the total weight of waste generated in households. In addition, the author has shown that the most biowaste were developed in four persons families in multi-family block of flats estates.

Waste packaging is a valuable source of secondary raw materials. The market value of this waste depends largely on the degree of their contamination. Waste segregation "at source" allows to obtain the highest efficiency of separation and least polluted (especially by organic waste). In the analyzed households one person generated about 37 kg of waste packaging per year. Share of packaging waste (glass, paper and cardboard, plastics) in household waste was significant and accounted approximately 22% (average per year). The largest share (by weight) was glass waste (mainly white) due to the properties of this type of packaging material. A group of plastic packag-

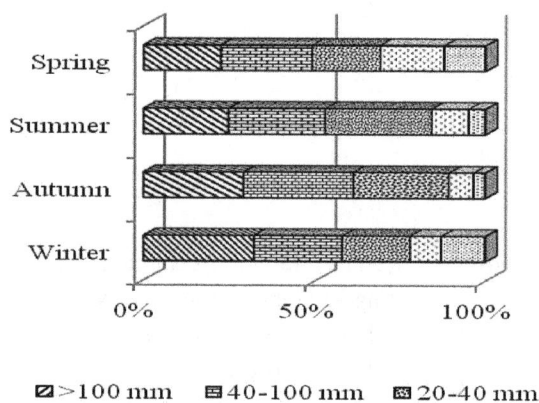

Figure 1. Share of size fractions of wastes from households, depending on season of the year

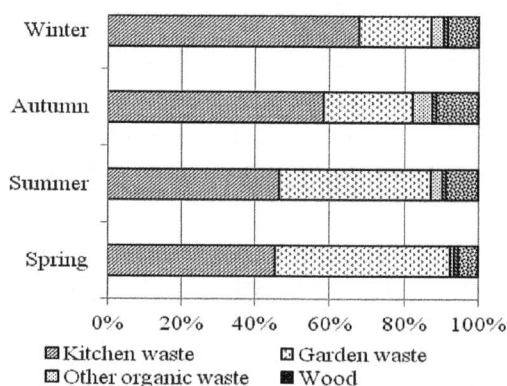

Figure 2. Share of morphological fractions of biodegradable wastes from households, depending on season of the year. The share of wood: spring and autumn: both 0,02%, summer and winter: both 0,04%

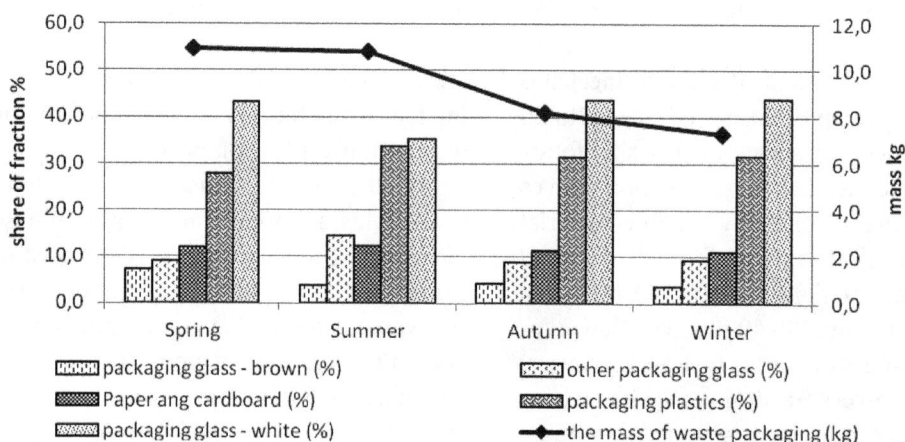

Figure 3. Waste packaging structure (%) and weight (kg per capita*year⁻¹) in household at suburban areas

ing is worth mentioning. Its share in the mass of packaging waste ranged (depending on season) from 28% to 34% (Figure 3). Plastic packaging waste even though their unit of weight is small occupy a very large volume. It is important in the aspect of waste management system designing in the community, the number of containers in the household and the organization of collection and transport of waste.

Analysis of morphological composition of size fractions showed that the largest mass in the fraction >100 mm constituted disposable diapers (about 30%). The second with respect to mass was the group of wastes from plastics (PL4). The fraction >100 mm contained over 14% of them including over 80% of plastic packaging. Paper and cardboard (mainly newspapers) formed about 9% of the mass of this fraction size.

Dominating in the second fraction of size 100-40 mm, were biowastes (47% mass) and glass waste - over 26% mass of the whole fraction. Over 80% of glass waste constituted packaging wastes of white (colourless) glass. The consecutive fractions (40-20 mm and 20-10 mm) almost wholly consisted of biowastes constituting 93.2% and 86.2% of wastes mass, respectively. The remaining groups did not exceed 2% share of the mass of the given fraction. The so-called "under-sieve" fraction, obtained after passing through sieves of the smallest mesh consisted up to almost 60% of ashes and 1/3 of biowastes (Table 1). The ashes can be traditionally used for sprinkling pathways on the household site during winter. From the analyses of morphological wastes (Table 2) it is concluded that irrespective of the season of the year, having the largest share in the mass are biowastes, then glass wastes and used disposable diapers. Due to heating season, ashes indicate seasonal variations. The results of extensive studies of municipal wastes in Wales [Burnley et al., 2007] vary slight. In these studies, the % age share of paper amounted to approx. 23%, share of biowastes was lower, slightly more than 35%, whereas more metal waste. The differences obtained in the results may be caused by different ways of accumulating domestic wastes. In the studies carried out in Spain and in the Balkans, comparing different systems of sorting wastes right at the place of their generation (households), the most effective turned out to be the 4-bag system (similar to that applied on the outskirts of Wrocław), separated in which was the kitchen-garden fraction. This enabled obtain-

ing 37% of biowastes [Gallardo et al., 2007; Vaccari, Di Bella, Vitali, Collivignarelli, 2013]. This, however, is significantly less than the result of our studies.

Thanks to separation of the "bio" fraction from other wastes, dry, clean raw material easy for mechanical segregation is obtained. From the studies of wastes conducted from the areas of suburban rural districts, it was ascertained that on sieves of mesh >100 mm, 97% of newspapers, over 60% of packing bags of foil and 75% of textiles are held back.

In fractions practically impossible for mechanical sorting such as 10-20 mm and <10 mm, they mainly comprise: ashes, inert and minor wastes as well as biowastes. Knowledge of material structures of individual fractions is important from the point of view of the possibility of utilizing the wastes and further procedures with them (Table 3).

The characteristics of wastes obtained in the households analyzed are similar to the data contained concerning municipal wastes of various Polish towns – Kraków (Cracow), Zgorzelec [Den Boer, Jędrczak, Kowalski, Kulczycka, Szpadt, 2010]. Although the data concerning studies in rural areas are from many years back [Skalmowski, Skalmowski, 2006], even then the characteristics of "wastes" from areas bordering with towns were similar to the characteristics of urban wastes.

The analysis of basic chemical components of biowastes showed that in case of utilizing them for production of compost, it would be necessary to modify (correct) the ratios C/N and C/P. The results obtained for C/N (between 13.7–18) and C/P (between 97.9–309.5) depending on the size fraction, do not ensure the proper process of composting [Sebastian, Szpadt, 1999; Czyżyk, Kozdraś, 2004]. The ratios recommended in literature for C/N and C/P of the components composted should not exceed 30:1, 100, respectively, moisture content up to 60% and reaction should be approximately neutral [Kasprzak, 1998]. To correct these ratios, it would be necessary to add sawdust, straw, hay to these wastes (for correcting C/N), and for correcting the ratio C/P – e.g. superphosphate (Table 4).

The analysis of chemical composition showed that the biowastes were characterized by very high moisture content and neutral pH (Table 4). Such high water content is due to the large proportion of kitchen wastes. The analysis of chemical components presented by Kumar et al. [Kumar et al. 2010] confirms high moisture content and

Table 3. Size structure, in individual material groups of wastes from suburban households (with the exception of OR1 01, 02, 03) in %

Size structure. in individual material groups of wastes (with the exception of OR1 01.02.03) (%)	Fraction (mm)					Total (%)
	>100	40–100	20–40	10–20	<10	
Non-treated wood	2.50	37.87	22.96	29.85	6.82	100.00
Treated wood	4.85	32.55	59.54	3.05	–	100.00
Glittering paper. wallpaper	61.54	38.40	0.06	0.00	–	100.00
Packaging paper and cardboard	49.25	49.72	0.98	0.05	–	100.00
Newspapers and magazines	97.66	2.34	0.00	0.00	—	100.00
Other paper and cardboard non–packaging	20.02	66.47	12.24	1.27	–	100.00
Packaging film	67.33	30.15	2.15	0.36	–	100.00
Non-packaging film	49.39	49.83	0.78	0.01	–	100.00
Other packaging plastic	59.32	36.62	3.57	0.49	–	100.00
Other non-packaging plastic	40.18	48.60	8.13	3.08	–	100.00
Packaging glass – white	20.59	77.23	2.12	0.06	–	100.00
Packaging glass – brown	29.12	70.88	0.00	0.00	–	100.00
Other packaging glass	57.72	40.39	1.76	0.14	–	100.00
Other non-packaging glass	68.65	28.57	1.69	1.08	–	100.00
Textiles – clothes	75.76	23.30	0.77	0.18	–	100.00
Textiles – other	55.11	40.42	4.32	0.15	–	100.00
Ferrous metal packaging	16.09	79.26	4.32	0.34	–	100.00
Non-ferrous packaging	46.92	49.01	3.29	0.65	0.13	100.00
Other ferrous metal	23.03	33.49	19.58	23.81	0.09	100.00
Other non-ferrous metal	34.33	48.61	10.53	3.38	3.15	100.00
Batteries	–	0.00	46.05	53.95	–	100.00
Other potentially hazardous	–	58.87	13.37	27.76	–	100.00
Multi component packaging	53.29	42.60	3.56	0.54	–	100.00
Non-multicomponent packaging	60.96	30.76	6.62	1.66	–	100.00
WEEE	12.44	49.47	13.47	24.62	–	100.00
Ashes	–	0.13	8.28	21.56	70.04	100.00
Disposable nappies	100.00	–	–	–	–	100.00
Identifiable clinical wastes	99.02	–	0.98	–	–	100.00
Other categories	22.90	25.00	17.31	34.17	0.62	100.00
Fine fraction	–	–	1.04	13.94	85.01	100.00
Inert waste	–	–	0.64	0.00	99.36	100.00

unfavourable ratio C/N in kitchen wastes which should be modified for the composting process. The contents of elements in biowastes obtained in our studies were similar to the data from literature [Boldrin, Christensen, 2010].

Composting biowastes is the most recommended method of their utilization. Rural areas have the highest possibility in this respect and this method is the cheapest and commonly used in rural households. The situation, however, changes when it concerns suburban areas. The households analyzed in the paper had neither been prepared nor were interested in composting on the area of the property where the biowastes were gen-

erated. Given as a reason was lack of space on the property, organizational difficulties, lack of interest in such a method and absence of the possibility of utilizing the compost obtained on the area of their property. In the context of the results obtained (53% of the total mass of wastes being biowastes), utilization of the biowastes in the place where they were generated would enable reducing the stream of wastes transferred. From the environmental point of view, it is very desirable not only because of decreasing the stream of municipal wastes but also related to the most recommended method of biowastes utilizing (for agricultural purposes). According to various authors

Table 4. Basic chemical parameters of biowastes determining their usefulness for composting (average for the 4 seasons)

Symbol	Unit	Fraction (mm)				
		>100	40–100	20–40	10–20	<10
C_{org}	[mg/g]	319.02	332.18	351.38	366.57	378.48
Organic matter content%	[%]	84.10	78.80	77.80	68.10	77.20
N_{ogK}	[mg/g]	19.12	18.45	22.62	26.66	25.51
P_{og}	[mg/g]	1.92	1.07	2.71	3.75	1.61
C/N	–	16.7	18.0	15.5	13.7	14.8
C/P	–	166.6	309.5	129.8	97.9	235.1
Na	[mg/g]	6.12	3.35	4.46	6.26	5.89
K	[mg/g]	18.96	21.75	16.49	13.25	10.26
Mg	[mg/g]	2.53	2.09	2.02	1.82	1.92
Ca	[mg/g]	47.32	48.46	43.80	26.98	39.49
Zn	[mg/g]	0.0265	0.0499	0.0579	0.0735	0.1206
Cu	[mg/g]	0.021	0.018	0.015	0.025	0.092
pH	–	6.06	6.36	6.37	6.18	6.33
Moisture content	[%]	86.2	76.8	91.3	83.4	73.2

[Burnley, 2007; What a Waste: A global review of solid waste management, 11.04.2014] depending on economical conditions of the society, the proportion of this group of wastes amounts from 20% to 75% of the total mass of waste.

It is therefore possible to reduce the mass of wastes directed for utilization by such a quantity in special installations and prevent pollution of the remaining wastes which must be transferred to the authorized party, in accordance with the prevailing regulations. Encouraging the residents for biowastes composting on their own property area enables to reduce significantly the costs of biowastes management system and it is very desirable for environmental reasons. The segregated biowastes from the whole stream enables obtaining other kinds of wastes in relatively clean condition facilitating further processing and making them more attractive commercially. Rural areas create very high possibilities for utilizing composts. If however there are regions in rural areas which differ considerably by their character from rural characteristics, detailed data should be obtained relating to wastes generated in specific types of structures and the arrangement or proportion of individual environments on the terrain of the district should be specified.

If the residents do not envisage the possibility of composting wastes in the vicinity of the household property, it is essential to study the chemical characteristics of biowastes (their various components) to assess the environmental benefits, management of such wastes and the possibility of their utilization.

Not all composition biowastes can be composted in domestic conditions and the usefulness of individual components for methods of biological utilization varies.

CONCLUSIONS

1. Municipal wastes generated in households on suburban areas of rural districts have the character of wastes generated in urban households.

2. Biodegradable wastes constitute over half the mass in the stream of wastes generated in suburban households.

3. Biowaste segregation and their composting by those generating them (residents) in the place where they are created reduces the stream of wastes transferred from individual households by about 50%. Separating the organic fraction of waste, which is suitable for composting and reusing on farm can effectively reduce the amount of waste going to landfill.

4. The share of waste packaging mass in the mass of waste generated in households in suburban areas accounted about 22%.

5. Waste segregation „at source", especially the separation of the organic fraction of waste stream, allows to obtain secondary raw materials of high quality parameters.

Acknowledgements

The studies were carried out within the framework of the long-term program for the years 2011–2015 "Standardization and monitoring of the environmental undertakings, agricultural technique and infrastructural solutions for safety and balanced development of agriculture and rural areas" – Resolution No. 202/2011 of the Council of Ministers dated 14 October 2011.

REFERENCES

1. Boldrin A., Christensen T.H. 2010. Seasonal generation and composition of garden waste in Aarhus (Denmark). Waste Management 30, 551–557.

2. Burnley S.J. 2007. A review of municipal solid waste composition in the United Kingdom. Waste Management 27, 1274–1285.

3. Burnley S.J., Ellis J.C., Flowerdew R., Poll A.J., Prosser H. 2007. Assessing the composition of municipal solid waste in Wales. Resources, Conservation and Recycling 49, 264–283.

4. Central Bureau for Statistics (GUS) 2013. Environment. Warsaw.

5. Central Bureau for Statistics (GUS) 2012. The results of National Census Population and Housing. Warsaw.

6. Central Bureau for Statistics (GUS) 2013. Population. Status and demographic and social structure. Warsaw.

7. Czyżyk F., Kozdraś M. 2004. Chemical properties and composting of sludge from rural wastewater treatment plant. Water -Environment- Rural Areas. 4, 2a (11), 559–569 (in polish).

8. Den Boer E., Jędrczak A., Kowalski Z., Kulczycka, Szpadt R. 2010. A review of municipal solid waste composition and quantities in Poland. Waste Management 30, 369–377.

9. Eisted R., Christensen T.H. 2011. Characterization of household waste in Greenland. Wast Management 31 (7), 1461–1466.

10. Gallardo A., Bovea M., Colomer F.J., Prades M. 2012. Analysis of collection systems for sorted household waste in Spain. Waste Management 32, 1623–1633.

11. Hanc A., Novak P., Dvorak M., Habart J., Svehla P. 2011. Composition and parameters of household bio-waste in four seasons. Waste Management 31, 1450–1460.

12. Hoornweg D., Bhada-Tata P. 2012. What a Waste, A Global Review of Solid Waste Management. Urban Development Series Knowledge Papers. (15), 3-116. From http://www-wds.worldbank.org/external/default/WDSContentServer/WDSP/IB/2012/07/25/000333037_20120725004131/Rendered/PDF/681350WP0REVIS0at0a0Waste20120Final.pdf [access: 14.04.2014].

13. Kotvicova J. 2010. Analysis of household waste composition and segregation in Blansko city. Infrastructure and ecology of rural areas. 8/2, 117-126. From http://www.infraeco.pl/pl/art/a_16052.htm?plik=879 (17.04.2014)

14. Kasprzak K. 1998. Theoretical and practical requirements for organic waste composting. [In:\ Municipal Enterprise and the Environment. IX Conference Poznań, Abrys, 125–238 (in polish).

15. Kumar M., Ou Y-L., Lin J-G. 2010. Co-composting of green waste and food waste at low C/N ratio. Waste Management 30, 602–609.

16. Sebastian M., Szpadt R. 1999. Individual composting of garden biowaste at home composting facility. In: Composting and disposal of compost. I Technical Sciences Conference. Puławy-Warsaw. IOŚ, IUNG, PTIE, 117–124 (in Polish).

17. Skalmowski K., Skalmowski A. 1996. Types of waste at rural areas and small towns. Ecological Guide for Local Government, 2 (in Polish).

18. Strzelczyk M. 2013. Structure and properties of municipal wastes from rural households in the aspect of the possibilities of their processing and legal regulations. Water-Environment-Rural Areas, Monography, 113, ITP Falenty (in Polish).

19. Szpadt R., Jędrczak, A. 2006. Research methodology to determine the composition of the sieve, morphological and chemical waste. (in Polish). From http://www.mos.gov.pl/g2/big/2009_04/0406c54fa55c86fb5c4548561bc4bce6.pdf [access: 11.04.2014].

20. Vaccari M., Di Bella V., Vitali F., Collivignarelli C. 2013. From mixed to separate collection of solid waste: Benefitsfor the town of Zavidovic´i (Bosnia and Herzegovina). Waste Management 33, 277–286.

9

THE PRESENT CONDITION OF SMALL WATER RETENTION AND THE PROSPECTS OF ITS DEVELOPMENT USING THE EXAMPLE OF THE PODLASKIE VOIVODESHIP

Joanna Szczykowska[1], Anna Siemieniuk[1]

[1] Faculty of Civil and Environmental Engineering, Bialystok University of Technology, Wiejska 45b, 15-351 Białystok, Poland, e-mail: j.szczykowska@gmail.com

ABSTRACT

The necessity and purposefulness of the investments related to water retention are justified mostly due to the preservation of the environment equilibrium as well as due to its farming, anti-flood, landscape and recreation aspects. Reasonable water management where various forms of retention are used gives large chances for the mitigation of the effects of unfavorable phenomena related to its insufficient amount. The creation of plans regarding the formation of reservoirs accumulating water is not necessarily synonymous with their realization. The reason of problems connected with the implementation of plans regarding the formation of new reservoirs lies mainly in financial measures and in problems with obtaining them. Water deficit in Poland is the reason for which the principles of its national usage need to be complied with. Realization of plans at both Voivodeship and municipality level that are focused on small retention will contribute to considerable increase in the retention capacity and will enable considerable increase in available resources in hydrographic catchments of both the characterized area and the entire country. The paper presents the characteristics of the present state and assumes the perspective development of small water retention in the Podlaskie Voivodeship using the example of the Podlaskie Voivodeship.

Keywords: small water retention, water reservoirs, small retention programs, Podlaskie Voivodeship.

INTRODUCTION

Small water retention means accumulation of water in small reservoirs as the consequence of retardation or deceleration of water flow in the situation of simultaneous conservation and diversification of the natural landscape. In low-lying areas small retention denotes mainly increasing the possibilities of retention and also counteracting drought and flood. In mountainous areas the aim of small retention is counteracting the effects of outflow of precipitation waters. The positive impact of small retention reactions which cause increase in the retention possibilities of small catchments on the water balance does not raise any doubts and is widely accepted. However, its role in shaping global water resources on the country level is slight, whereas as the type of non-steerable retention it has small significance for anti-flood protection. Its asset is the totality of local impacts which have considerable influence on farming, forestry and natural environment.

Apart from the improvement of the water balance of catchments, small retention objects perform also various economic functions, mostly on the local level, such as: small water power engineering, breeding of fish, source of water irrigations in farming and forestry and also in recreation. Not only do catchments serve accumulation of water, but they also have impact on increase in the retention of surface water in adjoining areas and perform numerous functions related to the environment and nature. The main functions include: improvement of water quality, reduction in the water erosion of soils and watercourses, increase in biological diversity and also the improvement of landscape values and microclimatic conditions. One of the possibilities of having rational water management of the region are development programs devoted to small retention for a certain area, Voivodeship or river catchment.

Small retention programs constitute the basis for undertaking complex investment activities in planning and realization of activities which can be carried out using a relatively small number of measures and the purpose of which is to construct various forms of small retention. The realization of the assumptions of small retention development programs is the task for Voivodeship self-governments and also for the Regional Water Management Boards and for the Boards of Melioration and Water Systems of particular regions. However, the activities related to the shaping of water resources are unfortunately not the investments that yield income in a direct way.

WATER RESOURCES OF PODLASKIE VOIVODESHIP

The resources of surface waters in the Podlaskie Voivodeship include running water of rivers, canals and, periodically, also of ditches as well as lakes, reservoirs, fish-breeding ponds, swamps and marshy areas. The river system in Podlaskie Voivodeship is well developed, yet its configuration is mainly the result of the landform and is related to the Narew river. The occurrence of groundwaters is, to a large degree, dependent on the structure and origin of rocks, hypsometry, deposition of scarcely permeable layers and on the granulation. The most hydrated areas are the river basins of Narew and Biebrza [1]. The Northern part of the Voivodeship is abounding with natural reservoirs of stagnant waters. The total number of lakes in the following three districts: Augustowski, Suwalski and Sejneński is 250. Water systems enrich ponds, canals and dam reservoirs, among which Siemianówka reservoir is the largest one [2]. The remaining area does not have large reservoirs of stagnant water. The largest canal in the Voivodeship is Augustowski canal (more than 100 km of length), which connects Augustowskie lake district with Biebrza river. At present, melioration facilities in Podlasie require large financial outlays in order to function properly. However, most facilities need to be modernized or exchanged for new ones [3].

Extreme phenomena, which are caused by water factors, such as flood or drought are of natural type. Rising watercourses or periods with small atmospheric precipitation and low flows are nothing new in nature. It is only in combination with man's activity related to improper water management and when natural changeability of water resources in space and time is taken into consideration that real dangers can appear [4]. The prevention of drought effects as well as anti-flood protection are the tasks for both governmental and local administration. One of the basic prevention activities focused on the protection from such dangers are retention reservoirs. They are becoming a crucial, yet frequently underestimated element both in the situation of drought and flood. The key purpose of forming retention reservoirs, irrespective of various assumptions, will always be the expansion of water resources. The most important tasks of small retention include also:

- formation of groundwaters by increasing their surface, which results also in larger dampness of habitats,
- regulation and control of water circulation in the environment,
- increase in the amount of water by means of storing it, i.e. protection and renewal of water resources,
- water supply for people,
- obtainment of energy,
- flood protection through taking over a certain capacity of flood wave and decelerating its rapid travel down the river,
- fire protection,
- usage as a watering place for both wild and farm animals,
- improvement of aesthetic, landscape and ecological qualities of the environment,
- stoppage of pollutions in typically farming catchments,
- recreation and decoration functions,
- breeding of fish or waterfowl, i.e. farming functions [5].

Poland's accession to the European Union has changed the approach to the formation of large and small water reservoirs. As the basic purpose there was taken water management at the source of forming water resources, i.e. in forests and on farmlands. It has crucial impact on the quality and quantity of these resources. The experience gained till that moment shows that they did not ensure good results at all times [6]. For that reason it is necessary to implement new methods in the management of water resources, whereas the shift in the approach to this issue was caused by Poland's entry into the EU. The new approach to water management ensues from the resolutions of the "Water Law" act as well as from the recommendations of the Framework EU Water Directive. Both documents focus on the necessity of water management at the level of a particular catchment as well as on the aspiration to achieve "good ecological status of waters" – as the main aim of the Framework Water Directive of the European Union [7].

The actions taken in connection with small retention constitute one of the elements that meet the requirements of the Water Directive. The obtainment of good status of waters and the improvement of water quality and natural amenities is possible only as regards the development of small retention.

THE PRESENT STATE OF SMALL RETENTION IN SELECTED MUNICIPALITIES OF THE VOIVODESHIP

In Podlaskie Voivodeship one of the largest water area in Poland, Siemianówka reservoir is located. The crucial aspect regarding retention structures is the economic profitability of projects. Numerous economic analyses conducted in this domain reveal considerable advantage of small investments, such as "small retention" over large retention reservoirs. In accordance with the data presented by the Coordination Centre for Environmental Projects the cost of storing 1 m³ of water in large reservoirs oscillates around 15–40 PLN, whereas in case of small retention reservoirs the cost is only 2–5 PLN. Not only does "small retention" have considerable advantage in economic terms, but it also has many merits in environmental aspect which is difficult to estimate [8].

In Podlaskie Voivodeship the total area of flooded areas amounts to approximately 175 km³. The number of households that are exposed to flood and the owners of which could possibly be evacuated is 654 and 88 buildings in the urban areas, i.e. 2670 people (230 from towns and 2440 from the countryside) [9]. In accordance with the prepared plans regarding the enlargement of small retention reservoirs in the Podlaskie Voivodeship, 12 reservoirs have been completed since 1995.

The 22 reservoirs described below include not only 12 reservoirs which have been completed in the recent years, but also other 10 reservoirs which were formed in the previous years. Retention reservoirs are not located in a uniform manner. Most of them are localized in the southern and central part of the Voivodeship . The approximate locations of retention objects in the Voivodeship were presented in Figure 1. Consecutive points on the map correspond with certain retention reservoirs located in the Voivodeship:

1. Complex of retention reservoirs in Siemiaty-cze.
2. Small water retention reservoir Grodzisk.
3. Water reservoir in Ciechanowiec.
4. Small water retention reservoir Otapy-Kier-snówek.
5. Small water retention reservoir Repczyce.
6. Water reservoir in the village named Lady.
7. Small water retention reservoir Trywieża.
8. Small water retention reservoir Narewka.
9. Water reservoir Siemianówka.
10. Small water retention reservoir Michałowo.
11. Small water retention reservoir Zarzeczany.

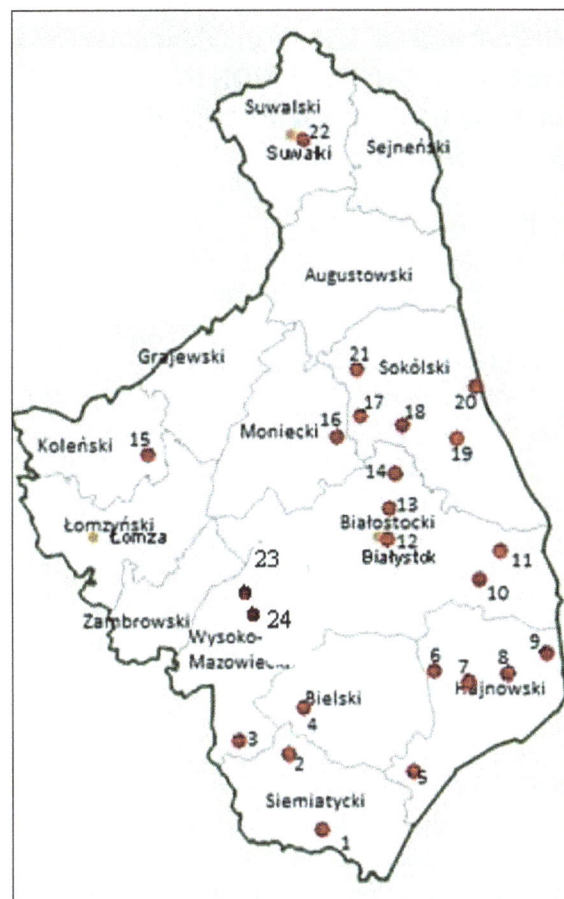

Figure 1. The map of Podlaskie Voivodeship presenting the locations of retention reservoirs

12. Dojlidy creek.
13. Wasilków creek.
14. Small water retention reservoir Czapielówka.
15. Small water retention reservoir Stawiski.
16. Small water retention reservoir Jasionówka.
17. Small water retention reservoir Korycin.
18. Small water retention reservoir Sitawka.
19. Creek in Sokółka.
20. Small water retention reservoir Kuźnica.
21. Small water retention reservoir Suchowola.
22. Arkadia creek in Suwałki.

Most of the aforementioned reservoirs were formed after 1995 because in that year there was signed the agreement regarding the plans of undertaking actions including the reconstruction of old and formation of new small retention reservoirs in particular Voivodeship s of Poland. It was at that time that in the Podlaskie Voivodeship the following reservoirs were either formed or renewed: Zarzeczany, Sitawka, Grodzisk, Jasionówka, Kuźnica, Repczyce, Korycin, Suchowola, Trywieża, Narewka, Michałowo and Otapy-Kiersnówek.

PERSPECTIVES OF THE DEVELOPMENT OF SMALL WATER RETENTION IN SELECTED MUNICIPALITIES OF THE VOIVODESHIP

In Podlaskie Voivodeship there has been observed in the recent years an increasing demand for the investments that contribute especially to larger attractiveness of farmland areas and to larger activity of municipalities in terms of tourism, which is undoubtedly caused by the formation of small retention reservoirs. So far it has been decided that in our Voivodeship 12 objects of small water retention will be formed or reconstructed. These include: reservoir "Suchowola" in the municipality Suchowola, "Bobra Wielka" in the municipality Nowy Dwór, "Szumowo-Olszanka" in the municipality Korycin, "Dzierzbia" in the municipality Stawiski, "Choroszcz" in the municipality Choroszcz, "Turość" in the municipality Turość Kościelna, "Kalinowo Solki" and "Wnory Wiechy" in the municipality Kulesze Kościelne, "Czyżew" in the municipality Czyżew- Osada, "Hajnówka" in the municipality Hajnówka, "Studziwody" in the municipality Bielsk Podlaski and "Chanie-Chursy" in the municipality Nurzec-Stacja (Figure 2).

Small water retention "Suchowola" will be located in Suchowola, in the municipality Suchowola, on the Olszanka river. The information obtained from the municipality office entails that this will be a water region having the area of 1ha and the depth of 1.5 m. The total cost of the investment is estimated at 1 500 000 PLN, the completion of the enterprise is planned for the year 2014.

Another planned investment is the small water retention reservoir "Bobra Wielka" which will be situated in Bobra Wielka, in the municipality Nowy Dwór, on the Biebrza river. The actions are planned to include the reconstruction of the retention reservoir in the location of the existing carp ponds. The water area is planned to have the water surface of 13.5 ha and the retention capacity of 130 000 m^3 of water. The completion date of the enterprise realization: September 2014.

A subsequent small retention reservoir will be situated in Szumowo-Olszynka, in the municipality of Korycin, on the Kumiałka river. The water area is supposed to have the total area of 6 ha. The completion of the enterprise realization was planned for the end of 2013, yet it was postponed for one year.

Figure 2. The map of Podlaskie Voivodeship presenting the locations of planned retention reservoirs

During the realization of the tasks aiming at the improvement of water management in the municipality Stawiski in 2010 it was decided to create a project involving the reconstruction of a small water retention reservoir on the Dzierzba river in Dzierzba (the eastern part of the municipality). The reservoir will be located approximately in the section 5.194–5.432 km of the river. The plan of the reservoir reconstruction entails the formation of a man-made instantaneous reservoir having the area of 2.35 ha and the maximum capacity of 46 650 m^3 and it is still in the process of realization.

Another small retention reservoir is planned for the town Choroszcz, in the municipality Choroszcz, its planned area will be 5.5 ha and the capacity – 99 000 m^3 of water, and it is intended to be formed on the Horodnianka river. However, the date of the enterprise completion has still not been specified due to the fact that there is lack of financial measures necessary for purchasing the land required for the investment.

Turośń Kościelna, in the municipality of Turośń Kościelna, is planned to be the location for forming a reservoir having the area of 8.2 ha and the capacity of 94 500 m^3 of water. The investment realization started in 2000 and the completion was scheduled for 2012, but it was postponed for the turn of the years 2014/2015.

With the object of improving the aesthetics and landscape amenities of the municipality Kulesze Kościelne there were created two projects regarding the reconstruction and organization of water reservoirs in the following villages: Wnory Wiechy and Kalinowo Solki. The village Kalinowo Solki is situated in the eastern part of the municipality Kulesze Kościelne. The plot on which the actions related to the investment will be conducted is in the possession of the municipality and occupies the area of 23 000 m^2, where the area of stagnant water is 9400 m^2. Presently there is located an instantaneous trench overgrown with plants and filled with a small amount of water. Another object designed for renovation is the reservoir in the village Wnory Wiechy located several kilometers south from the village Kulesze Kościelne. Similar to the village Kalinowo Solki, the investment area is the possession of the Municipality Office. The total area of the object is 0.55 ha, where the water surface area is 2500 m^2. It is difficult to specify precisely the completion date of the enterprise realization because presently tenders are being organized for renovation works.

Small retention reservoir will be situated in Czyżew-Osada, in the municipality of Czyżew-Osada along the right side of the Brok river. The water area is intended to have the area of 2.43 ha and the maximum retention capacity of 33 973 m^3 of water. The investment is in the process of realization.

In Hajnówka in the municipality Hajnówka there is a melioration ditch in the fragmentary catchment of the Leśna Prawa river which is planned to become the location for the formation of a small retention reservoir. The water area is supposed to have the area of 5,1 ha and the maximum retention capacity of 106 600 m^3 of water. The completion date of the enterprise realization has not still been specified.

What is more, for the town Bielsk Podlaski in its district Studziwody, in the municipality Bielsk Podlaski there is a plan to form another small retention reservoir – on the right side of the Biała river. The water area is supposed to have the water surface of 3.06 ha and the maximum retention capacity of 57 000 m^3 of water. The planned spatial concept of forming the retention reservoir was prepared and positively assessed in 2006, but due to the lack of funds the investment is unlikely to be realized.

A small water retention reservoir will be situated in the village Chanie – Chursy, in the municipality Nurzec Stacja in the catchment of Nurzec river. The water area is planned to occupy the area of 1.1 ha and have the retention capacity of 13 920 m^3 of water.

SUMMATION

The necessity and purposefulness of the investments related to water retention are justified mostly due to the preservation of the environment equilibrium as well as due to its farming, anti-flood, landscape and recreation aspects. Reasonable water management where various forms of retention are used gives large chances for the mitigation of the effects of unfavorable phenomena related to its insufficient amount.

Human activity, infrastructure development and elimination of natural water objects caused by earmarking areas for other purposes contributed to considerable acceleration of surface confluence and to reduction in natural retention.

The available resources of Podlaskie Voivodeship are sufficient to meet the basic needs of industry, municipal economy and farming, however, they are not equally set in time and space perspective. The activities regarding small water retention were undertaken mainly with focus on reservoir retention, whereas less attention was paid to melioration ditches or to increase in the natural soil retention. Retention reservoirs which are planned for the Voivodeship are supposed to perform a number of functions including mainly those related to farming, recreation and landscape. The more and more frequent phenomenon of drought causes deficiencies of water for both farming and population and small retention will enable its reserves in the periods of low amounts. It is of particular importance when into consideration is taken the fact that Podlasie is the area of primarily agricultural use.

So far there have been formed 11 plans regarding the construction and modernization of small retention objects. Most of them were set in the Southern part of the Voivodeship. The reason of such situation is deficiency of stagnant water in this area when compared with the Northern part.

In the following districts: Suwalski, Sejneński, Augustowski, Grajewski, Moniecki, Łomżyński and Zambrowski, in accordance with the statements of the municipality workers responsible for spatial management and water management, there is no need to modernize or construct small retention objects. At the same time in the following districts: Sokólski, Kolneński, Białostocki, Wysokomazowiecki, Bielski, Hajnowski and Siemiatycki it has already been planned to either renovate or construct new objects.

The creation of plans regarding the formation of reservoirs accumulating water is not necessarily synonymous with their realization. The reason of problems connected with the implementation of plans regarding the formation of new reservoirs lies mainly in financial measures and in problems with obtaining them.

In order to expand the available water resources of good quality it is necessary to manage properly the water resources in smaller hydrographic catchments. Complex retention of surface waters, mainly in rural areas, has many advantages as it enables more universal usage of water resources without causing changes in the natural water regime. Small retention introduces only corrections which improve water balance but do not cause violation of the ecosystem biological equilibrium.

Simultaneous improvement of water quality and landscape amenities aims not only at attracting tourists and focusing on agrotourism, but it also is aimed at improving the living standard of the local community.

Water deficit in Poland is the reason for which the principles of its national usage need to be complied with. Realization of plans at both Voivodeship and municipality level that are focused on small retention will contribute to considerable increase in the retention capacity and will enable considerable increase in available resources in hydrographic catchments of both the characterized area and the entire country.

CONCLUSIONS

1. In Podlaskie Voivodeship the establishment of 12 small retention reservoirs has been planned by the year 2015.
2. Realization of the investments causes expansion of the area of stagnant waters in the Voivodeship by approximately 47 ha.
3. Presently two objects are being realized: Turośń Kościelna (municipality Turośń Kościelna) and Czyżew Osada (municipality Czyżew Osada).
4. In the municipality Hajnówka (town Hajnówka) the realization of the investment came to a halt at the stage of coming into the possession of the lands allocated for the formation of the reservoir.
5. In the municipality Suchowola (town Suchowola) and Korycin (town Szumowo–Olszanka) there are continued works on preparing complete documentation.
6. For the following objects: Wnory Wiechy and Kalinowo Solki (municipality Kulesze Kościelne) tenders for renovation works are presently organized.
7. In the municipality Nowy Dwór, in the village of Bobra Wielka the works connected with the realization of the investment were delayed for technical reasons. There is concern whether the dyke left there and separating the upper and lower water area will stand the water pressure.

REFERENCES

1. Program małej retencji wodnej dla województwa białostockiego, Bipromel, Warszawa 1996, udostępnione przez Wojewódzki Zarząd Melioracji i Urządzeń Wodnych w Białymstoku.

2. Prognoza oddziaływania na środowisko regionalnego programu operacyjnego województwa podlaskiego 2007–2013, udostępnione przez Urząd Marszałkowski Województwa Podlaskiego.

3. Program ochrony środowiska województwa podlaskiego na lata 2007–2010, udostępnione przez Instytut Zrównoważonego Rozwoju

4. Mioduszewski W. 1999. Ochrona i kształtowanie zasobów wodnych w krajobrazie rolniczym, IMUZ, Falenty.

5. Szczykowska J. 2007. Mała retencja wodna na obszarze województwa podlaskiego, Samorząd w procesie rozwoju regionów Polski wschodniej, red. B. Plawgo, WSAP, Białystok.

6. Mioduszewski W. 1999. Mała retencja. Ochrona zasobów wodnych i środowiska naturalnego. Poradnik, Wydawnictwo IMUZ, Falenty.

7. Mioduszewski W. 1999. Ochrona i kształtowanie zasobów wodnych w krajobrazie rolniczym, Wydawnictwo IMUZ, Falenty.

8. Mioduszewski W. 1997. Mała retncja a ochrona zasobów wodnych. Gospodarka wodna, no. 3.

9. http://www.straz.bialystok.pl

10

MŚCIWOJÓW RESERVOIR – STUDY OF A SMALL RETENTION RESERVOIR WITH AN INNOVATIVE WATER SELF-PURIFICATION SYSTEM

Jolanta Dąbrowska[1], Olgierd Kempa[2], Joanna Markowska[1], Jerzy Sobota[3]

[1] Institute of Environmental Engineering, Wrocław University of Environmental and Life Sciences, Plac Grunwaldzki 24, 50-363 Wrocław, Poland, e-mail: jolanta.dabrowska@up.wroc.pl; joanna.markowska@up.wroc.pl

[2] Department of Spatial Economy, Wrocław University of Environmental and Life Sciences, Plac Grunwaldzki 24, 50-363 Wrocław, Poland, e-mail: olgierd.kempa@up.wroc.pl

[3] Institute of Building, Wrocław University of Environmental and Life Sciences, Plac Grunwaldzki 24, 50-363 Wrocław, Poland, e-mail: jerzy.sobota@up.wroc.pl

ABSTRACT

The study presents the characteristics of the Mściwojów Reservoir equipped with a unique pre-reservoir structure that supports the process of self-purification of waters. The authors present more than ten years of studies, focusing mainly on the issues of water quality and the concentration of phosphorus, which is considered as the main factor influencing water eutrophication process. Quite a high concentration of phosphates was noted in the outflow from the main reservoir, in spite of a lower concentration of these compounds in the water leaving the pre-reservoir. Basing on the conducted analyses, the catchment of the reservoir was qualified as group 4, being very prone to the movement and supply of material to the reservoir. The negative value of the retention coefficient of phosphorus obtained for the main part of the reservoir points to the existence of an internal source of phosphorus supply to the reservoir. During over 10 years of studies, new directions of the development of the rural areas were determined. Future works should be extended so as to cover all elements of the ecosystem of the reservoir. In a longer term it seems natural to extend the research works to cover the whole catchment of Wierzbiak River.

Keywords: surface water quality, eutrophication, dam reservoir, phosphorus retention.

INTRODUCTION

The retention reservoir Mściwojów was constructed in the end of the 1990's, in Przedgórze Sudeckie (Sudetian Foothills), as a result of the involvement of local community. The design of the object was a result of joint works of the employees of the Institute of Environmental Engineering of the former Agricultural University of Wrocław (now the Wrocław University of Environmental and Life Sciences) and of the design studio Water Service in Wrocław. The investor and administrator is the Lower Silesian Board for Amelioration and Water Management (DZMiUW). The Mściwojów Reservoir was designed mainly for the purposes of water supply for agriculture and flood protection. Presently, the reservoir is also used for fishing and recreational purposes.

CHARACTERISTICS OF THE MŚCIWOJÓW RESERVOIR

Mściwojów is a dam reservoir located on the Wierzbiak River (Figure 1), which is also supplied by the Zimnik watercourse. The shares of Wierzbiak and Zimnik in the total river supply is, respectively: 71% and 29% [Wiatkowski et al. 2006]. Wierzbiak is a right-side tributary of Kaczawa. The catchment area to the cross-section of the reservoir A = 47 km². This is a submontane area of an average inclination of 2.26% and forest coverage below 10%. It is a typical agricultural region, which is confirmed by the land usage structure. Agricultural lands account for over 80% of its area, of which more than 60% are arable lands. Due to quite good quality of soils (soils of the 1st, 2nd and 3rd class), the sowing structure is dominated by cereals followed by root plants.

The proper operation of this multi-purpose reservoir in a typically agricultural catchment was to be ensured by the innovative concept of usage, consisting in the separation of a pre-reservoir with a sedimentation tank (Figure 1).

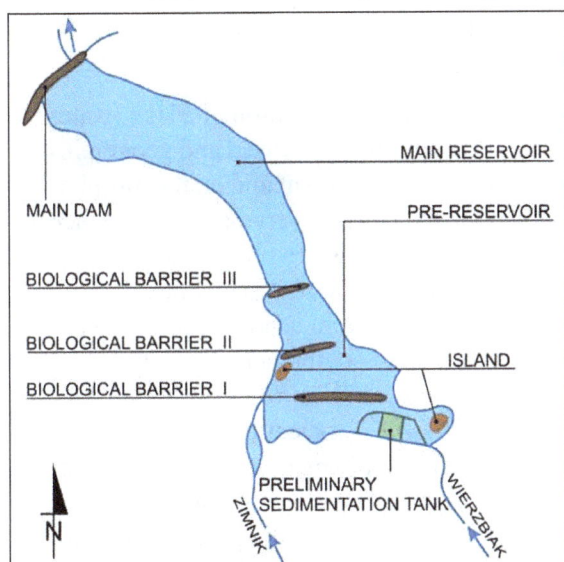

Figure 1. Mściwojów Reservoir

The main function of the reservoir is the storage of water for agricultural purposes, which means that the water should be retained during the season preceding plant vegetation. The operational capacity designated for securing water for agriculture is 700 000 m³.

The design of the reservoir also foresees the possibility to use the object for the production of electric energy. Due to a small volume of flow and low hydraulic head (approx. 3.5 m), the design of discharge facilities took into account the installation of the smallest turbo generators available at that time. The power generating function mentioned above has not been realised so far.

For the local community it is important to increase the attractiveness of the region, because, in the opinion of local inhabitants, this object should have a positive influence on economic growth through the development of its tourist and recreational functions as an additional, non-agricultural activity in the region. This task of the reservoir requires proper water management in the object and ensuring a constant level of elevation along with good water quality.

A detailed specification of the parameters of the reservoir and technical data is presented in Table 1 [Zbiornik... 1995].

Individual designed elements of the reservoir, such as the dam (Figure 2), the bowl and the sediment tank create a functional structure, integrated with the landscape of the Sudetian Foothills.

The bowl of the reservoir is the valley of Wierzbiak and Zimnik (Rakowiec). The average length of the reservoir along the axis is approx. 1600 (depending on the elevation), and the maximum approx. 2250 m. The area of normal elevation is limited by the isohypse 193.35 m above sea level and the maximum elevation by the isohypse 194.50 m above sea level. The length of the shoreline at normal elevation level is approx. 5.3 km, total length, including islands – 5.6 km.

The innovative structure, mutually exclusive functions and the resulting implications make this reservoir an object of interest in the field of scientific research.

In the years 2000–2007, Wiatkowski and Kasperek conducted studies on the quantity and quality of sediments. The results allow us to determine that the content of phosphorus in the res-

Table 1. Basic parameters of the Mściwojów Reservoir

Item	Parameter	Symbol	Unit	Amount
1	Mean annual flow (inflow to the reservoir)	SQ	m³/s	0.20
2	Mean annual outflow	ΣQ	mln m³	6.30
3	Surface of reservoir at normal storage level (193.35 m.a.s.l.)	F_z	ha	34.59
4	Maximum flood surface	$F_{z\,max}$	ha	57.07
5	Water surface elevation	H	m	7.50
6	Total capacity	V_c	mln m³	1.35
7	Useful capacity	V_u	mln m³	0.713
8	Dead storage	V_m	mln m³	0.024
9	Reserve storage obtained as a result of acceptable elevation	R_f	mln m³	0.61

Figure 2. Part of the earth dam and spillway of the Mściwojów Reservoir (photo: J. Dąbrowska)

ervoir sediments was not high, and the content of heavy metals remained in the range of the 1st and 2nd class of water purity. Moreover, the study confirmed that the sediment tank and pre-reservoir were effective in retaining the fine fraction of sediments [Wiatkowski 2006, Wiatkowski and Kasperek 2008]. Mokwa and Pikul [2006] analysed the influence of immersed plants on the concentration of the suspended sediments, confirming their beneficial influence on the deposition of sediments in the tank.

The reservoir has a positive influence on water management in the upper part of the Wierzbiak catchment, as it influences the equalisation of flow and the stabilisation of ground water level [Szafrański and Stefanek 2008].

MATERIAL AND METHODOLOGY

The source data for analysis were the results of research that has been conducted in the analysed area since 1999 by scientists from the Wrocław University of Environmental and Life Sciences, the University of Agriculture in Krakow, the Opole University, the University of Environmental and Life Sciences in Poznań. The design documentation of the reservoir has also been used, along with the results of field studies and analyses from the years 1999–2013, conducted by the authors of this article.

The evaluation of the catchment as a supplier of biogenic compounds was conducted with use of the method developed by Bajkiewicz-Grabowska [2002]. This method is a part of the comprehensive evaluation system of the catchment as a supplier of matter and the lake as the receiver. The evaluation system is based on the assumption that

the pace of natural eutrophication of a lake (water reservoir) depends on the physical and geographical structure of the catchment and on the morphometric parameters of the reservoir. The influence of the catchment on the reservoir is evaluated basing on the characteristics of the catchment.

The method takes into account the following parameters: the Ohle coefficient (total catchment of the reservoir divided by reservoir area), balance type of the lake – indicator of point sources of matter supply, morphometrics of the catchment in a form of: river network density, average inclination of the catchment and the percentage share of areas without drainage as well as the geological structure and land usage.

The phosphorus retention coefficient in the reservoir was also calculated [Bajkiewicz-Grabowska 2002] in order to diagnose the processes of deposition and release of phosphorus occurring in individual parts of the reservoir.

The presented research issues were divided as follows: design objectives of the innovative structure of the reservoir, the problem of water protection as the verification of design objectives and the correctness of operation, integration of the reservoir with the agricultural landscape and the resulting new development directions for the studied area, further research perspectives.

RESEARCH PROBLEMS

Innovative structure

The first analysed problem was the designing of a reservoir with an innovative structure. The structure of the bowl in the Mściwojów Reservoir is unusual for dam reservoirs, as it is divided into two operational parts. The first is the basic reservoir, being generally a reservoir of pure water, and the second part is the pre-reservoir – where sediments are deposited and biogenic substances are biologically removed. The division is realised by means of earth biological barrier III. The flow of water from the pre-reservoir and its even distribution in the main reservoir occurs through 5 spillways of the ordinate 192.50 m above sea level, situated in the barrier. The width of each spillway in the bottom is 3.0 m. The reservoir is equipped with two additional earth barriers. Biological barrier II is located at approx. 150 m distance from the dam, it has 2 spillways and it divides the pre-reservoir into two parts: sedimentation part and biological part. Barrier I is located near the sedimentation tank (Figure 1).

In front of the pre-reservoir, inside the bowl, a three-chamber earth tank is located. It performs an important role in the first stage of purification of water from Wierzbiak supplying the reservoir. In the first chamber, of an area of 6136 m² and bottom ordinate 192.85 m above sea level(depth – 1.1 m) sedimentation of coarser particles occurs. In the second, shallowest chamber, of an area of 4884 m² and bottom ordinate 193.10 m above sea level (depth – 0.4 m), which is overgrown by common reed (*Phragmites australis*) is the place where intense processes of biological removal of nitrogen and phosphorus occur. The third, deepest and last chamber, of an area of 4424 m² and bottom ordinate 192.10 (depth – 1.4 m) is the receiver of water that flowed through the two preceding chambers. This is where additional purification processes take place with the participation of living organisms [Dąbrowska 2010].

Quality of surface waters

The research on the influence of the sedimentation tank and the pre-reservoir on water quality tested first of all the degree of reduction of such biogenic compounds as nitrate nitrogen, nitrite nitrogen, ammonium nitrogen, total phosphorus and phosphates in specific parts of the reservoir. During the initial period of operation of the object – in the years 2000–2002 – the mean reduction in the concentration of nitrates in the pre-reservoir amounted to 66.5% and further reduction in the content was observed after the water flowed through the main reservoir [Wiatkowski et al. 2006, Czamara and Grześków 2008]. According to Wiatkowski [2006], the nitrites were reduced by 50% and their amount also decreased after leaving the main reservoir. The mean reduction in the content of phosphates flowing through the pre-reservoir and main reservoir was 52.8% in comparison to the total concentration of phosphates flowing into the reservoir [Wiatkowski et al. 2006]. The obtained reduction values are generally similar to data presented in literature by other authors describing the operation of water reservoirs [Lothar 2003, Kasza 2009, Bendorf and Putz 1987a, Bendorf and Putz 1987b, Putz and Bendorf 1988, Mazur 2010]. It is worth mentioning here that the values presented by individual authors referred to reservoirs of various area and depth. These objects also differed by the retention time and flow volume. For example, Mazur [2010] presents the values for a pre-reservoir of an average depth of 0.70 m and an area of 178 ha, and Lothar [2003] for reservoirs from 0.4 to 12 ha, whereas the area of the pre-reservoir in Mściwojów is 14 ha, and the average depth 1.5 m (max. 2.5 m). Basing on research conducted and published so far, the authors of the present study would like to emphasise the issue of changes in the concentration of phosphates in the pre-reservoir and main reservoir.

Table 2 presents the results of measurements of the concentration of phosphates in three observation periods. These are averaged values obtained by different authors. In the years 2006–2008 a quite high concentration of phosphates was noted on the outflow from the main reservoir, in spite of a lower concentration of these compounds in the water leaving the pre-reservoir. The

Table 2. Mean concentration of phosphates (mg $PO_4^{3-} \cdot dm^{-3}$) in the surface waters in the analysed area in the years 2000–2009

Test period	Wierzbiak	Zimnik	Pre-reservoir	Main reservoir	Main reservoir (outflow)	Publication
XI 2000 – X 2002	0.54	0.51	0.25	–	0.25	Wiatkowski et al. [2006]
XI 2006 – X 2008	0.61	0.77	0.44	0.45	0.61	Wiatkowski [2011]
V 2008 – V 2009	0.73	0.78	0.68	–	–	Dąbrowska, Markowska [2012]

problem of the increasing concentration of phosphates in the main part of the reservoir was also addressed in the study by Dąbrowska and Markowska [2012].

The study by Policht-Latawiec [2013], conducted from IV to IX 2012 showed an unusually low concentration of phosphates both in the inflows and in the reservoir. The mean values amounted, respectively, to 0.21 mg $PO_4^3 \cdot dm^{-3}$ for the waters of Wierzbiak, 0.19 mg $PO_4^3 \cdot dm^{-3}$ for the waters of Zimnik and 0.16 mg $PO_4^3 \cdot dm^{-3}$ in the main reservoir. Further research on water quality in the upcoming years may explain whether there is a constant decreasing trend in the concentration of phosphates and whether this results from the completion of the construction of the sewage network in the catchment.

The main objective of the Water Framework Directive is to achieve good water quality. The problem of contamination of water with phosphorus compounds has been widely discussed in literature and phosphorus compounds are considered a factor limiting the processes connected with eutrophication and algal bloom [van Puijenbroek et al. 2014, Bechmann et al. 2005, Larsson and Granstedt 2010]. Phosphorus is introduced into the water reservoir with inflowing waters, from atmospheric deposition, from inflows from the direct catchment, however, in certain conditions, it may be re-activated from bottom sediments. Literature emphasises mainly the fact that external sources, in particular the load carried by supplying water courses, has the strongest influence on the contamination of dam reservoirs, and that inflows from direct catchment of the reservoir account for a smaller share [Dojlido 1995, Dojlido and Woyciechowska 1996, Kasza 2009]. The method protecting the reservoir waters from phosphorus introduced from inflows are pre-reservoirs [Bendorf and Putz 1987a, Bendorf and Putz 1987b, Putz and Bendorf 1988, Lothar 2003]. Kajak [1995], referring only to pre-reservoirs, claimed that too long retention time leads to the development of zooplankton and the reintroduction of biogenic substances into the circulation. However, the same author points out that the internal load of phosphorus has a small significance in comparison to the external load. Kasza [2009] states that the inflow of phosphorus from bottom sediments is relatively low and accounts for 0.2 to 5.3% of the total external supply. He determined this fact basing on data from six analysed reservoirs. It is generally believed that the problem of re-activation of phosphorus from bottom sediments occurs mainly in the case of natural lakes. Its importance in river and lake ecosystems was widely documented by Bajkiewicz-Grabowska [2002]. Bartoszek [2007] presents a broad survey of the issues related to the release of phosphorus from bottom sediments, listing the factors influencing the course of this phenomenon, including: lowered oxygen content above the sediment layer, the existence of ions of iron and calcium, changes in pH, the participation of bacteria in the mineralisation of organic sediments, undulation etc. The author also points out that various mechanisms can contribute to the release of phosphorus, but in the case of water reservoirs in Poland this process has not been well recognised.

The question arises: what may be the reason of the high concentration of phosphates on the outflow from the main reservoir in Mściwojów, in spite of a lower concentration of these compounds in the water leaving the pre-reservoir.

In order to explain this phenomenon, the authors decided to evaluate the catchment as a supplier of biogenic matter and estimate the possibility of contamination of the reservoir with phosphorus from the sediments.

Evaluation of the catchment as a supplier of biogenic compounds

Bajkiewicz-Grabowska [2002] proposed a method of evaluating the catchment on a scale from 0 to 3 points. The final result is the qualification of the catchment to one of 4 groups of exposure to supply of matter to the reservoir. The group is determined basing on the calculation of the average value of the total points awarded.

The author of the method proposes 4 groups of exposure: group 1 – for the average value lower or equal to one, which is proof of a practical lack of possibility of matter supply to the reservoir; group 2 – for average values within the range (1.1–1.4), which suggests low exposure to the activation of the load deposited in the area and only a slight possibility of it being supplied to the reservoir; group 3 for average values within the range (1.5–1.9), which means average exposure to matter supply to the reservoir and group 4 for average values 2 and higher, where the catchment is characterised by high exposure to the activation of the load and its transportation to the reservoir.

Basing on available materials in form of maps (physical map in the scale of 1:25 000, geological map in the scale 1:50 000) and field studies conducted in the years 2000–2013 (geotechnical profiles, verification of land usage) the catchment of the Mściwojów Reservoir was evaluated in the aspect of matter supply.

Table 3 contains a list of the characteristics of the catchment that influence the eutrophication of the reservoir. Table cells containing numerical values of specific parameters or the results of descriptive evaluation for the catchment of the Mściwojów Reservoir are marked grey.

Basing on the conducted analyses and the obtained average value – 2.00 points, the catchment of the reservoir was qualified as group 4, being very prone to the movement and supply of the matter to the reservoir.

The Mściwojów Reservoir belongs to the group of medium-sized reservoirs [Radczuk and Olearczyk 2002] and, due to its size and location, it is particularly prone to matter supply from the area of its catchment, which is confirmed by the analysis conducted pursuant to the method proposed by Bajkiewicz–Grabowska [2002] for the evaluation of the catchment as matter supplier.

Calculation of the phosphorus retention coefficient in the reservoir

The next step in the course of explaining the increased concentration of phosphorus on the outflow from the reservoir was the determination of the possibility of contamination of the water in the main reservoir with phosphorus released from bottom sediments. In order to do so, the following formula [Bajkiewicz-Grabowska 2002] was used:

$$R = 1 - \frac{V_{odp} \cdot TP}{Ł_c}$$

where: R – phosphorus retention coefficient in the reservoir,

V_{odp} – annual outflow from the reservoir [m³],

TP – total phosphorus concentration in the reservoir [mgP·m³],

$Ł_c$ – total annual load introduced into the reservoir [mg].

Retention of phosphorus is usually determined with the use of the retention coefficient developed in 1974 by Dillon and Rigler and it is the key element of models used for predicting its concentration and the trophic status of the waters [Hejzlar et al. 2006].

Calculations were conducted for four variants: In variant I the phosphorus retention coefficient "R" was calculated for the whole reservoir. Basing on the field survey and analysis of the physical map 1:25 000, as well as the analysis of available literature [Behrendt and Dannowski 2005], it was assumed that the reservoir is prone to the influence of the area contamination from the direct catchment and to the deposition of phosphorus from atmosphere. Variant II encompassed the calculation of the "R" coefficient for the pre-reservoir supplied by a load of phosphorus introduced by Wierzbiak and Zimnik and,

Table 3. Catchment as the supplier of matter – evaluation of individual parameters

Characteristics	Number of points			
	0	1	2	3
Lake coefficient	<10	10 – 40	40 – 150 Mściwojów: 135.8	>150
Lake balance type	–	outflow	without outflow	flow–through
Density of river network	<5	0.5 – 1.0 Mściwojów: 0.83	1.0 – 1.5	>1.5
Mean slope of the catchment (%)	<5 Mściwojów: I=2.26%	5 – 10	10 – 20	>20
Drainless areas (%)	>60	45 – 60	20 – 45	<20 Mściwojów: ≅ 1%
Geological structure of the catchment	clayey, peaty	sandy, clayey	clayey, sandy	sandy
Land usage	forestry, agriculture and forestry, pasture, agriculture and forestry, pasture and forestry	forestry and agriculture, pasture and agriculture	agriculture, pasture, forestry and agriculture – developed land	forestry and agriculture – developed land, pasture and agriculture – developed land, agriculture – developed land

additionally, phosphorus originating from area sources in the direct catchment and atmospheric deposition onto the surface of the pre-reservoir. In the third variant, the main reservoir was separated, which was supplied with phosphorus by water from the pre-reservoir and the inflow from direct catchment along with atmospheric deposition of phosphorus. Variant IV was based on the assumption that the reservoir is only supplied by river load and atmospheric deposition.

The input data consist of the following:

- mean annual flow for Wierzbiak and Zimnik $SQ = 0.171$ m^3/s [Wiatkowski 2011],
- annual emission of phosphorus from area sources for the catchment of Kaczawa – 112.6 t P/year [Behrendt and Dannowski 2005],
- annual deposition of phosphorus from atmosphere in the catchment of Kaczawa – 1.2 t/year [Behrendt and Dannowski 2005],
- area of the catchment of Kaczawa – 2261 km^2,
- area of the direct catchment of the main reservoir – 3.5 km^2 (calculated),
- area of the direct catchment of the main reservoir – 2.0 km^2, (calculated),
- area of the pre-reservoir – 14.00 ha,
- area of the main reservoir – 20.59 ha,
- concentration of phosphates on the inflow from Wierzbiak 0.61 mg $PO_4 \cdot dm^{-3}$ [Wiatkowski 2011], share in river supply 71%,
- concentration of phosphates on the inflow from Zimnik 0.77 mg $PO_4 \cdot dm^{-3}$ [Wiatkowski 2011], share in river supply 29%,
- concentration of phosphates on the outflow from the pre-reservoir – 0.44 mg $PO_4 \cdot dm^{-3}$ [Wiatkowski 2011],
- concentration of phosphates on the outflow from the main reservoir – 0.61 mg $PO_4 \cdot dm^{-3}$, [Wiatkowski 2011],

The calculations take into account the conversion of phosphate phosphorus to total phosphorus. The following results were obtained:

- Variant I   – R = 0.25,
- Variant II   – R = 0.39,
- Variant III – R = -0.13,
- Variant IV   – R = 0.08.

The calculated values of coefficients suggest that the quality of waters flowing out of the Mściwojów Reservoir is influenced both by internal and external phosphorus supply. The negative value of the coefficient R = -0.13, considering the area contamination and atmospheric deposition shows the significant influence of the release of

phosphorus from bottom sediments on the circulation and transformations of this element in the reservoir. This process has not been analysed in this reservoir so far.

The difference resulting from the comparison of results obtained for variant I (R = 0.25) and IV (R = 0.08) proves that area contamination from direct catchment influences the water quality in the reservoir.

Long-term observations conducted by the authors on the discussed object prove that appropriate maintenance works are not performed here, neither in the sediment tank nor in the pre-reservoir, and that there is no properly managed environmental zone around the object, although it was foreseen in the design as a correctly planted and managed buffer zone capturing the contaminants from the direct catchment. The fact that the reservoir is being intensely exploited for fishing purposes – the supply of biogenic substances with bait and disadvantageous structure of fish species contribute to the deterioration of water quality. Recreational activities on the reservoir are conducted in a disorderly manner, without a sanitary base and waste management. However, cultivation of soil is maintained in the direct proximity of the water, so, in spite of the reduction of biogenic compounds in the pre-reservoir, the water is polluted by the discharge of surface waters from the arable land to the reservoir, even in average humidity conditions as occurred in the years 2006–2008. Only in the year 2008 precipitation was below the normal value and the noted deficit amounted to 5% for the south-western part of the Lower Silesian Voivodeship [Raport... 2009].

The analysis of land formation, soil conditions and field tests and observations shows that the analysed area is exposed to intense phenomena of water and wind erosion. In particular the transportation of wind erosion products from the fields directly to the waters of the reservoir on the western side (land inclination up to 2%) is noticeable, as there are no barriers for the wind in form of hills, buildings or trees. Water erosion is dominant on the eastern coast, with much higher inclinations (>8%).

All the listed factors influence the quality of water leaving the reservoir, which doubtlessly proves that not only technological solutions, such as pre-reservoirs and orderly water and wastewater management in the whole catchment area are important, but also rational management of the areas directly adjacent to the reservoir and located in its direct catchment. Works connected with the transformation of rural areas, promoting such arrangement of area structures that would perform various assigned economic functions, including the protection of water resources, should perform an important role in this aspect.

Works connected with the transformation of rural areas

New development directions for the analysed area were set by Wrocław University of Environmental and Life Sciences, emphasising the need to split monocultures and introduce tree planting in the fields [Kempa 2005], the necessity to balance the needs of agricultural production with the use of resources and assets of the natural environment. A concept of reorganisation of the agricultural production area and the introduction of new functions into the area was developed – the PROJECT "Blue Triangle" (Figure 4). It encompasses the area adjacent to the reservoir.

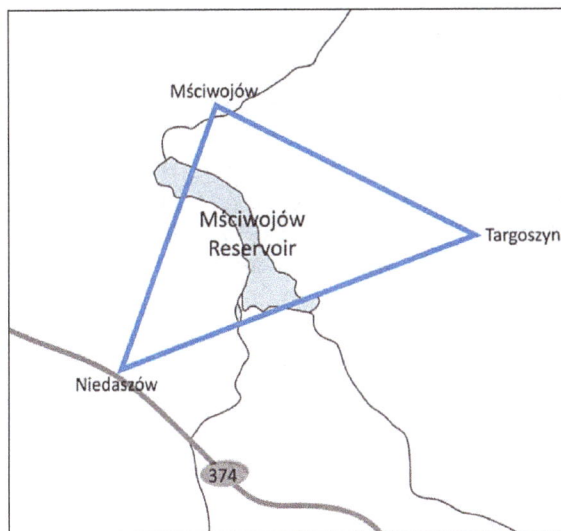

Figure 3. „Blue Triangle" - location

This concept is based on the principle that changes in the agricultural area should be introduced in a planned way and preceded by thorough studies.

A vital element of the concept is the proposal to use the palace and park complex in Targoszyn as a hotel base and to locate the Garden of the Nations here (this is an international project of the Wrocław University of Environmental and Life Sciences, which foresees the creation of a botanical garden on a plot of an area of over 18 ha), as a tourist attraction. Due to the proximity of Rogoźnica and the Museum of Gross-Rosen located there, this village can also provide a source of hotel and service facilities for potential organised trips.

Prospects of further research

The studies conducted for more than ten years were mainly aimed at the verification of the applied design solution and the correctness of the operation. The scope of future studies should encompass broader research on sediments, taking into account also interstitial water, a detailed analysis of taller plants overgrowing the area adjacent to the water course and the reservoir with reference to specific environmental zones, as well as communities of blue algae and algae, including phytoplanctone, periphyton and epiphyton as well as benthos organisms. The quantitative and qualitative aspects of ichtyofauna are also worth noting. Such research works should be aimed at the determination of the role of living organisms in the phosphorus circulation process. This will provide an opportunity to solve the problem of internal supply of phosphorus to the main reservoir.

Studies on the catchment of the reservoir should focus on model and simulation works, based on documented results of field studies. Such activities would result in some proposals, suggesting the optimal management of the catchment of the reservoir in the environmental and economic aspects.

In further perspective, it only seems natural to broaden the scope of research to encompass the whole catchment of Wierzbiak of an area of 273.4 km^2 according to the Hydrographical Map of Poland (MPHP) and to conduct such studies that would lead to the determination of principles of effective water management, taking into consideration flood protection and the threat of failure of hydrotechnical constructions [Sobota et al. 2009], and to the development of a dynamical water economic balance of the catchment that would encompass the hierarchy of needs of various users. This would lead to the creation of water management and environmental framework for spatial development, which would be able to rationally manage the environmental resources for the purposes of the development of economic activity, settlement, infrastructure, as well as the needs of the nature itself.

CONCLUDING REMARKS

The design of the reservoir is a result of several years of researchers' work. The over ten-year operation period of the reservoir presented in this article along with research carried out simultaneously constitutes a certain stage, whose end is the completion of the construction of sewage networks for settlements located in the upper part of the reservoir catchment. Wastewater from Goczałkowo, Kostrza, Rogoźnica, Żółkiewka and Wieśnica will be transported to the developed wastewater treatment plant in Strzegom, and then to the Strzegomka River (left-side tributary of Bystrzyca). This solution will certainly contribute to lowering negative influence of point contamination on the water quality in Wierzbiak, and thus in the Mściwojów Reservoir.

However, the reservoir will still be exposed to area contamination resulting from the nature of the catchment (class 4 of exposure). As shown by the conducted estimate calculations, in certain circumstances the contamination from direct catchment may eliminate the beneficial influence of the sediment tank and the pre-reservoir. This proves doubtlessly that any solutions aimed at the improvement of water quality directly connected with the water reservoir as an object of main focus, are, in a sense, of an ad-hoc nature, as the key to the solution is broader perspective of the problem.

The introduction of the reservoir into an agricultural area and the creation of a research base in its catchment had a positive influence on the local community. It led to the development of tourism and agro tourism, along with recreational opportunities at the reservoir. Projects conducted by academic entities create an opportunity for development for towns and villages located in the proximity of the reservoir. However, recreational activities at the reservoir have to be more orderly: a sanitary base and waste bins are needed in order to avoid potential threats to water quality.

Future works should be extended so as to cover all elements of the ecosystem of the reservoir. In a longer term it seems natural to extend the research works to cover the whole catchment of Wierzbiak River.

BIBLIOGRAPHY

1. Bajkiewicz–Grabowska E.: Obieg materii w systemach rzeczno-jeziornych. Uniwersytet Warszawski Wydział Geografii i Studiów Regionalnych. Warszawa, 2002.

2. Bartoszek L.: Wydzielanie fosforu z osadów dennych. Zeszyty Naukowe Politechniki Rzeszowskiej nr 240. Budownictwo i Inżynieria Środowiska, 2007, 5–19.

3. Bechmann M.E., Berge D., Eggestad H.O., Vandsemb S.M.: Phosphorus transfer from agricultural areas and its impact on the eutrophication of lakes – two long-term integrated studies from Norway. Journal of Hydrology, Vol. 304, Iss. 1–4, 2005, 238–250.

4. Behrendt H., Dannowski R.: Nutriens and Heavy Metals in the Odra River System. Emission from Point and Diffuse Sources, Their Loads, and Scenario Calculations on Possible Changes. Weissence Verlag, Berlin, 2005.

5. Benndorf J., Pütz K.: Control of eutrophication of lakes and reservoirs by means of pre-dams – I. Mode of operation and calculation of the nutrient elimination capacity. Wat. Res., 21, 1987a, 829-838.

6. Benndorf J., Pütz K.: Control of eutrophication of lakes and reservoirs by means of pre-dams – II. Validation of the phosphate removal model and size optimization. Wat. Res., 21, 1987b, 839-842.

7. Dąbrowska J.: Wpływ osadnika wstępnego z filtrem biologicznym na zmiany wartości wybranych parametrów fizykochemicznych wody. Infrastruktura i ekologia terenów wiejskich, nr 8/2, 2010, 5-13.

8. Dąbrowska J. Markowska J.: Wpływ zbiornika wstępnego na jakość wód retencjonowanych w zbiorniku Mściwojów. Nauka Przyroda Technologie. Dział Melioracje i Inżynieria Środowiska. Vol. 6, Iss. 2, 2012, 1-11.

9. Dojlido J.R.: Chemia wód powierzchniowych. Wydawnictwo Ekonomia i Środowisko. Białystok, 1995.

10. Dojlido J.R, Woyciechowska J., Świderska D.: Bilans ładunków zanieczyszczeń dopływających do Jeziora Zegrzyńskiego. Gospodarka Wodna nr 9, 1996, 273-276.

11. Grześków L, Czamara A.: Ocena skuteczności działania zbiornika wstępnego w Mściwojowie. Zeszyty Problemowe postepów Nauk Rolniczych. Zeszyt 528. Melioracje Wodne w Inżynierii Kształtowania Środowiska. Polska Akademia Nauk. Wydział Nauk Rolniczych, Leśnych i Weterynaryjnych, 2008, 361-372,.

12. Hejzlar J., Šámalová K., Boers P., Kronvang B.: Modelling Phosphorus Retention in Lakes and Reservoirs. Water, Air and Soil Pollution: Focus, Vol. 6, No 5-6, 2006, 487-494.

13. Kajak Z.: Eutrofizacja nizinnych zbiorników zaporowych. Procesy biologiczne w ochronie i rekultywacji nizinnych zbiorników zaporowych. Biblioteka Monitoringu Środowiska. PIOŚ, WIOŚ, Zes. UŁ. Łódź 1995, 33-41.

14. Kasperek R., Wiatkowski M.: Badania osadów dennych ze zbiornika Mściwojów. Przegląd Naukowy. Inżynieria i Kształtowanie Środowiska. SGGW. No 40, 2008, 194-201.

15. Kasza H.: Zbiorniki zaporowe. Znaczenie – Eutrofizacja – Ochrona. Akademia Techniczno-Humanistyczna w Bielsku Białej, Bielsko-Biała 2009.

16. Kempa O.: Ocena metod waloryzacji krajobrazu dla potrzeb prac urządzenioworolnych. Akademia Rolnicza we Wrocławiu, Wydział Inżynierii Kształtowania Środowiska i Geodezji, rozprawa doktorska, maszynopis, Wrocław 2005.

17. Larsson M., Granstedt A.: Sustainable governance of the agriculture and the Baltic Sea – Agricultural reforms, food production and curbed eutrophication. Ecological Economics, Vol. 69, Iss. 10, 2010, 1943-1951.

18. Lothar P.: Nutrient elimination in pre-dam: results of long term studies. Hydrobiologia 504. eds. Straskrabova V., Kenedy R.H., Lind O.T., Tundisi J.G., Hejzlar J. Reservoir Limnology and Water Quality, Kluwer, 200, 289-295.

19. Mazur A.: Influence of the pre-dam reservoir on the quality of surface waters supplying reservoir "Nielisz". Teka Kom. Ochr. Kszt. Środ. Przyr. – OL PAN, 7, 2010, 243-250.

20. Pikul K., Mokwa M:. Badania modelowe wpływu zmian koncentracji materiału unoszonego w wodach płynących. Infrastruktura i Ekologia Terenów Wiejskich, 4(2), 2006. Polska Akademia Nauk Oddział w Krakowie, Komisja Technicznej Infrastruktury Wsi, 2006, 119–128.

21. Pikul K., Mokwa M.: Wpływ osadnika wstępnego na proces zamulania zbiornika głównego. Przegląd Naukowy. Inżynieria i Kształtowanie Środowiska. SGGW. Zeszyt 40, 2008, 185-193.

22. Policht-Latawiec A.: Assessment of water inflowing, stored and flowing away form Mściwojów Reservoir. Geomatics, Landmanagement and Landscape No. 1, 2013, 107-115.

23. Pütz K., Benndorf J.: The importance of pre-reservoirs for the control of eutrophication of reservoir. Water Science and Technology, Vol. 37, Iss. 2, 1988, 317-324.

24. Radczuk L., Olearczyk D.: Małe zbiorniki retencyjne jako element poprawy bilansu wodnego zlewni użytkowanych rolniczo. Zeszyty Naukowe AR w Krakowie. Inżynieria Środowiska, 23, 2002, 139-148.

25. Raport o stanie środowiska w woj. dolnośląskim w 2008 r. WIOŚ, Wrocław 2009.

26. Sobota J., Gavardashvili G., Ayyub B. M., Bouranski E., Arbidze V.: Simulation of flood and mudflow scenarios in case of failure of the Zhinvali Earth Dam. International Symposium on Floods and Modern Methods of Control Measures. Red. Givi Gavardashvili, Alistair Borthwick, Lorenz King. Ministry of Education and Science of Georgia, Georgian Water Management Institute. Tbilisi, Georgia, 2008, 148-163.

27. Szafrański Cz., Stefanek P.: Wstępna ocena wpływu zbiornika Mściwojów na przepływy w rzece Wierzbiak i głębokości zwierciadła wody gruntowej w terenach przyległych. Rocznik Ochrony Środowiska, Wydawnictwo Środkowo-Pomorskiego Towarzystwa Naukowego Ochrony Środowiska, Vol. 10, 2008, 491-502.

28. Van Puijenbroek P.J.T.M., Cleij P., Visser H.: Aggregated indices for trends in eutrophication of different types of fresh water in the Netherlands. Ecological Indicators, Vol. 36, 2014, 456-462.

29. Wiatkowski M., Czamara Wł., Kuczewski K.: Wpływ zbiorników wstępnych na zmiany jakości wód retencjonowanych w zbiornikach głównych. Instytut Podstaw Inżynierii Środowiska Polskiej Akademii Nauk. Zabrze 2006.

30. Wiatkowski M.: Influence of Msciwojow pre-dam reservoir on water quality In the water reservoir dam and below the reservoir. Ecological Chemistry and Engineering. Vol. 18. No. 2, 2011, 123-134.

31. Zbiornik wodny „Mściwojów" na rzece Wierzbiak gmina Mściwojów woj. legnickie. Projekt budowlany urządzeń i obiektów hydrotechnicznych. Maszynopis. Instytut Inżynierii Środowiska AR, Wrocław 1995.

11

BIODEGRADATION OF DIESEL OIL IN SOIL AND ITS ENHANCEMENT BY APPLICATION OF BIOVENTING AND AMENDMENT WITH BREWERY WASTE EFFLUENTS AS BIOSTIMULATION-BIOAUGMENTATION AGENTS

Samuel Agarry[1], Ganiyu K. Latinwo[1]

[1] Ladoke Akintola University of Technology, Ogbomoso, Nigeria, e-mail: sam_agarry@yahoo.com; gklatinwo@lautech.edu.ng

ABSTRACT

The purpose of this study is to investigate and evaluate the effects of natural bioattenuation, bioventing, and brewery waste effluents amendment as biostimulation-bioaugmentation agent on biodegradation of diesel oil in unsaturated soil. A microcosm system was constructed consisting of five plastic buckets containing 1 kg of soil, artificially contaminated or spiked with 10% w/w of diesel oil. Biodegradation was monitored over 28 days by determining the total petroleum hydrocarbon content of the soil and total hydrocarbon degrading bacteria. The results showed that combination of brewery waste effluents amendment and bioventing technique was the most effective, reaching up to 91.5% of diesel removal from contaminated soil; with the brewery waste effluents amendment (biostimulation-bioaugmentation), the percentage of diesel oil removal was 78.7%; with bioventing, diesel oil percentage degradation was 61.7% and the natural bioattenuation technique resulted in diesel oil removal percentage be not higher than 40%. Also, the total hydrocarbon-degrading bacteria (THDB) count in all the treatments increased throughout the remediation period. The highest bacterial growth was observed for combined brewery waste effluents amendment with bioventing treatment strategy. A first-order kinetic model was fitted to the biodegradation data to evaluate the biodegradation rate and the corresponding half-life time was estimated. The model revealed that diesel oil contaminated-soil microcosms under combined brewery waste effluents amendment with bioventing treatment strategy had higher biodegradation rate constants, k as well as lower half-life times, $t_{1/2}$ than other remediation systems. This study showed that the microbial consortium, organic solids, nitrogen and phosphorus present in the brewery waste effluents proved to be efficient as potential biostimulation-bioaugmentation agents for bioremediation processes of soils contaminated with diesel oil.

Keywords: bioremediation, bioventing, biostimulation, bioaugmentation, brewery waste effluents, diesel oil, first-order kinetics.

INTRODUCTION

Petroleum-based products are the major source of energy for industry and daily life. Leakages and accidental spills occur regularly during the exploration, production, refining, transport, and storage of petroleum and petroleum products. The contamination of soil by crude oil and petroleum products has become a serious problem that represents a global concern for the potential consequences on ecosystem and human health [Onwurah et al., 2007]. Among petroleum products, diesel oil is a complex mixture of alkanes and aromatic compounds that are frequently reported as soil contaminants leaking from storage tanks and pipelines or released in accidental spills [Gallego et al., 2001]. The scale of hazards imposed on the natural environment depends on the surface of the area contaminated by the petroleum products, their chemical composition, and the depth at which pollutants occur [Wolicka et al., 2009]. The technology commonly used for soil remediation includes mechanical, burying, evaporation, dispersion, and washing. However, these technologies are expensive and can lead to incomplete decomposition of contaminants [Das and Chandra, 2011]. For this reason an increasing attention has been directed toward the research of new strategies and environmental-friendly technologies to be applied for the remediation of soil contaminated by petroleum hydrocarbons. Among these, bioremediation technology which involves the use of microorganisms to detoxify or remove pollutants through the mechanisms of biodegradation has been found to be an environmentally-friendly, noninvasive and relatively cost-effective option [April et al., 2000].

The activation of natural degradation potentials in environmental media is currently the challenge in the environmental research addressed to remediation methods. Ways to activate these potentials must consider that most degradation potentials are widely distributed among microorganisms (Alexander, 1999) but indigenous microbes are usually present in a very small number. Moreover, the degradative metabolism towards specific pollutants needs often to be induced. Possible ways to overcome these limitations include changes of physicochemical parameters (pH, T, electron donors or acceptor, etc.) as well as a "niche adjustment" by the inoculation of competent microorganisms into these systems (bioaugmentation). Thus, petroleum hydrocarbon bioremediation in soil can be promoted or activated through stimulation of the indigenous microbial population, by introducing nutrients and oxygen into the soil (biostimulation) [Seklemova et al., 2001] or through inoculation of an enriched microbial consortium into soil (bioaugmentation) [Richard and Vogel, 1999; Bento et al., 2005].

Recently, researchers have focused on the use of oxygen-release compounds (ORCs) and supply of air to promote the direct oxidation of pollutants and, at the same time, to increase aerobic microbial degradation [Tsai et al., 2009; Agarry et al., 2012; Zhang et al., 2014]. Injection of air through the method of bioventing is amongst the known bioremediation technologies for cleaning up the petroleum contaminated soil in the unsaturated zone. Bioventing has been defined as a method that stimulates indigenous microorganisms to biodegrade aerobically degradable compounds in soil by providing adequate oxygen or aeration [Leeson and Hinchee, 1997; Byun et al., 2006]. Biostimulation and bioaugmentation strategies has been widely used in remediation of organic contaminated soils in the past few years. Effective bioremediation of petroleum-contaminated soil using bioventing has been proved by some research [Moller et al., 1996; Kirsten et al., 2005; Byun et al., 2006; Morales et al., 2013; Thome et al., 2014].

Positive effects of nitrogen amendment using nitrogenous fertilizer on microbial activity and/or petroleum hydrocarbon degradation have been widely demonstrated [Brook et al., 2001; Margesin et al., 2007]. However, in developing countries, inorganic chemical fertilizers are costly as well as not sufficient for agriculture, let alone for cleaning oil spills. Therefore, it necessitates the search for cheaper and environmentally friendly options of enhancing petroleum hydrocarbon degradation. One of such options is the use of organic wastes effluents that could act as bulking agents and also as bacterial biomass suppliers. There are no adequate literatures on the potential use of these industrial-derived organic wastes effluents as biostimulating and bioaugmentation agents. However, few workers have investigated the potential use of solid organic wastes such sugarcane bagasse [Molina-Barahona et al., 2004], spent brewery grain [Abioye et al., 2009] and animal wastes like cow dung [Akinde and Obire, 2008; Singh and Fulekar, 2009], pig dung [Yakubu, 2007], poultry manure [Okiemen and Okiemen, 2005], goat dung [Agarry et al., 2010], and sewage sludge [Mao and Yue, 2010] as biostimulating agents in the cleanup of soil contaminated with petroleum hydrocarbons and were found to show positive influence on petroleum hydrocarbon biodegradation in a polluted environment. Nevertheless, the search for cost effective and environmentally friendly methods of petroleum hydrocarbon removal from contaminated sites still needs to be further investigated. However, to the best of our knowledge, there is a dearth of information on the use of brewery waste effluents for stimulation of autochthonous microflora of petroleum hydrocarbons-contaminated soils. Furthermore, the evaluation and comparison of bioventing and liquid organic wastes in petroleum hydrocarbon bioremediation has not been reported in the literature.

Therefore, the objectives of this study are to determine the biostimulation-bioaugmentation potential of brewery waste effluents as either alone and/or in combination with bioventing in enhancing the biodegradation of diesel oil as the target contaminant in soil. Also, to evaluate and compare, model and analyzed the degradation kinetics of diesel oil for natural attenuation, bioventing, and organic wastes effluents enhanced bioremediation.

MATERIALS AND METHODS

Sample collection

The main materials used for this study are: soil, brewery waste effluents and diesel oil. Soil samples were collected from an un-impacted zone of Ladoke Akintola University of Technology Agricultural Farm. About 2 kg of bulk surface and subsurface soil was collected from different areas on the sampling sites. The bulked composite soil was preserved by putting them in sterile polyethylene bag, properly sealed and kept in the refrigerator prior to further use. The brewery waste effluents were collected from Nigerian Breweries Plc., Ibadan, Nigeria. The diesel oil was obtained from a commercial petroleum products station, Ogbomoso, Nigeria.

Soil and brewery waste effluents characterisation

The soil sample was characterized for total organic carbon (TOC), total nitrogen (N), total phosphorus, moisture content, pH, and total hydrocarbon degrading bacteria (THDB) according to standard methods. Soil pH was determined according to the modified method of McLean (1982); total organic carbon was determined by the modified wet combustion method [Nelson and Sommers, 1982] and total nitrogen was determined by the semi-micro-Kjeldahl method [Bremner and Mulvaney, 1982]. Available phosphorus was determined by Brays No. 1 method

[Olsen and Sommers, 1982] and moisture content was determined by the dry weight method. The total hydrocarbon degrading bacteria (THDB) populations were determined by the vapor phase transfer method [Amanchukwu et al., 1989]. The physic-chemical and microbiological properties of the soil and brewery waste effluents are given in Table 1.

As shown in Table 1, brewery waste effluents contained significantly large amount of effective nitrogen, effective phosphorus and microorganism compared with the soil having low level of total nitrogen and phosphorus suggesting low nutrient level. These analyses indicate that brewery waste effluents can be added as nourishment for composting of soil. Moreover, the addition of brewery waste effluents can increase the microorganism density and the activity of soil. Organic component in brewery effluent (expressed as COD) is generally and easily biodegradable as it mainly consists of sugar, soluble starch, ethanol, volatile fatty acids, while the brewery solids (expressed as TSS) mainly consist of spent grains, kieselguhr, waste yeast and ('hot') trub [Inyang et al., 2012].

Isolation, Characterization, and Identification of Bacteria in Soil and Brewery waste effluents

The spread plate technique [APHA, 2005] using nutrient agar (Oxoid) was employed for the isolation of bacteria in soil and brewery waste effluents. The plates were incubated at 37 °C for 18 – 24 h. Pure bacterial isolates were characterized and identified using various criteria, as described by Krieg et al. (1994). Bacteria and fungi isolated from brewery waste effluents include *Pseudomonas* sp, *Acetobacter* sp, and *Flavobacterium* sp, *Aspergillus niger*, *Curvularia* sp and *Alternaria* sp, respectively. Meanwhile, the microorgan-

Table 1. Physical-chemical and microbiological characteristics of unimpacted soil and brewery waste effluents

Parameters	Soil	Brewery waste effluents
Total nitrogen (%)	0.06 ± 0.01	0.34
Total carbon (%)	0.45 ± 0.03	–
Total organic matter (%)	0.78 ± 0.02	–
Available phosphorus (%)	–	0.44
Biochemical Oxygen Demand (BOD) mg/l	–	30
Chemical Oxygen Demand (COD) mg/l	–	410
Total Suspended Solids (TSS) mg/l	–	65
pH	7	8.2
Bacterial count (cfu/g)	$0.3 \pm 1.6 \times 10^6$.	$0.2 \pm 5.0 \times 10^{10}$

isms present in the LAUTECH agricultural soil were identified to be made up of mainly *Bacillus* and *Pseudomonas* species.

Experimental design for the bioremediation of diesel oil-spiked soil

Five plastic buckets used as bioreactors were prepared for each soil treatment, designated as bioattenuation (treatment A), bioventing (treatment B), brewery waste effluents amendment (treatment C), combined brewery waste effluents amendment and bioventing (treatment D), and control experiment (treatment E). Each bioreactor contained 1 kg of soil, spiked with 150 ml of 100 g of diesel oil (10% *w/w*) and thoroughly mixed together to achieve complete artificial contamination. 10% spiking was adopted in order to achieve severe contamination as the concentration above 3%, oil has been reported to be increasingly deleterious to soil biota and crop growth [Osuji et al., 2005]. Two weeks after the contamination (to allow for aging), the different remediation treatments were applied. For bioreactors B and D, air compressor pump used by vulcanizer was used to allow for atmospheric air to be drawn in towards a perforated pipe in the centre. The bioreactor under treatment C was amended with 50 g of brewery waste effluents and the bioreactor under treatment D was amended with 50 g of brewery waste effluents and subjected to bioventing. The bioreactor under treatment A was neither amended with brewery waste effluent nor subjected to bioventing. Soil in the bioreactor used as control experiment was sterilized three times by autoclaving at 121 °C for 15 min. All the bioreactors with its contents were incubated at room temperature (28 ± 2 °C) for 28 days. Water content (moisture) of soil in each bioreactor was adjusted every week by addition of sterile de-ionized water to a moisture holding capacity of 50%. In order to avoid anaerobic conditions, contents of the bioreactor were aerated by mixing every 3 days. Samples were taken every week and analyzed for total petroleum hydrocarbon (TPH) (i.e. residual diesel oil) and total hydrocarbon–degrading bacteria (THDB), respectively. The experiments were carried out in triplicates.

Extraction and determination of total petroleum hydrocarbon

The extent of hydrocarbon utilization in the diesel oil was estimated gravimetrically and spectrophotometrically [Adesodun and Mbagwu, 2008]. Soil samples (approximately 10 g) was taken from each microcosm and put into a 50 mL flask and 20 mL of n-hexane was added. The mixture was shaken vigorously on a magnetic stirrer for 30 minutes to allow the hexane extract the oil from the soil sample. The solution was then filtered using a Whatman filter paper and the liquid phase extract (filtrate) diluted by taking 1 mL of the extract into 50 mL of hexane. The absorbance of this solution was measured spectrophotometrically at a wavelength of 400 nm HACH DR/2010 Spectrophotometer using n-hexane as blank. The total petroleum hydrocarbon in soil was estimated with reference to a standard curve derived from fresh crude oil of different concentration diluted with n-hexane. Percent degradation (D) was calculated using the following formula:

$$D = \frac{TPH_i - TPH_r}{TPH_i} \times 100 \qquad (1)$$

Where TPH_i and TPH_r are the initial and residual TPH concentrations, respectively.

Kinetic model analysis

Kinetic analysis is a key factor for understanding biodegradation process, bioremediation speed measurement and development of efficient clean up for a petroleum hydrocarbon contaminated environment. The information on the kinetics of soil bioremediation is of great importance because it characterizes the concentration of the contaminant remaining at any time and permit prediction of the level likely to be present at some future time. Petroleum hydrocarbon biodegradation rates are usually difficult to predict due to the complexity of the environment [Zhu et al., 2001]. Nevertheless, biodegradation rate of organic compounds by microorganisms is often described by the equation as follows [Wang et al., 2001]:

$$q = \frac{q_m c}{k + c} \qquad (2)$$

Where q is biodegradation rate, q_m is maximum specific biodegradation rate, c is the substrate concentration and k is half-saturation constant. If $c \leq k$; Eq. (2) can be reduced to:

$$q = \frac{q_m c}{k} \qquad (3)$$

Eq. (3) is a typical first-order model. The use of first-order kinetics in the description of biodegradation rates in environmental fate models is common because mathematically the expression can be easily incorporated into the model [Greene et al., 2000). Assuming $k_l = (q_m/k)$ and integrating Eq. (3), the following relation of substrate concentration to time can be obtained as given in Eq. (4):

$$\ln c = a + k_l t \qquad (4)$$

Estimation of biodegradation half-life times

The biological half-life is the time taken for a substance to lose half of its amount. Biodegradation half-lives are needed for many applications such as chemical screening [Aronson et al., 2006], environmental fate modeling [Sinkkonen and Paasivirta, 2000] and describing the transformation of pollutants [Dimitrov et al., 2007; Matthies et al., 2008]. Biodegradation half-life times $(t_{1/2})$ are calculated by Eq. (5) [Zahed et al., 2011; Agarry et al., 2013a]:

$$t_{1/2} = \frac{\ln 2}{k} \qquad (5)$$

Where k is the biodegradation rate constant (day^{-1}). The half-life model is based on the assumption that the biodegradation rate of hydrocarbons positively correlated with the hydrocarbon pool size in soil [Yeung et al., 1997].

Data analysis

The data were subjected to one-way analysis of variance (ANOVA) at 5% probability. The data analysis was performed using statistical package for social sciences, version 16.0 (SPSS Inc., Chicago, IL, USA).

RESULTS AND DISCUSSION

Diesel biodegradation

The biodegradation profile of diesel oil in soil subjected to bioventing and also amended with brewery waste effluents is shown in Figure 1.

It is observed that the percentage reduction in TPH was relatively fast within the first 14 days of remediation in all the soil microcosms subjected to bioventing and that amended with brewery waste effluents and rapidly continued up to the fourth week (day 28) when compared to that of the unamended soil microcosm (natural bioattenuation). At the end of day 14, there was 17%, 27.7% and 36.2% TPH reduction in soil microcosms B, C and D subjected to bioventing, amended with brewery waste effluents, and combined brewery waste effluents + bioventing, respectively; while 10.6% TPH reduction occurred in the un-amended soil microcosm A (natural bioattenuation). At the end of remediation period (day 28), the concentration of diesel oil (100,000 mg/kg) was reduced to 38,300, 21,300, and 8,500 mg/kg and correspondingly 61.7%, 78.7%, and 91.5% TPH reduction was achieved under bioventing, brewery waste effluents amendment (biostimulation-bioaugmentation), and brewery waste effluents amendment combined with bioventing (biostimulation-bioaugmentation + bioventing) treatments, respectively. This observation revealed that during the diesel oil biodegradation in soil, bioventing and amendment with brewery waste effluents individually resulted in a more effective bioremediation response than the natural bioattenuation.

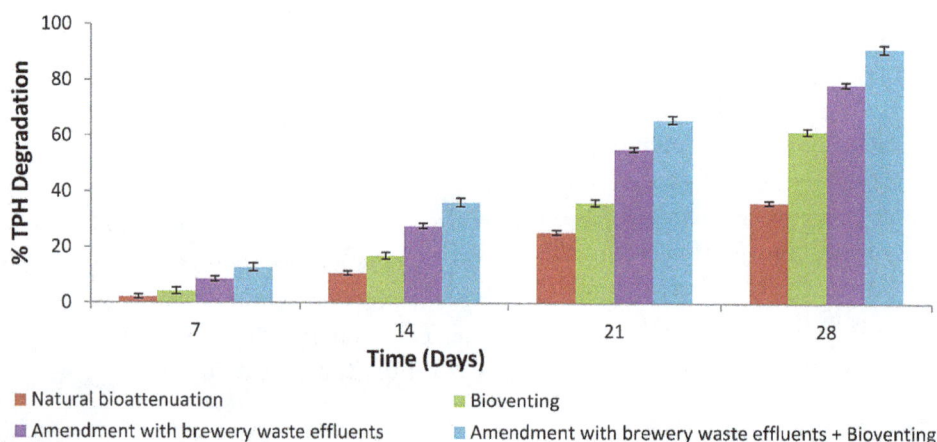

Figure 1. Time course for the biodegradation of diesel oil under natural bioattenuation, bioventing, brewery waste effluents amendment and combined brewery waste effluents amendment with bioventing. Bars indicate the average of triplicate samples while the error bars show the standard deviation

These observations may be due to the fact that bioventing increased the oxygen level in the soil required by the autochthonous microorganisms while the brewery waste effluents increased the nutrients level as well as the microbial load or density in the soil thus acting as biostimulation and bioaugmentation agent. Similar observations have been reported using bioventing technique [Morales et al., 2013; Thomé et al., 2014] as well as using mixture of cow, goat and poultry dungs (which acts as biostimulation-bioaugmentation agent) [Agarry and Jimoda, 2013b], and brewery spent grain [Abioye et al., 2012] Nevertheless, amendment with brewery waste effluents elicited a higher biodegradation of diesel oil than the bioventing treatment. However, Morales et al. (2013) reported that that bioventing strategy enhanced the bioremediation of diesel oil more than bioaugmentation. Generally, in this work, the combination of brewery waste effluents amendment and bioventing (biostimulation-bioaugmentation + bioventing) treatment strategy showed relatively greater %TPH degradation than the bioventing and brewery waste effluents amendment treatments respectively used alone during the whole period of remediation. Moller et al. (1996) and Thome et al. (2014) have respectively reported that biostimulation-bioventing elicited higher diesel oil biodegradation than biostimulation and bioventing respectively used alone.

Figure 2 shows the growth profiles of the total hydrocarbon degrading bacteria (THDB) in microcosms due to natural attenuation, bioventing, and brewery waste effluents amendments treatment methods.

Generally, it is seen that the microbial (THDB) counts increased from day 0 to day 28 in each of the treatment microcosms. For natural attenuation (microcosm A), the THDB count increased from 0.2 ± 1.00 to $0.1 \pm 2.54 \times 10^{10}$ cfu-g^{-1}, while it increased from 0.3 ± 1.10 to $0.2 \pm 4.10 \times 10^{10}$ cfu-g^{-1}, 0.1 ± 1.08 to $0.2 \pm 6.80 \times 10^{10}$ cfu-g^{-1}, and 0.3 ± 1.20 to $0.2 \pm 9.80 \times 10^{10}$ cfu-g^{-1} for bioventing (microcosm B), brewery waste effluents amendment (microcosm C), and brewery waste effluents amendment with bioventing (microcosm D), respectively. This corresponded to a growth increase of 364%, 530%, and 717% for soil microcosms B, C and D subjected to bioventing, brewery waste effluents amendments and brewery waste effluents amendment with bioventing, respectively. The percentage THDB growth in the unamended soil (natural attenuation) is 154%. This showed that the soil microcosms subjected to bioventing, brewery waste effluents amendments and brewery waste effluents amendment with bioventing enhanced the microbial growth rate which accounted for the higher microbial counts observed in all the soil microcosms subjected to bioventing and brewery waste effluents amendments either alone or in combination than the unamended soil microcosm (natural attenuation). The higher microbial count in soil microcosms subjected to bioventing and brewery waste effluents amendments either alone or in combination may be due to increased oxygen level and high nutrient level which stimulated increase in microbial population and activities thus leading to high energy (carbon) demand by the oil-degrading microbes.

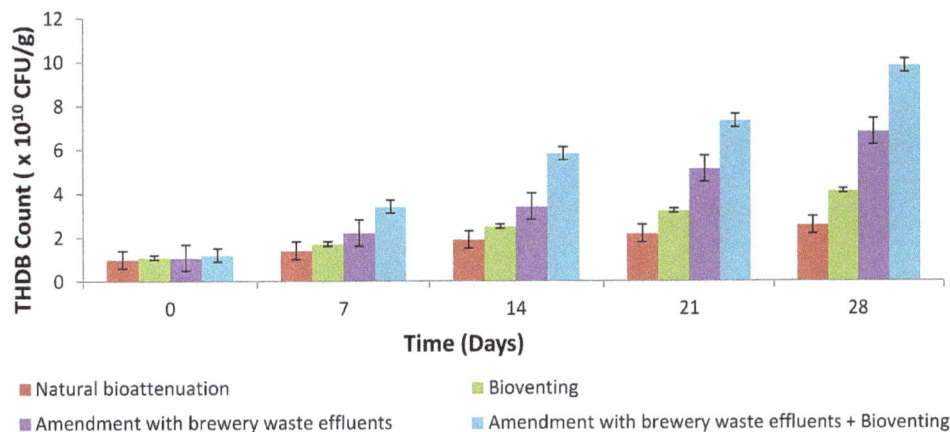

Figure 2. Time course for the growth of total hydrocarbon degrading bacteria (THDB) in diesel oil contaminated soil under natural bioattenuation, bioventing, brewery waste effluents amendment and combined brewery waste effluents amendment with bioventing treatments. Bars indicate the average of triplicate samples while the error bars show the standard deviation

Evaluation of Biodegradation Kinetics and Half-Life

First-order kinetic model equation (Eq. 4) fitted to the biodegradation data was used to determine the rate of biodegradation of diesel oil in the various remediation treatments which is illustrated in Figure 3.

The biodegradation data fitted well to the first-order kinetic model with high correlation coefficient (R^2) that lies between 0.90 and 0.98. The half-life times of diesel oil biodegradation was calculated using Eq. 5. The biodegradation rate constants (k) and half-life times ($t_{1/2}$) for the different remediation treatments are presented in Table 2.

It is to be noted that the higher biodegradation rate constants, the higher or faster is the rate of biodegradation and consequently the lower is the half-life times. Table 2 shows that the biodegradation of diesel oil in soil under combined brewery waste effluents amendments and bioventing treatment strategy had a higher k (0.108 day^{-1}) and lower $t_{1/2}$ (6.4 days) than that under brewery waste effluents amendment alone ($k = 0.069$ day^{-1} and $t_{1/2} = 10$ days), bioventing ($k = 0.043$ day^{-1} and $t_{1/2} = 16.1$ days), and natural attenuation ($k = 0.016$ day^{-1} and $t_{1/2} = 43.3$ days), respectively. Therefore, the value of the kinetic parameter showed that the degree of effectiveness of these treatment strategies in the clean-up of soil contaminated with diesel oil is in the following order: brewery waste effluents amendment with bioventing > brewery waste effluents amendment > bioventing > natural bioattenuation. Nevertheless, these observations indicate that bioventing and brewery waste effluents amendment alone or in combinations enhanced TPH reduction.

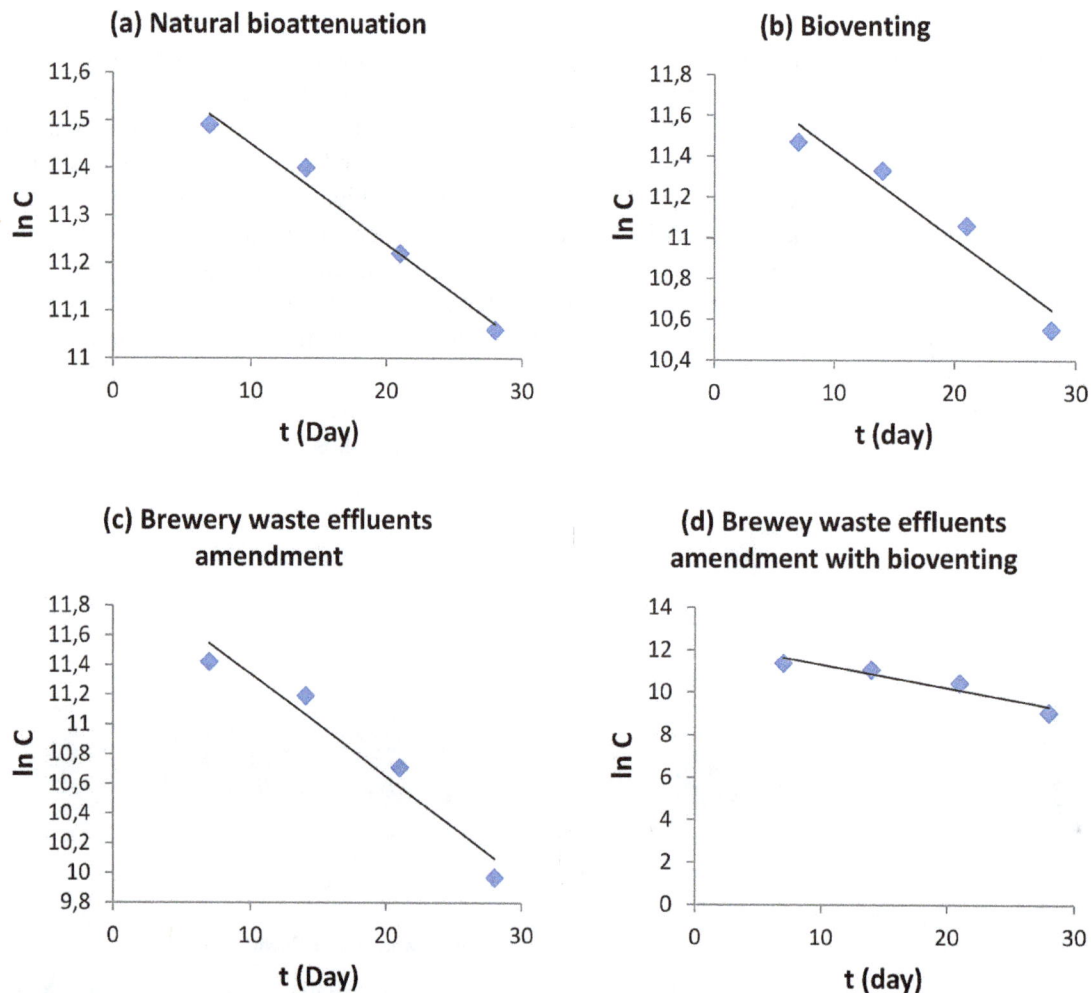

Figure 3. First-order kinetic model fitted to the diesel oil biodegradation data under (a) natural bioattenuation, (b) bioventing, (c) brewery waste effluents amendment, and (d) combined brewery waste effluents amendment with bioventing treatments

Table 2. The biodegradation rate constants (k) and half-life ($t_{1/2}$) time of diesel biodegradation in the various treatments

Microcosm code	Soil treatment	k (day^{-1})	R^2	$t_{1/2}$ (days)
A	Natural bioattenuation	0.016	0.983	43.3
B	Bioventing	0.043	0.929	16.1
C	Amendment with brewery waste effluents	0.069	0.947	10
D	Amendment with brewery waste effluents + Bioventing	0.108	0.909	6.4

CONCLUSIONS

From this present study, it can be concluded that the reduction of diesel oil in the contaminated soil indicates the presence of diesel-degrading microbial communities; and that the rate of diesel oil biodegradation in soil could be enhanced by bioventing and amendment with organic waste effluents that could serve as both biostimulation and bioaugmentation agents, respectively. The soil treatment under combined brewery waste effluents amendments and bioventing exhibited the highest degree of biodegradation with the highest biodegradation rate constant ($k = 0.108$ day^{-1}) and lowest half-life time ($t_{1/2} = 6.4$ days)) and the soil treatment under natural bioattenuation the least degradation with the lowest biodegradation rate constant ($k = 0.016$ day^{-1}) and highest half-life time ($t_{1/2} = 43.3$ days). Thus, the use of bioventing and biostimulation/bioaugmentation to enhance diesel oil biodegradation in the soil could be one of the severally sought bioremediation strategies of remediating natural ecosystem (environment) contaminated with petroleum hydrocarbons.

REFERENCES

1. Abioye P.O, Abdul Aziz A. and Agamuthu P. 2009. Enhanced biodegradation of used engine oil in soil amended with organic wastes. Water Air Soil Poll., 173–179.

2. Adesodun J.K. and Mbagwu J.S.C. 2008. Biodegradation of waste-lubricating petroleum oil in a tropical alfisol as mediated by animal droppings. Bioresource Technol., 99 (13), 5659–5665.

3. Agarry S.E., Owabor C.N., Yusuf R.O. 2012. Enhanced bioremediation of soil artificially contaminated with kerosene: Optimization of biostimulation agents through statistical experimental design. J. Pet. Environ. Biotechnol. 3, 120.

4. Agarry S.E., Aremu M.O., Aworanti O.A. 2013a. Kinetic modelling and half-life study on bioremediation of soil co-contaminated with lubricating motor oil and lead using different bioremediation strategies. Soil and Sediment Contam. An Int. J. 22 (7), 800–816.

5. Agarry S.E. and Jimoda L.A. 2013b. Application of Carbon-Nitrogen Supplementation from Plant and Animal Sources in In-situ Soil Bioremediation of Diesel Oil: Experimental Analysis and Kinetic Modelling.

6. Agarry S.E., Owabor C.N. and Yusuf R.O. 2010. Bioremediation of soil artificially contaminated with petroleum hydrocarbon mixtures: Evaluation of the use of animal manure and chemical fertilizer. Bioremediation J., 14 (4), 189–195.

7. Akinde S.B. and Obire O. 2008. Aerobic heterotrophic bacteria and petroleum-utilizing bacteria from cow dung and poultry manure. World J. Microbiol. Biotechnol., 24, 1999–2002.

8. Amanchukwu C.C., Obafemi A., Okpokwasili G.C. 1989. Hydrocarbon degradation and utilization by a palmwine yeast isolate. FEMS Microbiol. Lett. 57, 51–54.

9. APHA 1985. Standard Methods for Examination of water and wastewater. American Public Health Association Washington DC.

10. April T.M., Foght J.M. and Currah R.S. 2000. Hydrocarbon degrading filamentous fungi isolated from flare pit soils in northern and western Canada. Canadian J. Microbiol., 46 (1), 38–49.

11. Aronson D., Boethling R., Howard P., Stiteler W. 2006. Estimating biodegradation half-lives for use in chemical screening. Chemosphere 63, 1953–1960.

12. Bento F.M., Camargo F.A., Okeke B., Frankenberger Jr. T.W. 2003. Bioremediation of soil contaminated by diesel oil. Braz. J. Microbiol. 34 (Suppl. 1), 65–68.

13. Bremner J.M., Mulvaney C.S. 1982. Total nitrogen determination. In Method of Soil Analysis, vol. 2, ed. A.L. Page, R.H. Miller, and D.R. Keeney, pp. 595. Madison, WI: American Society of Agronomy.

14. Brook T.R., Stiver W.H., and Zytner R.G. 2001. Biodegradation of diesel fuel in soil under various nitrogen addition regimes. Soil Sediment Contamination 10, 539–553.

15. Das N. and Chandran P. 2011. Microbial degradation of petroleum hydrocarbon contaminants: an overview. Biotechnol. Res. Int., 1–13.

16. Dimitrov S., Pavlov T., Nedelcheva D., Reuschenbach P., Silvani M. et al. 2007. A kinetic model for predicting biodegradation. SAR QSAR Environ. Res. 18, 443–457.

17. Gallego J.R., Loredo J., Llamas J.F., Vazquez F. and Sanchez J. 2001. Bioremediation of diesel-contaminated soils: evaluation of potential in situ techniques by study of bacterial degradation. Biodegradation, 12, 325–335.

18. Greene E.A., Kay J.G., Jaber K., Stehmeier L.G., Voordouw G. 2000. Composition of soil microbial communities enriched on a mixture of aromatic hydrocarbons. Appl. Environ. Microbiol. 66, 5282–5289.

19. Inyang U.E., Bassey E.N., and Inyang J.D. 2012. Characterization of brewery effluent fluid. J. Eng. Appl. Sci. 4, 67–77.

20. Kirsten S., Hung L., Zytner R.G. 2005. Optimization of nitrogen for bioventing of gasoline contaminated soil. J. Environ. Eng. Sci. 4 (1), 29.

21. Krieg N.R., Holt J.G., Sneath P.H.A., Stanley J. T. and Williams S.T. 1994. Bergey's Manual of Determinative Bacteriology, 9th ed., Williams and Wilkins, Baltimore.

22. Lee T.H., Byun I.G., Kim Y.O., Hwang I.S., Park T.J. 2006. Monitoring biodegradation of diesel fuel in bioventing processes using in situ respiration rate. Water Science & Technology, 53 (4/5), 263

23. Mao L. and Yue Q. 2010. Remediation of diesel-contaminated soil by bioventing and composting technology. International Conference on Challenges in Environmental Science and Computer Engineering, pp. 3–6.

24. Margesin R., Hammerle M. and Tscherko D. 2007. Microbial activity and community composition during bioremediation of diesel-oil-contaminated soil: Effects of hydrocarbon concentration, fertilizers, and incubation time. Microbiol Ecol., 53, 259–269.

25. Matthies M., Witt J., Klasmeier J. 2008. Determination of soil biodegradation half lives from simulation testing under aerobic laboratory conditions: a kinetic model approach. Environ. Poll. 156, 99–105.

26. McLean E.O. 1982. Soil pH and lime requirement in methods in soil analysis: Chemical and microbiological properties. Part II, ed. C.A. Black. Madison, WI: American Society of Agronomy.

27. Molina-Barahona L., Rodriguez-Vázquez R., Hernández-Velasco M., Vega-Jarquin C., Zapata-Pérez O., Mendoza-Cantú A. and Albores A. 2004. Diesel removal from contaminated soils by biostimulation and supplementation with crop residues. Appl. Soil Ecol., 27, 165–175.

28. Møller J., Winther P., Lund B., Kirkebjerg K., and Westermann P. 1996. Bioventing of diesel oil-contaminated soil: Comparison of degradation rates in soil based on actual oil concentration and on respirometric data. J. Ind. Microbiol. 16 (2), 110–116.

29. Morales M., Maria A., Munoz S., Claudia Quintero P., Silvia L. 2013. Evaluation of natural attenuation, bioventing, bioaugmentation and bioaugmentation-bioventing techniques, for the biodegradation of diesel in a sandy soil, through column experiments. Gestion y Ambiente, 16(2), 83–94.

30. Nelson D.W., Sommers L.E. 1982. Determination of organic carbon. [In:] Method of Soil Analysis, vol. 2, ed. A.L. Page, R.H. Miller, and D.R. Keeney, 539. Madison, WI: American Society of Agronomy.

31. Okiemen C.O. and Okiemen F.E. 2005. Bioremediation of crude oil polluted soil. Effect of poutry droppings and natural rubber processing sludge application on biodegradation of petroleum hydrocarbon. Environ. Sci. 1(1), 1–8.

32. Olsen S.R., Sommers L.E. 1982. Determination of available phosphorus. [In:] Method of Soil Analysis, vol. 2, ed. A.L. Page, R H. Miller, and D.R. Keeney, 403. Madison, WI: American Society of Agronomy.

33. Onwurah I.N.E., Ogugua V.N., Onyike N.B., Ochonogor A.E. and Otitoju O.F. 2007. Crude oil spills in the environment, effects and some innovative clean-up biotechnologies. Int. J. Environ. Res., 1, 307–320.

34. Osuji L.C., Egbuson E.J.G. and Ojinnaka C.M. 2005. Chemical reclamation of crude-oil-inundated soils from Niger Delta. Nigeria. Chem. Ecol., 21(1), 1–10.

35. Richard J.Y. and Vogel T.M. 1999. Characterization of a soil bacterial consortium capable of degrading diesel fuel. Int. Biodeter. Biodegrad., 44, 93–100.

36. Seklemova E., Pavlova A. and Kovacheva K. 2001. Biostimulation based bioremediation of diesel fuel: field demonstration. Biodegradation, 12, 311–316.

37. Singh D. and Fulekar M.H. 2009. Bioremediation of benzene, toluene and o-xylene by cow dung microbial consortium. JABs 14, 788–795.

38. Sinkkonen S., Paasivirta J. 2000. Degradation half-life times of PCDDs, PCDF sand PCBs for environmental fate modeling. Chemosphere 40, 943–949.

39. Thomé A., Reginatto C., Cecchin I. and Colla L. 2014. Bioventing in a Residual Clayey Soil Contaminated with a Blend of Biodiesel and Diesel Oil. J. Environ. Eng., 10.1061/(ASCE)EE.1943-7870.0000863, 06014005.

40. Tsai T.T., Kao C.M., Surampalli R.Y., Chien H.Y. 2009. Enhanced Bioremediation of Fuel-Oil Contaminated Soils: Laboratory Feasibility Study. J. Environ. Eng. 135, 845–853.

41. Wang J.L., L.P. Han, H.C. Shi, and Y. Qian 2001. Biodegradation of quinoline by gel immobilized Burkholderia sp. Chemosphere 44, 1041–1046.

42. Wolicka D., Suszek A., Borkowski A. and Bielecka A. 2009. Application of aerobic microorganisms in bioremediation in situ of soil contaminated by petroleum products. Bioresour. Technol., 100, 3221–3227.

43. Yakubu M.B. 2007. Biodegradation of Lagoma crude oil using pig dung. Afri. J. Biotechnol., 6, 2821–2825.

44. Yeung P.Y., Johnson R.L., Xu J.G. 1997. Biodegradation of petroleum hydrocarbons in soil as affected by heating and forced aeration. J. Environ. Quality 26, 1511–1576.

45. Zahed M.A., Abdul Aziz H., Isa M.H., Mohajeri L., Mohajeri S., Kutty S.R.M. 2011. Kinetic modeling and half life study on bioremediation of crude oil dispersed by Corexit 9500. J. Hazard Mater. 185, 1027–1031.

46. Zhang S., Wang X., Zhu R., Li H., Wang P., Yang J., Lin K., Gu J., and Liu Y. 2014. Aerobic biodegradation of trichloroethylene by a bacterial community that uses hydrogen peroxide as sole oxygen source. http://www.paper.edu.cn/download/downpaper/201406-149.

47. Zhu X, Venosa A.D., Suidan M.T., Lee K. 2001. Guidelines for the Bioremediation of Marine Shorelines and Freshwaters Waterlands. US Environmental Protection Agency Office of Research and Development National Risk Management Research Laboratory.

12

THE ROLE OF PLANTS IN EXPERIMENTAL BIOLOGICAL RECLAMATION IN A BED OF FURNACE WASTE FROM COAL-BASED ENERGY

Kazimierz H. Dyguś[1]

[1] Faculty of Ecology, University of Ecology and Management, 12 Olszewska Str., 00-792 Warsaw, Poland, e-mail: dygus@wseiz.pl

ABSTRACT

In the model experiment, an assessment of the role of plants in the reclamation of the bed of combustion waste from coal-based power plants fertilised with compost and sewage sludge. The bed of combustion waste was stored in cylindrical containers with a diameter of 80 cm (0.5 m² of surface) and the height of 100 cm. The first stage of the experiment was carried out in 2006–2007. Then the bed was fertilised with four types of compost and sewage sludge, and then seeded with four species of grasses and white mustard. The second stage was undertaken in 2011–2013. In 2011, mixture of four species of grasses and white mustard was seeded on the same bed. It was assumed that the continuation of research in the second stage, whose results of are presented in this paper, will show a broader spectrum of vegetation changes, what will accurately track the process of biological reclamation of the bed of combustion waste. The aim of the study was to evaluate the effectiveness of reclamation in the experiment, based on the percentage estimates of the coverage of species and crop yields. During the three-year (second stage) experiment 78 species of self-seeding plants belonging to 19 taxa in the rank of families and 11 syntaxonomic groups were recorded. The most numerous were the families: aster family, grass family, papilionaceous family, goosefoot family and cabbage family. Among the syntaxonomic groups the dominating species belonged to the class *Stellarietea mediae*, *Molinio-Arrhenatheretea* and *Artemisietea vulgaris*. Among the forms of life hemicryptophyte and therophytes were the most represented. Highest total yields of plants were found in model containers with Complex compost and Radiowo compost and the model of sewage sludge. Based on the estimated models in each degree of coverage of species and crop yield, the highest reclamation efficiency was demonstrated in the models of reclamation of composts Complex and Radiowo, as well as in the model of sewage sludge. The lowest efficiency was demonstrated in models of composts ZUSOK and plant composts.

Keywords: model experiment, biological reclamation, flora and vegetation, crop plant, furnace waste of coal-based power plant, compost, sewage sludge.

INTRODUCTION

Landfills of coal-based power plants, due to their fine grain structure manifest high susceptibility to wind and water erosion. Technical ways of fixing this type of landfills do not eliminate the nuisance of dust. Properly selected and sown plant species can form continuous vegetation cover on the surface of landfills, and thus protect the landfill from the erosive action of wind and water, migration of heavy metals into groundwa-

ter and improve the aesthetics of the landscape [Siuta 2005, Antonkiewicz, Radkowski 2006].

Biological reclamation of landfill furnace, with the use of deposits of soil formation and vegetation, is more and more often used and improved [Hryncewicz et al. 1972, Żak 1972, Wysocki 1984, 1988, Siuta 2002, Dyguś et al. 2012]. For this purpose, there have been many experimental studies (model, lysimeter, field) [Kozłowska 1995, Siuta et al. 1997, Siuta 2005, Siuta, Kutla 2005, Siuta et al. 2008, Klimont 2011].

Methods of biological reclamation of land resources require large amounts of humus and significant financial resources. Initiating the process of soil formation to obtain appropriate habitat conditions for the growth and formation of plant cover is a cheaper solution. For this purpose, organic fertilizers are used, due to their high nutrient and cariogenic substance content. They are mainly sewage sludge, compost, municipal waste, peat, etc. The introduction of these substances onto the surface layer of furnace ashes initiates biological life and soil formation process, facilitating conditions for the development of plants. Organic matter forms an absorbing complex for nutrients and water [Gilewska 1999, Gilewska, Przybyła 2011, Polkowski, Sulek 1999, Siuta 2005, Siuta 2007, Siuta et al. 2008, Klimont 2011].

Appropriately conducted reclamation generates favourable conditions for the formation of the vegetation cover consisting of agronomically introduced and spontaneously grown reclamation plants [Góral 2001]. The plants which are most useful in reclamation Góral [2001] enumerates species of the families papilionaceous family and crossed cabbage family, in case of the use of sewage sludge; on the other hand, when composts are used a similar role can be played by segetal weeds of the family from goosefoot family and some taxa from the group of ruderal plants.

The aim of the study was to evaluate the effectiveness of reclamation on an experimental bed of combustion waste fed with several types of nutrients. This assessment was obtained on the basis of the percentage coverage of species and estimate crop yields and ecological characteristics of plant species.

MATERIALS AND METHODS

The richness of species and plant yield was studied on the model bed ash from the heat and power plant Kawęczyn. The deposits were stored in cylindrical containers with a diameter of 80 cm (0.5 m² area) and a height of 100 cm (Figure 1).

Model deposits had the following chemical composition:
- main ingredients: SiO_2 – 48,5%; Al_2O_3 – 4,8%; Fe_2O_3 – 2,8%; CaO – 2,8%; MgO – 2,7%; K_2O – 2,1%; TiO_2 – 1,1%; Na_2O – 0,8%; MnO_2 – 0,5%;
- heavy metals, in mg/100g: Zn – 90; Cu – 59; Pb – 50; Ni – 42; Co – 17; Cd – 15; Cr – 46.

Figure 1. The construction and dimensions of an experiment container

The containers with waste deposits were fertilized with following substances:
- experimentally produced plant compost from green municipal waste (kr) [Madej 2007, Madej et al. 2010];
- „Radiowo" compost from non-selectively collected municipal waste (kRa);
- compost from municipal waste processing plant – Zakład Unieszkodliwiania Stałych Odpadów Komunalnych „ZUSOK" (kZ);
- AG – COMPLEX compost made from green municipal waste from Warsaw (kC) [Opaliński 2007];
- sewage sludge from Radzymin plant (O);
- Mineral fertiliser N, P_2O_5, K_2O (NPK).

For one variant of the experiment there were six containers, two for each of the three doses of nutrients (single, doubled and tripled) (Figure 4–6).

The scheme of the experiment and reclamation dose of fertilizers is presented in Table 1, and their characteristics in Table 2 [Siuta et al. 2008].

The first stage of the experiment was carried out in 2005 and 2006 [Siuta et al. 2008]. In 2007–2010, plant vegetation and soil formation proceeded without any interference.

Prior to the second stage of the experiment (2011–2013) ground (dry) mass of the plant was removed, leaving the root part. In the spring of 2011. mixture of four grass species was seeded: perennial ryegrass var. Stadion, smooth meadow-grass var. Evona, tall fescue var. Starlett, red fescue var. Maxima and one species of dicotyledonous plant –

Table 1. Reclamation dose of fertilizers

Substances introduced	Doses					
	dm³/0.5 m²			m³/ha		
Sewage sludge (O)	5.0	10.0	15.0	100	200	300
Plant compost (kr)	5.0	7.5	10.0	100	150	200
Radiowo compost (kRa)	5.0	7.5	10.0	100	150	200
ZUSOK compost (kZ)	5.0	7.5	10.0	100	150	200
Complex compost (kC)	5.0	7.5	10.0	100	150	200
NPK	g/0.5 m²			kg/ha		
Nitrogen (N)	7.5	10.0	12.5	150	200	250
Phosphorus (P_2O_5)	3.5	4.5	6.0	70	90	120
Potassium (K_2O)	5.0	7.5	10.0	100	150	200
Total NPK	16.0	22.0	28.5	320	440	570

Table 2. Properties of organic fertilizers used in the experiment

Properties	Type of fertilizer				
	Radiowo compost (kRa)	Plant compost (kr)	ZUSOK compost (kZ)	Complex compost (kC)	Sewage sludge (O)
Organic substances content in d.m., %	25.8	42.8	24.0	74.4	37.7
Organic carbon content in d.m., %	12.8	19.3	11.9	39.2	21.2
Nitrogen content in d.m., %	1.1	2.1	1.0	5.8	1.4
Phosphorus content in d.m., %	0.6	0.8	0.5	1.5	0.6
Potassium content in d.m., %	0.5	1.0	0.8	0.3	1.3
C : N	10.8	9.2	11.9	6.1	15.1
pH	7.8	7.2	7.9	7.1	7.9
Fresh substances, g/dm³	807	330	790	950	490
Dry substances, g/dm³	472	219	504	185	350
Water content, %	40.1	31.7	36.9	80.9	40
Nitrogen content, gN/dm³	5.2	4.6	5.0	11.0	n.o.

n.o. – not determined.

white mustard (Figure 2). After sowing the plants the model deposits were fed with a complex mineral fertilizer (Azofoska) containing 13.6% of nitrogen, 6.4% P_2O_5, 19.1% K_2O 4.5% MgO and 23.0% SO_3 with low Cu, Fe, Mn, Mo and Zn content.

In the growing seasons of Stage II of the experiment detailed observations of the flora were carried out. The listed plant species were analyzed for floristic and phytosociological, taxonomic and ecological characteristics. Three-year dynamics of plants changes in containers during the multi-variant experiment were expressed as the percentage share of cover crops in different experimental models. Based on the species composition of plants, systematic, syntaxonomic, ecological, geographical and historical groups were distinguished, and the forms of life by Raunkiaer and environmental indicators according to Ellenberg were described. Over the three years of the experiment, each month during the vegetation period the species inventoried and their percentage of coverage was determined.

Three times a year (in June, August and October) plants crops were harvested (Figure 3). The collected plant biomass was dried in a drying

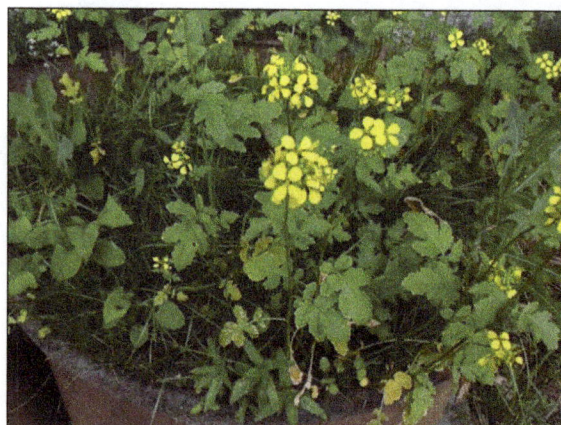

Figure 2. A container with vegetating grass and white mustard

Figure 3. The row of containers with grass and white mustard (*Sinapis alba*) during the first year of experience (on the right); containers on the left after plants crop harvesting

oven SLW STD INOX 240 at 75 °C, to obtain a dry mass, and then weighed to 0.0001 g on a laboratory scale Radwag XA 310.

The flora underwent ecological analysis (Table 3). Taxonomic data was compiled according to "The key to determine of vascular plants in Polish lowlands" [Rutkowski, 1998]. The names of the syntaxonomic groups according to Matuszkiewicz was used [2001]. Classification and share of flora life forms according to Raunkiaer was used [Zarzycki et al. 2002]. Geo-historical analysis of the plants was based on the studies by Rutkowski [1998] and Mirek et al. [2002]. The average coverage of plants with the models of fertilization were presented in a Braun-Blanquet scale (1964), including the modifications by Westhoff and van der Maarel (1978).

Latin names of vascular plants was used according to Mirek et al. [2002].

RESULT AND DISCUSSION

Ecological analysis of the flora

In 2011–2012, 67 species of plants were found in the containers, among them as many as 62 species which settled spontaneously. In 2013, the number of spontaneous species rose to 78. The floristic composition of plants has proven useful in assessing the effectiveness of reclamation sites. The abundance of plants species and their coverage on the surface of the containers were analyzed. The number of species and their coverage are important indicators of the reclamation efficiency of the applied fertilizers [Gutkowska,

Pawluśkiewicz 2006, Dyguś, Madej 2012, Dyguś et al. 2014]. It is worth mentioning that in the process of biological reclamation, proper selection of plants is important [Majtkowski et al. 1999, Nowak 2006]. The flora inventoried in 2011–2012 belonged to 17 taxa, in the rank of families, among which the dominant species were members of aster family (*Asteraceae*), goosefoot family (*Chenopodiaceae*) and cabbage family (*Brassicaceae*). During the 2013 growing season 19 taxa in the rank of the family were reported and their families dominance structure was partially changed. Still the species of aster family dominated, but the species of the grass family (*Poaceae*) and papilionaceous family (*Fabaceae*) became subdominants (Table 4). This is due to the predominance of anemochory in these families that produce large amounts of light, volatile seeds, allowing them to spread over long distances. The most numerous botanical types include goosefoot (*Chenopodium*), knotweed (*Polygonum*), sowthistle (*Sonchus*), bluegrass (*Poa*), orache (*Atriplex*), clover (*Trifolium*), plantain (*Plantago*) and sedge (*Carex*). This is a typical distribution of botanical taxa in the spontaneous formation of the vegetation cover in the early stages of reclamation of industrial and municipal landfills [Gutkowska, Pawluśkiewicz 2006, Rostański 2006, Dyguś et al. 2012].

Phytosociological and syntaxonomic analysis of flora of the years 2011–2013 proved the presence of 11 groups in the rank of classes. Among the distinguished groups of the studied flora more than 40% of the species belonged to nitrophilous communities of cultivated fields (class *Stellarietea mediae*). Two further groups of species were associated with the meso- and eutrophic communities of meadows (Class *Molinio-Arrhenatheretea*) and anthropogenic ruderal habitats (Class *Artemisietea vulgaris*). Other groups were represented by negligible number of syntaxonomic species (Table 5).

The spectrum of life forms of vascular plants vascular plants was partially changed. Among the identified species in 2011–2012 a clear dominance of annual plants (therophytes) was demonstrated, which accounted for almost half of the species of the studied flora. A relatively high proportion (39%) in the flora were perennials – hemicryptophytes. However, in 2013, hemicryptophytes clearly dominated (45%), with a simultaneous tendency to reduce the share of therophytes. Throughout the observation period there was a

Table 3. Ecological analysis of vegetating plants in experimental containers on the bed of combustion waste fed with fertilizers

No.	Species	Fż	Ecological indicators							Gt	Gg-h	Ge	Gs	Average coverage of plant in fertilization models					
			W	T	O	G	H	S	M					NPK	kRa	kC	kr	kZ	O
1	2	3	4							5	6	7	8	9					
1	*Acer negundo – s*	M	3–4	3–4	3–5	3–4	1–2	–	–	Ac	Kn	Ll	Sp	+	+	+			
2	*Achillea millefolium*	H	2–3	3–4	3–4	4	1–2	1	–	Ast	Ap	Ł	M–A				+	+	1
3	*Agrostis stolonifera*	H	4	3–4	3–5	2–4	1–2	1	1	Po	Ap	R	M–A	+		+	+	+	+
4	*Anthriscus sylvestris*	H	3	4–5	4	4	2	–	–	Api	Ap	R	Av				+		+
5	*Artemisia vulgaris*	H	3	4	4–5	4	2	–	1	Ast	Ap	R	Av	+	+	1	1	+	+
6	*Atriplex patula*	T	3	3–5	4–5	4–5	1–2	1	1	Che	Ar	Sg	Sm				+		
7	*Atriplex prostrata*	T	4	4	4	4–5	1–2	2	–	Che	Kn	R	Sm				+		
8	*Atriplex tatarica*	T	2	2–3	5	3–5	1–2	1	–	Che	Kn	Sg	Sm				+	+	
9	*Bidens frondosa*	T	3	4	4–5	2–4	2	1	–	Ast	Kn	R, Nm	Bt	+	1	+	+	1	+
10	*Bidens tripartita*	T	4–5	4	4–5	5	2	–	–	Ast	Ap	Nw	Bt	+	1	+	+	+	+
11	*Calamagrostis epigejos*	G,H	3	3	3	3	1	1	–	Po	Ap	R, O	Ea	+		+			
12	*Capsella bursa–pastoris*	H,T	3	4	4	4	1–3	–	–	Bra	Ar	Sg	Av			1	+	+	1
13	*Chamomilla recutita*	T	2–3	4	5	–	–	–	–	Ast	Ar	R	Sm						
14	*Chenopodium album*	T	3	4–5	4	3–5	2	–	–	Che	Ap	Sg	Sm	2a	2a	2a	2a	2a	2a
15	*Chenopodium glaucum*	T	4	4	4–5	2–5	2	1	1	Che	Ap	Nw	Bt, Sm			+		+	
16	*Chenopodium hybridum*	T	3	4	4–5	3–5	1–3	–	–	Che	A	R	Sm		+	+	+	+	
17	*Chenopodium murale*	T	2–3	3	4–5	3–5	1–2	–	–	Che	Ar	R	Sm		1	1		+	+
18	*Chenopodium polyspermum*	T	3	3–5	3–4	4–5	2–3	–	–	Che	Ap	Sg	Sm		+	+			
19	*Chenopodium urbicum*	T	3–4	4–5	4	2–4	2	–	–	Che	Ar	R	Sm			+			
20	*Cirsium arvense*	G	2–3	3–4	3–5	3–5	2	–	–	Ast	Ap	R	Sm		+	+	+	+	1
21	*Cirsium oleraceum*	H	4–5	4	4–5	4–5	2–3	–	–	Ast	Ap	Ł	M–A	+		+	+		
22	*Crisium vulgare*	H	3	3	3–4	3–5	1–2	–	–	Con	Ap	R	Av	+	+	2m			+
23	*Convolvulus arvensis*	G, H, li	2–3	3	3–5	4–5	2	1	1	Ast	Ap	R	Sm				+		+
24	*Conyza canadensis*	T, H	2–3	3	3–4	3–4	2	–	–	Ast	Kn	R	Sm	1	2m	2a	2m	1	1
25	*Dactylis glomerata*	H	3	4–5	4–5	4	2	1	–	Po	Ap	Ł	M–A		+	+	+	+	+
26	*Daucus carota*	H	3	4	4–5	4	2	1	–	Api	Ap	R	M–A				+		+
27	*Descurainia sophia*	T	3	4	4	3–4	2	–	–	Bra	Ar	Sg	Av	+	+		+	+	1
28	*Elymus repens*	G	3	3–4	3–5	4	1–2	1	–	Po	Ap	R	Sm	+	+		+	+	+
29	*Epilobium montanum*	H	3	4	4	4	2	–	–	Ona	Ap	Ll	Av						
30	*Erigeron annuus*	H, T	3	3	3–4	2–4	2	–	–	Ast	Kn	R	Av	+			+	+	+
31	*Erigeron ramosus*	H	3	3	3–4	2–4	2	–	–	Ast	Kn	R	Av	+		+		+	
32	*Erysimum cheiranthoides*	T	2	2–3	3	2–4	1–2	–	–	Bra	Ar	Sg	Av						
33	*Fallopia convolvulus*	T, H	3	3–4	3–4	2–5	2	–	–	Pol	Ar	Sg	Sm				+	+	1
34	**Festuca arundinacea var. Starlett**	H	3–4	4	4	4–5	2	2	–	Po	Up, Ap	R	M–A	2a	2a	2a	2a	2a	2b
35	**Festuca rubra var. Maxima**	H	2–4	3	4	3–4	3	2	1	Po	Up, Ap	Ł	M–A	2a	2m	2a	2a	2a	2a

Cont. table 3.

N	Species	LF								Fam											
36	*Geranium pyrenaicum*	T	3	3-4	3-4	4	2	-	-	Ger	Kn	R	Av	2a		2a		+		+
37	*Impatiens parviflora*	T	3	4	4	3-4	2	-	-	Bal	Kn	Li	Q-F	+		+	+	+	2a	2m
38	*Lamium maculatum*	H	4	4	4	3-4	2	-	-	Lam	Ap	Li, Z	Av	+	+		2a	+		+
39	*Lolium perenne* var. **Stadion**	H	3	4	4	4	2	1	1	Po	Up, Ap	Ł	M-A	2a	2a	2a	2a	+	2a	2b
40	*Lotus corniculatus*	H	3-4	3-4	3-5	4	2	1	1	Fab	Ap	Ł	M-A						+	
41	*Lycopus europaeus*	H, Hy	5	4	4-5	4-5	2-3	1	-	Lam	Ap	Nw	Ag	+	+		+			
42	*Matricaria maritima* ssp. *inodora*	H, T	3	4	4	3-4	2	-	-	Ast	Ar	Sg	Sm	1	+	+	+	+	+	
43	*Medicago falcata*	H	2-3	3-4	5	2-5	2	-	-	Fab	Ap	O	F-B	1	2m	2m	+	1		1
44	*Medicago lupulina*	H, T	2-3	3-4	3-5	2-4	1-2	-	-	Fab	Ap	Mks	Sm	2a	2a	2a	2a	2a	2m	2m
45	*Oxalis fontana*	G	3	3-4	3-4	3-4	2	-	-	Oxa	Kn	Sg	Sm						+	1
46	*Phleum pratense*	H	2-3	3-4	4-5	1-3	2	-	-	Po	Ap	Ł	M-A	+	+	+	+	+	+	+
47	*Plantago intermedia*	H, T	4	3-4	3-4	3-5	2	-	-	Pla	Ap	Nm	I-N	1		+	+	+	+	+
48	*Plantago lanceolata*	H	2-4	3-4	4	4	3	1	-	Pla	Ap	Ł	M-A		+	+		+	+	+
49	*Plantago major*	H	3-4	4-5	3-5	3-5	2-3	1	-	Pla	Ap	R	Sm	1		+	+	+	1	
50	*Poa angustifolia*	H	2-3	3	4-5	3-5	2-3	-	-	Po	Ap	Mks	F-B, M-A	+	+	+	+	+	+	+
51	*Poa annua*	H, T	3	4	4	3-5	2	-	-	Po	Ap	R	Sm	+	+	+	+	+	+	+
52	*Poa compressa*	H	2	3	5	2-5	2	-	-	Po	Ap	R	F-B	+	+			+	+	
53	*Poa pratensis* var. **Evona**	H	3	4	4	4	2	1	-	Po	Up, Ap	Ł	M-A	2a	2a	2b	2a	2a	2b	2b
54	*Polygonum aviculare*	T	3	3-4	4-5	2-5	1-2	-	1	Pol	Ap	R	Sm	+	2a	+	1	1	1	+
55	*Polygonum lapathifolium*	T	3-4	4-5	4	2-4	2-3	-	-	Pol	Ap	Nw	Bt				+		+	+
56	*Polygonum persicaria*	T	3	3-4	4	3-4	2	-	-	Pol	Ap	Sg	Sm						+	
57	*Quercus robur – s*	M	3-4	3-4	4	4	2	-	-	Fag	Ap	Ll	Q-F				+			+
58	*Raphanus raphanistrum*	T	3	3	3	2-4	2	-	-	Bra	Ar	Sg	Sm	1	1		+	+	+	+
59	*Rumex acetosa*	H	3-4	4	4	4	2	-	-	Pol	Ap	Ł	M-A				+		+	
60	*Rumex acetosella*	G, H, T	2	2	2-3	2-4	1-2	-	-	Pol	Ap	Mks	K-C		+	+	+	+		
61	*Rumex crispus*	H	3-4	4	4	3-4	2	1	-	Pol	Ap	R	Sm				+	+		
62	*Silene vulgaris*	H, C	3	3	4-5	2-4	2	-	-	Car	Ap	Z	Sm							+
63	*Sinapis alba*	T	bd	bd	bd	bd	bd	bd	bd	Bra	Up, Kn	Sg	Sm	2b	3	2b	+	+	2b	1
64	*Sinapis arvensis*	T	3	4	3-4	3-4	2	-	-	Bra	Ar	Sg	Sm	+	+	+	+		+	+
65	*Sisymbrium loeselii*	H, T	2	3	4-5	3-4	2	-	-	Bra	Kn	R	Sm	1	1	1	+	1	1	
66	*Sisymbrium officinale*	T	3	4-5	4-5	2-4	2	-	-	Bra	Ar	Sg	Av	+	+	+		+	+	+
67	*Solidago canadensis*	G, H	3-4	4	4	2-4	2	-	-	Ast	Kn	R	Av	2a	2a	2a	2a		+	1
68	*Solidago gigantea*	G, H	3-4	4	-	-	-	-	-	Ast	Kn	R	Av	1		+	+		+	1
69	*Sonchus arvensis*	G, H	3-4	3-4	3-5	3-5	2	-	-	Ast	Ap	Sg	Sm	2a	2a	1	2a	1	2a	2a
70	*Sonchus asper*	T	3	4	4	4	2	-	-	Ast	Ap	Sg	Sm	+	+		+	+	+	
71	*Sonchus oleraceus*	H, T	3	4	4	4	3	-	-	Car	Ar	Ł	Sm	1	+	+	1	+	+	1
72	*Stellaria media*	T, H	3-4	4-5	3-5	3-5	2	-	-	Car	Ap	Sg	Sm	2m	1	2m	2a	+	2m	2a
73	*Taraxacum officinale*	H	3	4	4-5	4-5	2	1	1	Ast	Ap	Ł	M-A	1	2a	1	1	1	2a	1

Cont. table 3.

74	Trifolium arvense	T	2	1–2	3–5	1–3	2	–	–	Fab	Ap	Mp	K–C	2a			1	1
75	Trifolium dubium	T	3	4	4	4	2	–	–	Fab	Ap	Ł	M–A			+	1	
76	Trifolium hybridum	H	4	4	4	4–5	3	–	–	Fab	Ap	R	M–A			1	1	+
77	Trifolium repens	C, H	3–4	4	4	4	2	1	–	Fab	Ap	Ł	M–A	2m	2a	2a	1	1
78	Tussilago farfara	G	3–4	3–4	4–5	4–5	1–2	–	–	Ast	Ap	R	Av	2m	2m	2m		
79	Veronica persica	T	3	4–5	3–5	3–5	2	–	–	Scr	Kn	Sg, R	Sm					+
80	Vicia cracca	H	3	4	4–5	4	2	–	–	Fab	Ap	Ł	M–A	+	+		+	
81	Vicia hirsuta	T	3	3–4	3–4	2–4	2	–	–	Fab	Ar	Sg	Sm		+	+		
82	Vicia sativa	T	3	3–4	4	3–4	2	–	–	Fab	Kn	Sg	Sm				+	
83	Viola arvensis	T	3	3–4	3–4	2–4	2	–	–	Vio	Ap	Sg	Sm	+				

Comments:

1. No. – number.

2. Species – the Latin name of the species.

3. Fz – type of growth: C – herbaceous chamaephyte, H – hemicryptophyte, G – geophyte, T – therophyte, Hy –hydrophyte and helophyte, li – liana.

4. Ecological indicators: W – soil moisture value, T – trophy value, O – deposit acidity value, G – soil granulometric value, H – organic matter content value, S – value of resistance to NaCl content in soil, M –value of resistance to increased heavy metal content in the soil, (–) – not available.

5. Gt – taxonomic groups in the rank of families: Ac – Aceraceae, Ast – Asteraceae, Che – Chenopodiaceae, Po – Poaceae, Bra – Brassicaceae, Con – Convolvulaceae, Api – Apiaceae, Ona – Onagraceae, Pol – Polygonaceae, Ger – Geraniaceae, Bal – Balsaminaceae, Lam – Lamiaceae, Fab – Fabaceae, Oxa – Oxalidaceae, Pla – Plantaginaceae, Scr – Scrophulariaceae, Car – Caryophyllaceae, Fag – Fagaceae, Vio – Violaceae.

6. Gg-h – geographical and historical genres: Ap – apophytes, Ar – archeophytes, Kn – kenophytes, A – presumably anthopophytes, Up – cultivated plants.

7. Ge – ecological groups of species: R – ruderal, Sg – segetal, Ł – meadow species, Ll – broadleaf forests, O –tall herb fringe communities, Mp – sandy grasslands, Mks – xerothermic grasslands, Z – bush, Nm – silt plants.

8. Gs – syntaxonomic groups: Av – Artemisietea vulgaris, Sm – Stellarietea mediae, M-A – Molinio-Arrhenatheretea, F-B – Festuco-Brometea, Ea – Epilobietea angustifolii, Sp – Salicetea purpureae, Q-F – Querco-Fagetea, K-C – Koelerio glaucae-Corynephoretea canescentis, I-N – Isoëto-Nanojuncetea, Bt – Bidentetea tripartiti, Ag – Alnetea glutinosae.

9. The average of plant coverage in various models expressed by Braun-Blanquet scale (1964), taking into account the modification by Westhoff and Van der Maarel (1978): NPK – fertilizer N, P₂O₅, K₂O without organic fertilization, kRa – Radiowo compost from municipal waste, kC – Complex compost from municipal and plant waste, O – sludge from municipal wastewater treatment. kZ – ZUSOK compost made from municipal and plant waste, kr – plant compost made from grass,

10. Other abbreviations: na - not available; s – seedling.

Species highlighted in bold were sown in 2011 (4 species of grasses and white mustard *Sinapis alba*).

Table 4 Share of families in the flora in the third year of the model experiment

Families	Flora total		Number and percentage of species in experimental models												
			NPK		kRa		kC		kr		kZ		O		
	number	%	number	%	number	%	number	%	number	%	number	%	number	%
Asteraceae – aster family	19	22.9	14	35.0	13	33.3	15	29.4	12	24.0	16	29.1	12	22.2
Poaceae – grass family	12	14.5	11	27.5	10	25.6	11	21.6	10	20.0	10	18.2	10	18.5
Fabaceae – popilionaceous family	10	12.1	5	12.5	3	7.7	6	11.7	5	10.0	6	10.9	6	11.1
Chenopodiaceae – goosefoot family	9	10.8	1	2.5	3	7.7	6	11.7	5	10.0	5	9.1	2	3.7
Brassicaceae – cabbage family	8	9.6	2	5.0	7	18.0	4	7.8	4	8.0	8	14.5	8	14.8
Polygonaceae – buckwheat family	7	8.4	2	5.0	2	5.1	2	3.9	6	12.0	4	7.3	2	3.7
Plantaginaceae – plantain family	3	3.6	1	2.5	–	–	2	3.9	2	4.0	3	5.5	2	3.7
Lamiaceae – mint family	2	2.4	1	2.5	–	–	1	2.0	–	–	–	–	1	1.9
Apiaceae – carrot family	2	2.4	–	–	–	–	–	–	2	4.0	–	–	2	3.7
Caryophyllaceae – pink family	2	2.4	1	2.5	–	–	1	2.0	1	2.0	1	1.8	2	3.7
Convolvulaceae – morning–glory family	1	1.2	–	–	–	–	–	–	1	2.0	–	–	1	1.9
Onagraceae – evening–primroses family	1	1.2	–	–	–	–	–	–	–	–	–	–	1	1.9
Geraniaceae – geranium family	1	1.2	–	–	–	–	–	–	1	2.0	–	–	1	1.9
Oxalidaceae – sorrel family	1	1.2	–	–	–	–	–	–	–	–	1	1.8	1	1.9
Balsaminaceae – balsamina family	1	1.2	1	2.5	–	–	1	2.0	1	2.0	1	1.8	1	1.9
Scrophulariaceae – figwort family	1	1.2	–	–	–	–	–	–	–	–	–	–	1	1.9
Violaceae – violet family	1	1.2	–	–	–	–	1	2.0	–	–	–	–	–	–
Aceraceae – maple family	1	1.2	1	2.5	1	2.6	1	2.0	–	–	–	–	–	–
Fagaceae – beech family	1	1.2	–	–	–	–	–	–	–	–	–	–	1	1.9
Total	83	100.0	40	100.0	39	100.0	51	100.0	50	100.0	55	100.0	54	100.0

Comments:

NPK – fertilizer N, P_2O_5, K_2O without organic fertilization,
kRa – Radiowo compost,
kC – Complex compost,
kr – plant compost,
kZ– ZUSOK compost,
O – sludge from municipal wastewater treatment.

Table 5. Participation syntaxonomic groups in the rank of classes in the third year of the experiment

Syntaxonomic group (class)	Number of species	%
Stellarietea mediae	35	42.2
Molinio-Arrhenatheretea	18	21.7
Artemisietea vulgaris	15	18.1
Bidentetea tripartiti	4	4.8
Festuco-Brometea	3	3.6
Koelerio-Corynephoretea	2	2.4
Querco-Fagetea	2	2.4
Isoëto-Nanojuncetea	1	1.2
Epilobietea angustifolii	1	1.2
Salicetea purpureae	1	1.2
Alnetea glutinosae	1	1.2
Total	83	100.0

significant share (12%) of geophytes, i.e. plants hiding their buds in the soil, often with storage organs (Table 6).

Evaluation of the effectiveness of reclamation in a model experiment based on plant cover

In the last year of the experiment in all models of fertilization, in terms of coverage, grass seeded at the beginning of the experiment dominated (Table 3 – Column 9, Figure 6 and 7). In contrast, among the self-seeding plants their presence varies between models, for example, in NPK model papilionaceous plants dominated, and in models with Radiowo and ZUSOK compost and sewage sludge aster plants had the greatest coverage. The role of the so-called reclamation plants should

Table 6. The spectrum of life forms of plants according to Raunkiaer in the third year of the experiment

Form of of living	Number of species	%
Hemicryptophytes (H)	37	44.6
Therophytes (T)	33	39.8
Geophytes (G)	10	12.0
Megafanerophytes (M)	2	2.4
Green chamephytes (C)	1	1.2
Total	83	100.0

also be noted. They have a high biomass level and are annuals in most cases (therophytes). Their increased participation in the coverage can be an indicator of appropriate initiation of the course of biological reclamation of landfills. It was quite clearly observed in a model of Complex compost. No less important is the role of papilionaceous plants and aster plants in the process of reclamation. Their large share facilitates the creation of the vegetation cover in most models, and therefore, the course of rehabilitation, because many species of these groups are perennials.

Experimental model with NPK without organic fertilization (NPK). In the first year of the experiment (2011). Almost half of the containers fertilized with NPK blend were occupied by the sown 4 species of grasses. In contrast, coverage of white mustard (*Sinapis alba*) was estimated at approx. 10%. Over 20 other species of dicotyledonous plants inhabited spontaneously, incl.: white clover (*Trifolium repens*), Canadian fleabane (*Conyza canadensis*), fat-hen (*Chenopodium album*). In 2012, the share of grasses decreased to approx. 10%, and white mustard was not found (annual plant). In comparison to 2011 an increase in the share of small balsam (*Impatiens parviflora*), false London rocket (*Sisymbrium loeselii*), fat-hen (*Chenopodium album*), corn sow-thistle (*Sonchus arvensis*), field clover (*Trifolium arvense*) was recorded. The total number of species increased to 30. In the third year of the experiment in containers supported with fertilizer NPK (without organic fertilization) two species of grasses, sown in 2011, were dominant: Evon smooth meadow-grass var. Evon (*Poa pratensis var. Evon*) and red fescue var. Maxima (*Festuca rubra var. Maxima*). The free surfaces were usually occupied by self-seeding black medic (*Medicago lupulina*). Most species belonged to therophytes, annual plants overwintering in a form

of seeds. In terms of the share of families, most species belonged to aster family, grasses and papilionaceous plants. In contrast, floristic richness remained at a similar level as in 2012.

Based on the floristic observations moderate effectiveness of NPK fertilizer on the reclamation of a bed of coal combustion waste was proved. This is evident by relatively low coverage with hemicryptophytes – turf and cluster plants, mainly grasses.

Experimental model with Radiowo compost (kRa). In this model, in the first year of the experiment, 26 species of plants were recorded. The sown grasses and white mustard constituted almost ¾ of the container surfaces, approx. 50 and 25% respectively. In the second year, the number of species did not change much, but there was a qualitative change in the structure of flora. Grass cover decreased to approx. 20% in favour of some species of dicotyledonous plants. These were mainly: puff-ball (*Taraxacum officinale*), corn sow-thistle (*Sonchus arvensis*), nettle-leaved goosefoot (*Chenopodium murale*), fat-hen (*Chenopodium album*), false London rocket (*Sisymbrium loeselii*), hedge mustard (*Sisymbrium officinale*), wild radish (*Raphanus raphanistrum*), white clover (*Trifolium repens*). A total of 30 species were recorded. In the third year of the experiment in the containers with Radiowo compost (kRa) there was again an increased grass cover up to almost 50%, mainly involving perennial ryegrass var. Stadium (*Lolium perenne var. Stadion*). The number of species did not change. In the third year of observations the cover of grasses in the containers with Radiowo compost reached 50%. In addition, the surfaces of containers were covered with goosefoot plants, cabbage plants and papilionaceous plants. They were both annuals (therophytes) and perennial plants (hemicryptophytes, geophytes, rhizomous plants).

The share of plants of these taxa manifests good effectiveness of reclamation on the bed. The importance of the process of biological reclamation plant of the papilionaceous plants and cabbage plants was also noted by Góral [2001].

Experimental model with Complex compost (kC). In the first year of the experiment, 25 plant species were found in the model bed. The dominant share in the coverage surface of the containers were sown with grass species. They covered approx. 60% of the containers, and white

Figure 4. General view of containers with plants as on 30.05.2011

Figure 5. General view of containers with plants as on 31.05.2012

Figure 6. General view of containers with plants as on 30.05.2013

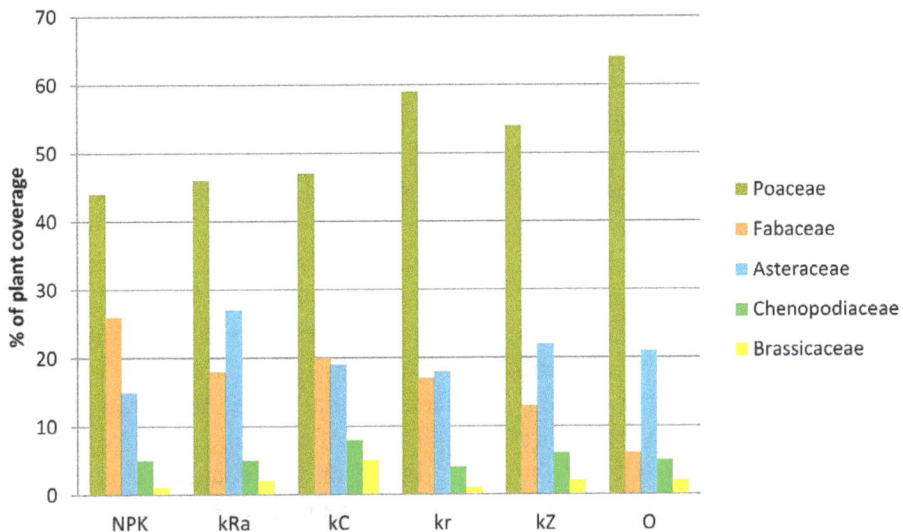

Figure 7. Percentage of dominant families in the creation of vegetation cover on the surface of experimental models after three years of the experiment

mustard approx. 20% of the area. Other areas were spontaneously entering inhabited by several dicotyledonous plant species. In 2012, the number of species increased to 34, and at the same time quantitative composition of plants changed radically with the participation of dicotyledonous plant species. The sown grass covered only approx. 10% of the area, without white mustard. A clear dominance of spontaneous dicotyledonous plant species was also recorded. Among them, in this model dominated: fat-hen (*Chenopodium album*), small balsam (*Impatiens parviflora*), Canadian fleabane (*Conyza canadensis*), white clover (*Trifolium repens*). More than 20 other species of small or occasional participation were recorded. In 2013, another spread of grasses (50% of the coverage) seeded at the start of the experiment was observed. There were also some self-seeded grasses, incl.: rougle cock's-foot (*Dactylis glomerata*) and couch grass (*Elymus repens*). In total, this model 45 species of self-seeding plants were recorded. Among them, the highest surface coverage in the containers was occupied by the same species as in 2012. These were mainly therophytes: goosefoot plants, papilionaceous plants and synanthropic plant species. The domination of therophytes at this stage of recovery is not evidence of its high efficiency. Due to the relatively high proportion of grasses and goosefoot plants can one can predict for this model the intensification of soil formation in the following years and, consequently, even greater increase in the efficiency of reclamation of the applied deposit.

Experimental model with plant compost (kr). In 2011, 22 species of plants were recorded. The species of grass clearly dominated (approx. 70% coverage). In addition to the grasses there were mainly chickweed (*Stellaria media*), coltsfoot (*Tussilago farfara*) *and* Canadian fleabane (*Conyza canadensis*). In 2012, the cover of grasses decreased to 10%. There was a significant expansion of dicotyledonous plant species, of which the majority were: fat-hen *Chenopodium album*, corn sow-thistle *Sonchus arvensis,* Canadian goldenrod *Solidago canadensis*, Canadian fleabane (*Conyza canadensis*). In the growing season 39 species were recorded in the containers in this model. In the following year, the third year of the experiment, more than 50% of the containers were grown with grass. For the rest of the surface more than 30 species of flowering plants were recorded, including goosefoot plants

family: goosefoot plants (incl. fat-hen *Chenopodium album* and several species of oraches *Atriplex*), cabbage plants (wild mustard *Sinapis arvensis,* false London rocket *Sisymbrium loeselii,* lady's purse *Capsella bursa-pastoris),* papilionaceous plants (white clover *Trifolium repens,* bastard clover *Trifolium hybridum,* black medic *Medicago lupulina)* and synatropic plans, mainly from aster family (Canadian goldenrod *Solidago canadensis,* Canadian fleabane *Conyza canadensis,* corn sow-thistle *Sonchus arvensis,* puff-ball *Taraxacum officinale,* coltsfoot *Tussilago farfara* and other).

The effectiveness of recultivation in this model was partially facilitated by the inhabitation of perennial dicotyledonous (hemicryptophytes, geophytes and herbaceous chamaephytes). However, the relatively low cover of grasses at this stage of recultivation proves its slow pace.

Experimental model of compost ZUSOK (kZ). Most of the surface of the containers in this model in 2011 were overgrown with grass species (approx. 80%) with minor contributions from white mustard. In addition, several individual self-seeding dicotyledonous plants were reported. In the next year of experiment, grasses clearly subsided, inhabiting only 20% of the area. 41 species of dicotyledonous plants were observed. Among them the greatest coverage was recorded for: small balsam *(Impatiens parviflora),* puffball *(Taraxacum officinale),* fat-hen (*Chenopodium album),* false London rocket (*Sisymbrium loeselii*) and corn sow-thistle (*Sonchus arvensis).* In the last year of the experiment a large share of grass cover was observed again. These were mainly grasses applied either at the beginning of the experiment, but also self-seeding species, for example: timothy grass (*Phleum pratense*), creeping bent (*Agrostis stolonifera*), annual meadowgrass (*Poa annua*), couch grass (*Elymus repens*). Among the dicotyledonous plants there were also those present in other models, typical reclamation plants. These were the plants belonging to the goosefoot family orache and goosefoot), cabbage (sisymbrium, radish and tansy mustard), papilionaceous (clover trefoil and medic). Ruderal plants, segetal weeds and meadow species had a high share of coverage.

Based on the identified floristic composition one can assume that the effectiveness of ZUSOK compost in the initial process of biological reclamation on the bed of combustion waste was merely satisfactory.

Experimental model of sewage sludge (O). In the first year of the experiment the surfaces of containers fertilized with sewage sludge were covered with sown grass species (more than 80% coverage) and occasionally by white mustard. The remaining 16 species were dicotyledonous plants, and among them was the greatest share common chickweed (*Stellaria media*). The following year, the share of grasses decreased almost threefold. These surfaces overgrown with 39 self-seeding dicotyledonous plant species, of which the predominant was: small balsam (*Impatiens parviflora*), corn sow-thistle (*Sonchus arvensis*), sickle medick (*Medicago falcata*), fat-hen (*Chenopodium album*), november goldenrod (*Solidago gigantea*) and common chickweed (*Stellaria media*). In the third year of observations grass cover increased again to over 60%. During the three years of experiment in containers fertilized with sewage sludge a fairly good growth of vegetation cover was demonstrated.

It can be assumed that further stages of reclamation should be more efficient for the presence of (even with a small coverage) up to several species of papilionaceous and cruciferous family which, together with grasses, play a major role in the process of reclamation of furnace waste.

Evaluating the effectiveness of reclamation in a model experiment based on the crop yield

Based on the results of the plant yield an attempt was made to assess the efficiency of the reclamation models with fertilization. Each model was analyzed in terms of average yield of all the variants of fertilization during the three-year experiment. The size of crops in a given model was suggested by the level of effectiveness of reclamation model.

Dry matter yields of plants, in of experimental area are shown in Table 7 and graphs 7 and 8. These data show the size of the crop yields obtained from different models of containers fertilized with three different variants (doses) of fertilizer during the three-year study.

Model N, P$_2$O$_5$, K$_2$O (NPK) without organic fertilization of deposits assumed as a zero object (test), for the comparison of the effectiveness of all other types of organic fertilization. In 2011–2012, in variants 1 and 2 (single and a double dose of NPK) relatively high yields of plants were obtained. In the variant 3 (triple dose of NPK) plant

yields were much lower than in variants 1 and 2. However, in the last (third) one, the yield of the experiment in all variants of fertilization were found to be relatively lower (Table 7). Average total yields of plants for the period 2011–2013 from all the variants of fertilization in the analyzed model followed a clearly declining trend (Figure 8). It can be considered that the efficiency of reclamation on the basis of the yield obtained in this model was relatively low (Figure 9).

In **the model of "Radiowo compost" (kRa)** no clear relationship between the yields of plants and the size of reclamation dose of compost was showed. In each year of the study, the average yield of plants, regardless of the dose of fertilization with compost, had aligned values, but at the same time it was decreasing in subsequent years

Table 7. Dry matter yields obtained in 2011–2013

Option number	Fertilizer dose	Years			Σ
		2011	2012	2013	
Plant compost (model kr)					
1	10 dm³	682	625	481	1788
2	15 dm³	887	442	323	1652
3	20 dm³	901	427	384	1712
ZUSOK compost (model kZ)					
1	10 dm³	318	575	337	1230
2	15 dm³	389	539	334	1262
3	20 dm³	572	607	337	1516
Radiowo compost (model kRa)					
1	10 dm³	911	687	544	2142
2	15 dm³	905	611	466	1982
3	20 dm³	835	751	656	2134
Sewage sludge (model O)					
1	10 dm³	416	578	415	1409
2	15 dm³	1103	721	545	2369
3	20 dm³	1285	652	544	2481
Complex compost (model kC)					
1	10 dm³	1024	985	1021	3030
2	15 dm³	1077	826	674	2577
3	20 dm³	1735	1077	960	3372
N, P₂O₅, K₂O (model NPK)					
1	32 g	860	432	299	1591
2	44 g	784	522	317	1623
3	57 g	427	304	298	1029
Loam (model Z)					
1	10 dm³	b.d.	768	1012	b.d.
2	15 dm³	b.d.	741	902	b.d.
3	20 dm³	b.d.	863	1171	b.d.
b.d. – not available					

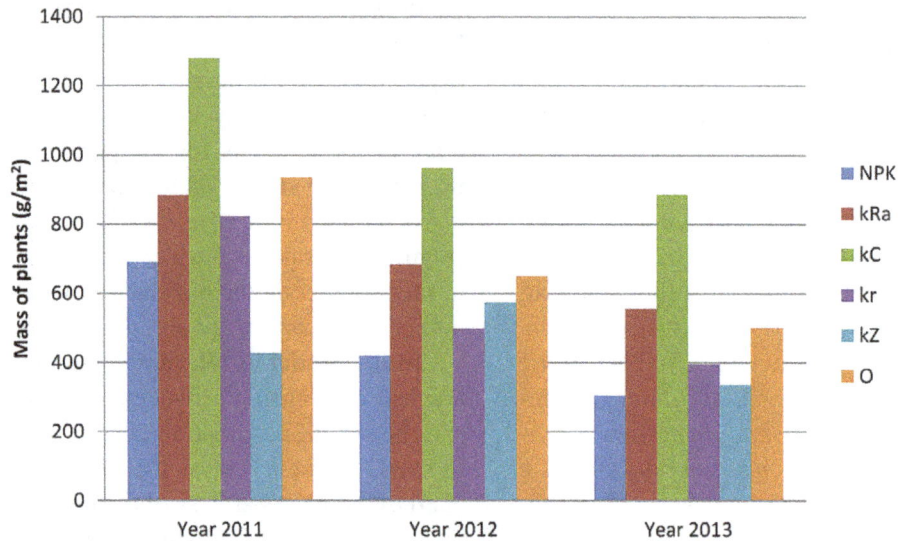

Figure 8. Average amounts of harvested plants in different forms of fertilizing in 2011–2013

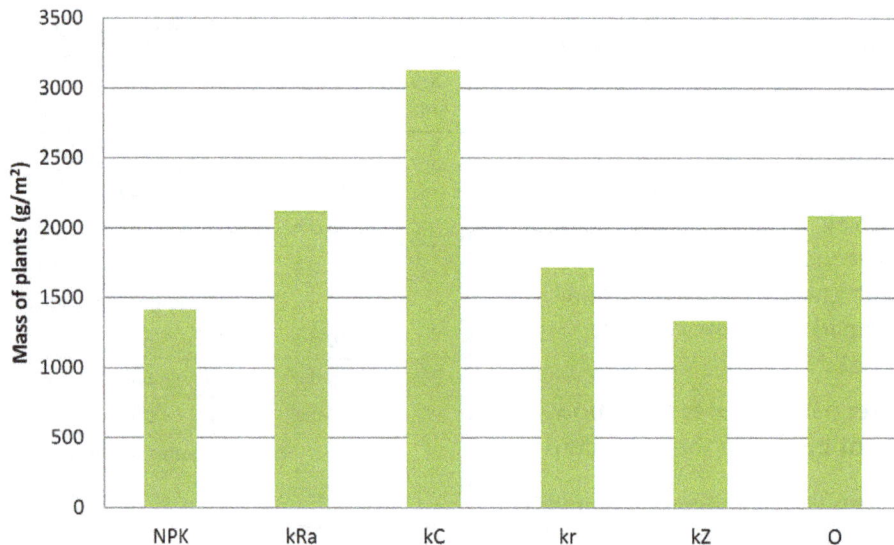

Figure 9. Evaluation of the effectiveness of reclamation on the basis of the yield of plants in containers in multi-variant experiment in bed with combustion waste

(Table 7). During the three years of the experiment in the model with Radiowo compost the total obtained crop yields decreased (Figure 8). The effectiveness of reclamation in this model, on the basis of the yield was high (Figure 9).

Three-year experiment in **"Complex compost" (kC) model** showed a significantly higher crop yields than in other models of fertilization. Mean total plant yield of three years and all variants of fertilization in this model were moderately decreasing (Table 7, Figure 8). Reclamation efficiency, on the basis of the results obtained in this model was very high (Figure 9).

In the first year of the experiment in **a model with plant compost (kr)** the highest yield of the plants was shown in variants 2 and 3. However,

in the next two years significantly higher yields in the first variant and lower in variants 2 and 3 were recorded (Table 7). During the three years of experiment in the model plant with compost the aggregated crop yields decreased (Figure 8). Based on the estimation of the cumulative yield of reclamation efficiency in this model proved to be low, but the yield slightly exceeded the test model without organic fertilization – NPK (Figure 9).

In **the model with ZUSOK compost (kZ)** yields in 2011 were lower than in the other models. But in 2012, there was an increase in yields. However, in the third year of the experiment yields again decreased clearly (Table 7). Average total yields of plants for the period 2011–2013 from all variants of fertilization in this model were relatively aligned (Figure 8). The resulting low yield suggests low efficiency of reclamation in this model (Figure 9).

In 2011, **the model with sewage sludge (O)** gave approx. threefold higher yield in variants 2 and 3, as compared with the first variant. A similar trend occurred in the next two years, but with lower yielding (Table 7). During the three years of experiment in the model of sewage sludge average yields of all variants decreased (Figure 8). The effectiveness of reclamation in this model, on the basis of the yield proved to be high (Figure 9).

CONCLUSION

1. The experiment on the model bed of combustion waste showed that an important role in the course of reclamation is played by the presence of appropriate forms of plant life. In the initial stage of reclamation annuals (therophytes) are important as they stimulate the process of soil formation. In further phases of reclamation they ensure the highest efficiency of grasses and dicotyledonous plants, especially biennial, and perennial (hemicryptophytes, geophytes, herbaceous chmaephytes), which perpetuate the ground.

2. During the three-year experiment a vital role in improving reclamation on the experimental fields was played (except grass) by plants of the families: aster family, goosefoot family and papilionaceous family, which gathered a large amounts of biomass, which is important in the process of reclamation.

3. Based on the estimates of the degree of coverage of species and yield reclamation in the studied models, the highest efficiency was demonstrated in containers with Complex compost and Radiowe compost, as well as in the model with sewage sludge. The lowest efficiency has been demonstrated in the models of ZUSOK composts and plant compost.

4. Weather conditions (particularly high temperatures, reduced humidity) and the depletion of nutrients in the substrate had a large influence on the experience and efficiency of reclamation in the bed of combustion waste. This illustrates the dynamics of yield decline in subsequent years of the experiment.

5. Impromptu experimental model (ex situ) of biological reclamation of furnace waste can-

not refer to real systems (in situ – dumps, landfills, etc.). Its functionality is fragmented and sometimes limiting accurate conclusions. Therefore, multiple iterations and multifaceted testing is required.

REFEENCES

1. Antonkiewicz J, Radkowski A. 2006. Przydatność wybranych gatunków traw i roślin motylkowych do biologicznej rekultywacji składowisk popiołów paleniskowych. Annales UMCS, Sec. E, 2006, 61, 413–421.

2. Braun-Blanquet J. 1964. Pflanzensoziologie, Grundzüge der Vegetationskunde 3. Aufl. Springer, Vienna-Nev York, 865.

3. Dyguś K.H., Madej M. 2012. Roślinność wielowarian-towego doświadczenia modelowego na złożu odpadów paleniskowych energetyki węglowej. Inżynieria Ekologiczna, 30, 227–240.

4. Dyguś K.H., Siuta J., Wasiak G., Madej M. 2012. Roślinność składowisk odpadów komunalnych i przemysłowych. Wyd. Wyższej Szkoły Ekologii i Zarządzania, Warszawa, pp. 134.

5. Dyguś K.H., Wasiak G., Madej M. 2014. Dynamika zmian roślinności w doświadczeniu modelowym ze złożem odpadów paleniskowych energetyki węglowej. Inżynieria Ekologiczna, 40, 100–121.

6. Gilewska M. 1999. Utilization of sewage sludge in the reclamation of post-mining soil and ash disposal sites. Roczniki AR Poznań, 310. Melioracje i Inżynieria Środowiska, 20/II, 273–281.

7. Gilewska M., Przybyła Cz. 2011. Wykorzystanie osadów ściekowych w rekultywacji składowisk popiołowych. Zesz. Prob. Post. Nauk Roln. PAN, 477, 217–222.

8. Góral S. 2001. Roślinność zielna w ochronie i rekultywacji gruntów. Inżynieria Ekologiczna, 3, 161–178.

9. Gutkowska A., Pawluśkiewicz B. 2006. Kształtowanie zadarnienia i składu florystycznego zbiorowisk trawiastych pod wpływem zabiegów pratotechnicznych na składowisku popiołu EC Siekierki. Annales UMCS, Sec. E, 61, 249–255.

10. Hryncewicz J., Balicka N., Giedrojć B., Małysowa E. 1972. Badania nad utrwalaniem i zagospodarowaniem hałdy popiołowej w elektrowni „Halemba". XIX Zjazd Naukowy PTGleb., Puławy.

11. Kozłowska B. 1995. Zastosowanie osadu ściekowego do biologicznego zagospodarowania składowisk odpadów paleniskowych. Zesz. Prob. Post. Nauk Roln. PAN, 418, 859–868.

12. Klimont K. 2011. Rekultywacyjna efektywność osadów ściekowych na bezglebowym podłożu

wapna poflotacyjnego i popiołów paleniskowych. Problemy Inżynierii Rolniczej, 2, 165–176.

13. Madej M. 2007. Zieleń miejska źródłem surowca do produkcji kompostu. Praca doktorska, WSEiZ, Warszawa, pp. 140.

14. Madej M., Siuta J., Wasiak G. 2010. Zieleń Warszawy źródłem surowca do produkcji kompostu. Cz. II. Skład chemiczny masy roślinnej z różnych powierzchni zieleni warszawskiej. Inżynieria Ekologiczna, 23, 22–36.

15. Majtkowski W., Głażewski M., Schmidt J. 1999. Roślinność trawiasta składowiska fosfogipsów w Wiślince koło Gdańska. Fol. Uniw. Agric. Stein. 197, Agricultura 75, 207–210.

16. Matuszkiewicz W. 2001. Przewodnik do oznaczania zbiorowisk roślinnych Polski. Wyd. Naukowe PWN, Warszawa.

17. Mirek Z., Piękoś-Mirkowa H., Zając A., Zając M., 2002. Flowering Plants and pteridophytes of Poland a checklist. W. Szafer Institute of Botany, PAS, Kraków.

18. Nowak W. 2006. Rekultywacja biologiczna hałdy fosfogipsu w zakładach chemicznych „Wizów" S. A. Zesz. Nauk. Uniw. Przyr. we Wrocławiu. Ser. Rolnictwo, 88, 545, 195–203.

19. Opaliński R. 2007. Rekultywacyjna efektywność kompostu Complex na odpadach paleniskowych w doświadczeniu lizymetrycznym. Praca magisterska. WSEiZ, Warszawa, pp. 78.

20. Polkowski M, Sułek St. 1999. Kompostowanie masy roślinnej ze strefy bezleśnej przy Zakładach Azotowych Puławy. Kompostowanie i użytkowanie kompostu. IOŚ, IUNG, PTIEkol. Warszawa, 71–74.

21. Rostański A. 2006. Spontaniczne kształtowanie się pokrywy roślinnej na zwałowiskach po górnictwie węgla kamiennego na Górnym Śląsku. Wydawnictwo Uniwersytetu Śląskiego, Katowice.

22. Rutkowski L. 1998. Klucz do oznaczania roślin naczyniowych Polski niżowej. Wyd. Naukowe PWN. Warszawa.

23. Siuta J. 2002. Przyrodnicze użytkowanie odpadów. IOŚ, Warszawa.

24. Siuta J. 2005. Rekultywacyjna efektywność osadów ściekowych na składowiskach odpadów przemysłowych. Acta Agrophysica, 5(2), 417–425.

25. Siuta J. 2007. System uprawy i kompostowania roślin na składowisku odpadów posodowych w Janikowie z zastosowaniem osadów ściekowych. Inżynieria Ekologiczna, 19, 38–58 + 6 fot.

26. Siuta J., Dyguś K.H. 2013. Plony i chemizm roślin wielowariantowego doświadczenia na modelowym złożu odpadów paleniskowych energetyki węglowej. Inżynieria Ekologiczna, 35, 7–31.

27. Siuta J., Kutla 2005. Rekultywacyjne działanie osadów ściekowych na złożach odpadów paleniskowych w energetyce węglowej. Inżynieria Ekologiczna, 10, 58–69.

28. Siuta J., Wasiak G., Chłopecki K., Mamełka D. 1997. Rekultywacja efektywności osadu ściekowego na bezglebowych podłożach w doświadczeniu lizymetrycznym. II Konf. Przyrodnicze użytkowanie osadów ściekowych. Puławy-Lublin-Jeziórko, 135–154.

29. Siuta J., Wasiak G., Madej M. 2008. Rekultywacja efektywności kompostów i osadów ściekowych na złożu odpadów paleniskowych w doświadczeniu modelowym. Ochrona Środowiska i Zasobów Naturalnych, 34, 145–172 + 26 fot.

30. Westhoff V., van der Maarel E. 1978. The Braun-Blanquet approach. [In:] Classification of plant communities (ed. R.H. Whittaker), 287–297. Junk, The Hague.

31. Wysocki W. 1984. Reclamation of Alkalien Ask Piles USEPA Cincinnati. Ohio.

32. Wysocki W. 1988. Rekultywacja składowisk odpadów elektrowni węglowych. Sozologia i Sozotechnika, 26, AGH Kraków.

33. Zarzycki K., Trzcińska-Tacik H., Różański W., Szeląg Z., Wołek J., Korzeniak U. 2002. Ecological indicator values of vascular plants of Poland. W. Szafer Institute of Botany, PAS, Kraków.

34. Żak M. 1972. Wpływ powłok asfaltowych przeciwdziałających wtórnemu pyleniu składowisk popiołów lotnych na wegetację roślin. XIX Zjazd Naukowy PTGleb., Puławy.

13

OPERATING DIFFICULTIES OF SMALL WATER RESERVOIR LOCATED IN WASILKOW

Anna Siemieniuk[1], Joannna Szczykowska[1], Józefa Wiater[1]

[1] Department of Technology in Engineering and Environmental Protection, Faculty of Civil and Environmental Engineering, Białystok University of Technology, Wiejska 45A, 15-351 Białystok, Poland, e-mail: j.szczykowska@gmail.com

ABSTRACT

When considering the issue of the functioning of small water reservoirs, the attempt to assess changes in trophy of small retention reservoir located in Wasilkow, Podlasie, before and after remediation, was carried out. Water samples tests were carried out once a month from April 2007 to March 2008, from April 2009 to March 2010 (before remediation), and from April 2013 to March 2014 (after removal of silt). Prior to works related to the reservoir remediation, a gradual increase in the number of tested contaminants and disturbances in the seasonal occurrence of nitrogen and phosphorus compounds were observed. Advanced eutrophic processes in Wasilków reservoir occur probably due to the supply of large amounts of humic and biogenic substances from the catchment, because a significant percentage of its area is covered by forests and agricultural lands. The development of the trophic status of the reservoir is largely influenced by the amount of phosphorus and total nitrogen supplied to the reservoir; the least affected by chlorophyll "a". Comparing the analyses performed in 2007/2008 and 2009/2010, a slight, but growing trend of average trophic levels of water in the basin Wasilków was found. Studies conducted in 2013/2014 revealed a significant decrease in the concentrations of all analyzed pollutants, and hence lower TSI values. It can be concluded that the reclamation associated with the removal of sediments brought the expected results.

Keywords: small retention reservoirs, contamination, biogenic compounds, trophy.

INTRODUCTION

Water retention is the capacity of a catchment to retain water. It depends on the terrain and covering vegetation, yet human activity is also very significant. Despite the good natural conditions, water retention and its storage can be very limited. In contrast, meeting the needs of all water users is based primarily on water storing in its proper quantity and quality.

River dams, as a result of artificial water damming, contribute to increased sedimentation of substances brought by the river into the storage reservoir. However, water retention in a reservoir is conducive to the intensity of metabolic processes, in particular, under conditions of excessive loads of phosphorus and nitrogen supplied to the reservoir [1]. Speaking of lake and reservoir resistance to pollution, the processes of self-purification and the capacity of the ecosystem, exhibiting as the decomposition of the influent contaminants are taken into account. It should be understood as the ability to precipitate the digested pollutants in sediments [2]. Then, in a very short period of time, the high trophic status of the ecosystem with all its consequences can be achieved. Appearing losses, deficits, and the total lack of oxygen in the bottom water layers make bottom sediments begin to release previously stored contaminants, which accelerate the degradation process [3, 4].

Considering the issue of the functioning of small water reservoirs, an attempt to assess trophy changes in a small retention reservoir located in Wasilkow, Podlasie, before and after remediation, was undertaken.

MATERIAL AND METHODS

Studies upon the changes in water trophy were carried out on small retention water reservoir located in Podlasie in Wasilkow on the river Supraśl.

The reservoir in Wasilkow was completed in 1968, the volume of the waters contained is 150 thousand m³, and the surface of the reservoir is 12 hectares, while the catchment area supplying the reservoir is 1448.2 km². The Supraśl river catchment above the water reservoir in Wasilkow is covered mainly by woodlands and to a lesser extent, the areas used for agricultural purposes. The lake was formed by the damming of river Supraśl for the water intake in Wasilkow. Maximum depth of the reservoir is 3 m. Reservoir formerly served for recreational functions, but its water quality, bottom, and banks state deteriorated from year to year. Up to 2010, the reservoir was gradually silting, the maximum thickness of sediments averaged 1.4 m, the reservoir was also in 80% grown by emergent vegetation. Therefore, the reclamation was performed.

For many years, the community was looking for funds to cope with this problem. Half of the finances was allocated from the local funds, and the second half from The Regional Board for Melioration and Water Structures in Bialystok. About six hectares of overgrown land was cleared and more than 80 thousand cubic meters of sediments were transported. The de-silting cost was over 4 million PLZ.

Three measurement-control points were selected in the reservoir Wasilkow for testing – they were located within the inlet (first point) and outflow (third point) of river Supraśl, as well as in the central part of the reservoir (second point). The selection and placement of measurement and control points on the reservoir was dictated by the ability to capture the changes taking place in the study object. They were used to assess the trophic status of the tested waters. Collected waters were subject to the following determinations: total nitrogen, total phosphorus, and chlorophyll "a". Tests of water samples collected from the surface layer of the coastal zone, were carried out once a month from April 2007 to March 2008, from April 2009 to March 2010 (before remedia-

tion), and from April 2013 to March 2014 (after de-silting the reservoir). All determinations were performed in accordance with current methodology [5].

The trophic state of Wasilków reservoir was also evaluated according to the concentration criteria and based on the calculated trophy indices TSI by Carlson and Kratzer and Brezonik. Trophy indices were calculated after Carlson trophic TSI (Chl) in mg/dm³, TSI (TP) in μg/ dm³, according to Kratzer and Brezonika TSI (TN) in mg/dm³ as well as the overall trophy of the reservoir TSI was determined using the mean value obtained from the three calculated indicators: TSI (Chl), TSI (TP), TSI (TN). The water transparency measured by Secchi disk was skipped, because it is of little importance in dam reservoirs, when assessing the trophic status. Water inflow during heavy rainfall brings a large amount of suspensions, which cause turbidity of retained water. The reduction in transparency due to this fact is not related to the development of phytoplankton [6, 7]. The values of trophic status indicators to assess according to Carlson are as follows: TSI < 40 – oligotrophy, 40–50 mesotrophy, 50–70 eutrophy, > 70 hypertrophy [8]. The assessment of the eutrophication degree is made on the basis of average annual values biogenic indicators for flowing waters [9].

RESULTS AND DISCUSSION

The analyzes showed marked variability in the concentration of total nitrogen in individual years of research. Prior to works related to reservoir de-silting, average annual total nitrogen concentrations ranged from 2.34 to 2.62 mg N/dm³, while after the reclamation, this value decreased to 0.94 mg N/dm³. In 2007/2008 and 2009/2010, there was no clear seasonal changes in this indicator, instead a gradual increase in the average content of contaminants expressed in total nitrogen quantities, was monitored. Probably silts, from which nitrogen compounds could be released were the disturbing factor in such situation. River outwash was probably delivered in the form of humic substances along with waters of the Supraśl river, whose catchment is mostly wooded above the test reservoir. During the reservoir use, growth of the organic matter always occurs, and the lack of adequate amount of oxygen required for its rapid degradation leads to an excessive deposition of sediments on the bottom of the reservoir. This situation favors oxygen deficiency, a condition for anaerobic processes that emit poisonous gases - hydrogen sulfide and methane, toxic amines, and other degradation products of amino acids. When there is more sediment, the reservoir gets shallow and is converted into a muddy pond, which was the case in the Wasilków reservoir. Another factor that could affect the seasonality of nitrogen compounds occurrence could be their inflow from the catchment, that is of the forest-agriculture character. In 2013/2014, variations in the amount of total nitrogen: a decrease was observed in the summer time. It was affected by developing autotrophic organisms absorbing nitrogen compounds and building them into cell structure in spring and summer [10]. Undoubtedly, it was influenced by renewal of the system by removing silt from the reservoir bottom.

Similar results were achieved in the case of phosphorus, for which no clear seasonal changes in its concentration in the test water were recorded in 2007/2008 and 2009/2010. The average annual total phosphorus concentrations ranged from

0.99 to 1.22 mg P/dm^3. Also in this case, the reason may be the accelerated oxygen consumption in the reservoir waters by sediments accumulated at the bottom of the reservoir. Anaerobic conditions affect the release of phosphorus from bottom sediments into the water column. This may be an internal source of water enrichment in biogenic substances accumulated in the sediment. Reducing the amount of average annual total phosphorus to 0.27 mg P/dm^3 and seasonality of its incidence was observed in 2013/2014, which may indicate the restoration of the proper reservoir functioning.

When observing the content of chlorophyll "a" in waters of the reservoir Wasilków, regardless of the study years, seasonal variations in its content can be observed. The increase in the concentration of chlorophyll "a" is observed in spring, while in summer and fall, it slightly decreases to almost complete disappearance during winter. Average concentration of chlorophyll "a" in Wasilków reservoir ranged from 20.89 to 21.88 mg/dm^3 during the study years before de-

Table 1. Waters testing results for reservoir Wasilków in 2007/2008

Wasilków 2007/2008		First point	Second point	Third point	Mean value
Tested parameter	Unit	min.–max mean	min.–max mean	min.–max mean	min.–max mean
Total nitrogen	mg·N/dm³	0.75 – 3.55 2.15	0.82 – 3.85 2.34	0.94 – 4.10 2.52	0.84 – 3.83 2.34
Total phosphorus	mg·P/dm³	0.14 –1.27 0.70	0.21 – 1.83 1.02	0.53 – 1.96 1.24	0.29 – 1.69 0.99
Chlorophyll „a"	µg/dm³	12.24 – 24.5 18.37	14.31 – 26.83 20.57	16.9 – 30.54 23.72	14.48 – 27.29 20.89

Table 2. Waters testing results for reservoir Wasilków in 2009/2010

Wasilków 2009/2010		First point	Second point	Third point	Mean value
Tested parameter	Unit	min. – max mean	min. – max mean	min. – max mean	min. – max mean
Total nitrogen	mg·N/dm³	0.81 – 3.95 2.38	0.89 – 4.25 2.57	1.04 – 4.78 2.91	0.91 – 4.33 2.62
Total phosphorus	mg·P/dm³	0.19 – 1.79 0.99	0.29 – 2.03 1.16	0.58 – 2.46 1.52	0.35 – 2.09 1.22
Chlorophyll „a"	µg/dm³	14.04 – 26.1 18.57	14.98 – 27.03 21.01	17.1 – 32.00 24.55	15.37 – 28.38 21.88

Table 3. Waters testing results for reservoir Wasilków in 2013/2014

Wasilków 2013/2014		First point	Second point	Third point	Mean value
Tested parameter	Unit	Min. – max mean	Min. – max mean	Min. – max mean	Min. – max mean
Total nitrogen	mg·N/dm³	0.01 – 1.25 0.63	0.19 – 1.85 1.02	0.21 – 2.08 1.15	0.14 – 1.73 0.94
Total phosphorus	mg·P/dm³	0.01 – 0.42 0.21	0.02 – 0.53 0.28	0.03 – 0.60 0.32	0.02 – 0.52 0.27
Chlorophyll „a"	µg/dm³	1.04 – 2.14 1.59	1.98 – 3.03 2.51	2.01 – 3.30 2.66	1.68 – 2.82 2.25

silting of the reservoir, to 2.25 mg/dm³ after its reclamation.

The study also evaluated the trophic status of Wasilków reservoir: the overall trophic level in years preceding the reclamation of the reservoir ranged from 76.92 to 78.62 and on that grounds, it could be ranked among the hypertrophic, while the overall TSI decreased to 58.98 in the last year of the research, qualifying it to eutrophic one.

Comparing the analysis performed in 2007/2008 with those in 2009/2010, a slight, but growing trend in average trophy levels of water was found in Wasilków reservoir. In the studies carried out in 2013/2014, a significant decrease in the concentrations of all the studied pollutants, and hence lower TSI values, can be seen. It can be concluded that the reclamation associated with the removal of sediments brought the expected results.

The development of the trophic status of the reservoir is largely affected by the amount of total phosphorus delivered to the water. Also, higher concentrations of total nitrogen have impact on the growth of the reservoir fertility. Chlorophyll "a" had the least effect on the poor trophy status. A reduced amount of biogenic compounds in the waters of reservoir Wasilków is currently observed. Unfortunately, despite the diminished quantities of biogenic substances and chlorophyll "a" in waters of the reservoir, a high trophy degree of its waters was recorded in the last years of the study.

Advanced eutrophication processes in reservoir Wasilków occur probably due to the supply of large amounts of humic and biogenic substances from the catchment, because a significant percentage of its area is covered by forests and agricultural lands. Organic compounds inflowing from these areas will be subject to biodegradation, and their intensive decomposition will occur in the water column. Some compounds present in the form of suspensions will accumulate at the bottom, forming bottom sediments. And again, after many years of use, the accumulation of large amounts of outwashes containing a variety of organic compounds, will take place, and they will be subject to chemical, biochemical, and biological transformations.

In addition, water quality in the reservoir may be worsen by not entirely ordered wastewater management in the catchment. Bringing mineral compounds of phosphorus and nitrogen into the reservoirs, accelerates the natural aging and, in extreme cases, a rapid degradation [11], which undoubtedly was the case in Wasilkow reservoir.

From the user's point of view, it is desirable that reservoirs accumulated waters of good quality and stable physicochemical properties as a result of reclamation processes, or the new aqueous environment was characterized by high capacity for regeneration. Concentration of municipal, industrial, and transport pollution sources determines that as a rule, quality of water in such objects, often shallow, with a small area, so the small capacity, is unsatisfactory. In addition, such condition can worsen the runoff from the catchment of both humic and biogenic substances. However, despite this, and despite the high costs, a lot of benefits can be found. Reclamation of small water reservoirs seems to be worth considering, against keeping them in a form of degraded wastelands.

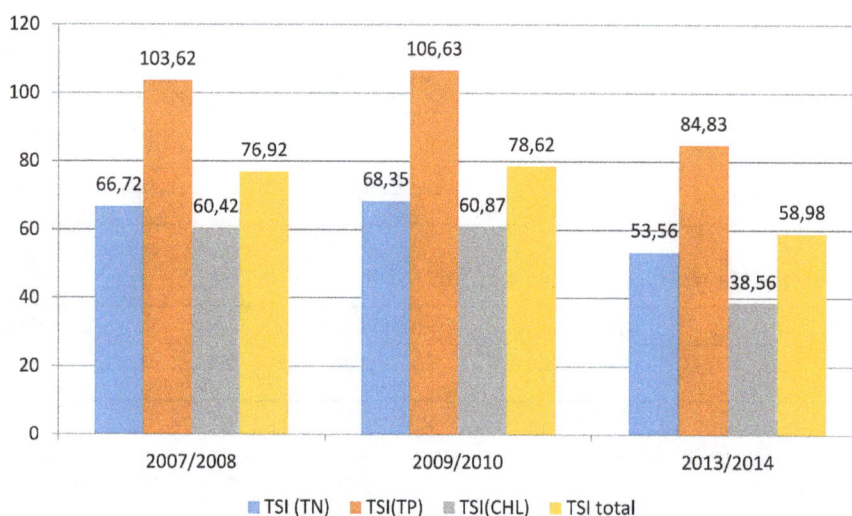

Figure 1. The annual average trophy level in Wasilków reservoir in particular years of the research

CONCLUSIONS

1. Before reclamation of the reservoir, a gradual increase in the examined pollution indicators and disturbances in the seasonal occurrence of nitrogen and phosphorus compounds was observed.

2. Advanced eutrophication processes in Wasilków reservoir occur probably due to supply of large amounts of humic and biogenic substances from the catchment, because a significant percentage of its area is covered by forests and agricultural lands.

3. Development of the trophic status of the reservoir is largely affected by the amount of total phosphorus and nitrogen supplied to the water, chlorophyll „a" had a weaker impact.

4. Reclamation of the reservoir associated with the removal of silt, brought the expected results, including a change of the reservoir trophy.

REFERENCES

1. Dunalska J. 2001. Effect of limited hypolimnetic with drawal on the content of nitrogen and phosphorus in the waters of Kortowskie Lake. Natural Sciences, 9, 333–346.

2. Ripl W., Wolter K.-D. 2005. The assault on the quality and value of lakes. [In:] O'Sullivan P.E. & Reynolda C.S. (eds.) The lakes handbook. Volume 2. Part I – General Issues. Chapter 2. Blackwell, Oxford, 25–61.

3. Mientki Cz. 1986. Wpływ usuwania wód hipilimionu na układy termiczne i tlenowe oraz zawartość związków azotu i fosforu w wodzie Jeziora Kortowskiego. Acta Acad. Agricult. Tech. Olst. Suppl. A 14, 1–53.

4. Nurnberg G.K. 1995. Quantifyinganoxia in lakes. Limnol. Oceanogr. 40, 100.

5. Hermanowicz W. et al. 1999. Fizyczno-chemiczne badanie wody i ścieków. Wydawnictwo Arkady, Warszawa.

6. Szczykowska J., Siemieniuk A., Wiater J. 2013. Sezonowe zmiany stanu troficznego zbiorników retencyjnych. Ekonomia i Środowisko, 2, 107–116.

7. Szczykowska J., Siemieniuk A., Wiater J. 2013. Problemy ekologiczne zbiorników małej retencji na Podlasiu. Ekonomia i Środowisko, 4, 234–244.

8. Carlson R.E. 1977. A tropic state index for lakes. Limnology and Oceanography, 22.

9. Rozporządzenie Ministra Środowiska z dnia 23 grudnia 2002 r. w sprawie kryteriów wyznaczania wód wrażliwych na zanieczyszczenia związkami azotu ze źródeł rolniczych. Dz. U. 2002 r. Nr 241 poz. 2093.

10. Kajak Z. 2001. Hydrobiologia – Limnologia. Ekosystemy wód śródlądowych. Wydawnictwo PWN, Warszawa.

11. Bartsch A.F. 1972. Role of phosphorus in eutrofication. EPA-R3-1972.

14

COMPARISON OF MORPHOLOGICAL TRAITS AND MINERAL CONTENT IN *EUCOMIS AUTUMNALIS* (MILL.) CHITT. PLANTS OBTAINED FROM BULBS TREATED WITH FUNGICIDES AND COATED WITH NATURAL POLYSACCHARIDES

Piotr Salachna[1], Agnieszka Zawadzińska[1]

[1] Department of Horticulture, Faculty of Environmental, Management and Agriculture, West Pomeranian University of Technology in Szczecin, Papieża Pawła VI 3, 71-459 Szczecin, Poland, e-mail: piotr.salachna@zut.edu.pl; agnieszka.zawadzinska@zut.edu.pl

ABSTRACT

Eucomis autumnalis is an attractive ornamental species from the South Africa, commonly used in natural medicine. Plant protection programs, particularly those concerning plants grown for phytotherapeutics, are focused on prophylactic treatments that facilitate a limited use of pesticides negatively affecting the environment. Polysaccharides, such as chitosan and sodium alginate are exemplary non-toxic and biodegradable substances used for hydrogel coatings. The aim of this study was to investigate the effects of treating *E. autumnalis* bulbs with fungicide or coating with natural polysaccharides on the morphological traits and content of minerals in the leaves and bulbs. Prior to planting, the bulbs were divided into three groups: (I) untreated bulbs (control); (II) bulbs treated with Kaptan and Topsin fungicides; (III) bulbs coated with oligochitosan and sodium alginate. Bulb coating was found to exert a stimulating effect on plant height, number and length of leaf, greenness index (SPAD), number of flowers per inflorescence, fresh weight of the aboveground part and fresh weight of bulbs. The leaves and bulbs of plants grown from coated bulbs contained more nitrogen, potassium and boron. Treating the bulbs with fungicides positively affected the number of leaves, greenness index and fresh weight of the aboveground part.

Keywords: pineapple lily, oligochitosan, sodium alginate, macronutrients, micronutrients.

INTRODUCTION

Synthetic fungicides used in plant protection have a negative impact on the environment, and therefore, environmentally-friendly solutions based on non-toxic and biodegradable substances are sought for. Some examples of environmentally-safe compounds are natural polysaccharides, of which chitosan and its derivatives are particularly promising [1]. Chitosan is a polymer obtained by N-deacetylation of natural chitin. It is a linear polymer containing D-glucosamine (GlcN) and N-acetyl-D-glucosamine (GlcNAc) molecules linked by β-1,4-glycosidic bonds [2]. It is used in agriculture and horticulture, mainly as an elicitor of resistance to certain pathogenic infections and as a growth stimulator [3]. In plants, chitosan can stimulate seed germination [4], accelerate flowering [5], increase chlorophyll content [6], and plant tolerance to stress [7], and improve yield quantity and quality [8]. Another natural polymer that can be potentially used in plant cultivation is sodium alginate. It is composed of β-1,4 linked β-D-mannuronic acid and α-L-guluronic acid. On the industrial scale, alginates are extracted from brown algae, mainly *Laminaria* and *Lessonia* [9]. Sodium alginate, particularly in the form of depolymerized oligosaccharide, was reported to have a stimulating effect on the growth in several plants, such as mint [10], fennel [11], and potato [12]. Natural polysaccharides can be used for coating formation on the surface of plant organs [13].

Eucomis autumnalis (Mill.) Chitt., also known as "pineapple lily", is an original bulbous plant of Asparagaceae family, native to southern Africa. The plants have very decorative raceme inflorescences composed of numerous, star-shaped, greenish-white and sweetly scented flowers. This species can be grown as garden plants, cut flowers and flowering potted plants, with high demand by the European and North American markets [14]. Due to antibacterial and antifungal compounds, the bulbs of *E. autumnalis* are used in southern African traditional medicine [15]. The anti-inflammatory activity in *E. autumnalis* plant depends on age, season, fertilization and growth conditions [16]. The subject literature lacks data on *E. autumnalis* cultivation in the Central European climatic conditions, which is a problem for the producers of ornamental and medicinal plants interested in wider use of this species. So far, no information has been published on the use of natural polysaccharides for encapsulation of *E. autumnalis* bulbs before planting. Therefore, the aim of this study was to evaluate growth, development and the content of minerals in *E. autumnalis* plants grown from the bulbs treated with traditional fungicides and from the bulbs coated with oligochitosan and sodium alginate.

MATERIAL AND METHODS

Plant material and treatments

The study was conducted at a research plot of the West Pomeranian University of Technology in Szczecin (53°25' N, 14°32' E; 25 m asl.). Study material were *E. autumnalis* bulbs, 14–16 cm in circumference, obtained from Dutch plantations. Prior to planting, the bulbs were divided into three groups: (I) non-treated control bulbs; (II) bulbs treated for 30 min with fungicide suspension 1.0% (w/v) Kaptan 50 WP (active ingredient: Captan) and 0.7% (w/v) Topsin M 500 SC (active ingredient: thiophanate-methyl); (III) bulbs coated as described by Startek et al. [17] in 0.2% (w/v) oligochitosan and 1% (w/v) sodium alginate (Sigma Aldrich). Oligochitosan (M_w 48 000 g·mol^{-1}; DD 85%) was purchased from Center of Bioimmobilisation and Innovative Packaging Materials (Szczecin, Poland). The products have been obtained using the free radical degradation process [18].

Growth conditions and plant measurements

The bulbs were planted on 15th April 2012 and 17th April 2013, into polyethylene boxes 60×40×19 cm, filled with deacidified peat (Kronen, Poland), pH 6.0, with mean content of macronutrients amounting to 11 mg·dm^{-3} N-NO$_3$, 39 mg·dm^{-3} P, and 13 mg·dm^{-3} K. The substrate was supplemented with multicomponent fertilizer Hydrocomplex (Yara International ASA, Norway) at a dose of 5 g·dm^{-3} that contained 12% N, 11% P$_2$O$_5$, 18% K$_2$O, 2.7% MgO, 8% S, 0.015% B, 0.2% Fe, 0.02% Mn, and 0.02% Zn. Each box contained 8 bulbs, planted at a spacing of 10×10 cm. The plants were grown in an unheated plastic tunnel, under natural photoperiod. Air temperature was controlled with vents that were opened when the temperature exceeded 20 °C. The plants were cultivated according to agrotechnical standards developed for *Eucomis* by the International Flower Bulb Centre [19]. When the plants reached the full bloom stage, the following parameters were evaluated: plant height, leaf number and length, inflorescence length, and number of flowers per inflorescence. Leaf greenness index that highly correlates with chlorophyll content was determined using Chlorophyll Meter SPAD-502 (Minolta, Japan). This evaluation is based on determining the quotient of difference in light absorption by a leaf at 650 nm and 940 nm. The result is given in dimensionless units called SPAD. Mean SPAD value was calculated based on four readings of four leaves from each plant. The plants were dug out in both years of the study on 20th October and the aboveground part and bulbs were weighted.

Determination of mineral content

The content of macro- and micronutrients in the bulbs and leaves obtained at full anthesis was analyzed in a laboratory. To this end, 8 plants from each variant were randomly dug out, and the maternal bulbs without roots and fully developed leaf blades were separated. Chemical analyzes were carried out in dried at 105 °C material, according to the standards of an accredited laboratory of the Chemical and Agricultural Station in Szczecin. The mineralized plant material was used for the determination of nitrate nitrogen by means of a colorimetric method, potassium and calcium by means of the atomic emission spectrometry (Solaar S AA spectrometer), phosphorus content was evaluated based on Baton method on Marcel s 330 PRO spectrophotometer at a wavelength $\lambda = 470$ nm, and magnesium and micronutrient (B, Cu, Fe, Mn, Zn) content were measured by means of the flame atomic absorption spectroscopy using Solaar S AA spectrometer [20].

Statistical analysis

The study was designed as a univariate experiment involving randomized sub-blocks, with four replications comprising 8 plants each. The results were verified statistically by means of one-way analysis of variance (ANOVA), using FR-ANALWAR software developed by Professor Franciszek Rudnicki of the University of Science and Technology in Bydgoszcz. Significance of mean values variation was assessed by Tukey`s multiple comparison test at p = 0.05.

RESULTS AND DISCUSSION

Statistical analysis of the experimental results revealed that the treatment of *E. autumnalis* bulbs before planting significantly affected most of the evaluated traits (Table 1). It was found that the plants obtained from the bulbs coated with oligochitosan and sodium alginate were the highest (35.2 cm), had the longest leaves (27.9 cm), the highest number of flowers per inflorescence (75.0), and the greatest fresh weight of the aboveground part (126 g) and dug out bulbs (43.5 g). Additionally, a positive effect of coatings containing natural polysaccharides was observed with reference to the number of leaves and their greenness index expressed in SPAD units, by respectively 15.0% and 10.5%, as compared to the control plants. Previous studies [5, 17] showed positive effects of coating freesia (*Freesia hybrida*) bulbs with polysaccharides on the morphological characteristics of plants, flowering and corms yield, with type of polymer and molecular weight of chitosan significantly affecting these traits. Several studies

suggested that chitosan stimulates the vegetative growth and the developing of roots and shoots and improves the efficiency of nutrient and water uptake, which can be beneficial for plant growth [21, 22]. Apart from oligochitosan, the coatings investigated in our study included also sodium alginate, which is another biostimulator exerting positive effects on plant metabolism. According to Khan et al. [23], treating opium poppy plants (*Papaver somniferum* L.) with oligo-alginates at the doses of 0.02 to 0.1 mg·ml[-1], improved root and shoot length, dry weight, and total content of chlorophylls, carotenoids and alkaloids, including codeine and morphine. In fennel (*Foeniculum vulgare* Mill.), oligo-alginates positively affected the activity of nitrate reductase, plant growth and development, seed yield, and the content of assimilation pigments, proline and essential oils [24].

The plants derived from the bulbs treated with a mixture of Kaptan and Topsin had significantly more leaves, higher greenness index and greater mass of the aboveground part, by respectively 10.0%, 11.0% and 11.6%, as compared to the control (Table 1). A positive impact of Captan 50 WP, a formulation containing the same active substance as Kaptan used in this study, was also found in some ornamental geophytes. Piskornik et al. [25] showed that, apart from protective properties, the fungicide affected also the quality and yield of *Ixia hybrida* L. flowers. The plants grown from bulbs treated with 2% Captan solution produced the highest number of flowers. Furthermore, the yield of daughter bulbs derived from plants grown from treated bulbs was the most abundant and of the highest quality [25]. Klimek et al. [26] also reported that buttercup (*Ranunculus asiaticus* L.) bulbs treated with 2% Captan 50 WP for 30 minutes produced plants with more flowers and of greater mass than non-treated control bulbs.

Table 1. Effect of fungicides and polysaccharides on morphological traits of *Eucomis autumnalis* (means for 2012–2013)

Traits	Treatment of bulbs before planting		
	non-treated (control)	dressed with fungicides	coated with polysaccharides
Height of plant (cm)	30.7 [b]	33.7 [ab]	35.2 [a]
Number of leaves	5.00 [b]	5.50 [a]	5.75 [a]
Lenght of leaf (cm)	22.2 [c]	24.1 [b]	27.9 [a]
Greenness index of leaves (SPAD)	39.0 [b]	43.3 [a]	43.1 [a]
Length of inflorescence (cm)	14.5 [a]	17.4 [a]	16.3 [a]
Number of flowers per inflorescence	64.0 [b]	66.0 [b]	75.0 [a]
Fresh weight of aboveground part (g)	90.5 [c]	101 [b]	126 [a]
Fresh weight of bulb (g)	36.1 [b]	40.7 [ab]	43.5 [a]

Values in the same row followed by the same letter do not differ significantly.

The way of treating *E. autumnalis* bulbs prior to planting significantly affected plant content of two macronutrients, nitrogen and potassium. No relationships were observed between bulb treatment and the content of phosphorus, magnesium and calcium in the leaves and bulbs (Table 2). The bulbs coated with chitosan and sodium alginate produced plants whose leaves contained the highest amount of nitrogen and potassium. The lowest content of nitrogen and potassium in the leaves was found in the plants grown from control bulbs. Nitrogen is an essential element of amino acids and, as a component of these and many other compounds, it is involved in almost all biochemical reactions in the plant tissues. Potassium is a key ion, the concentration of which determines the value of osmotic potential [27]. Increased leaf content of macronutrients as an effect of chitosan treatment was also observed by Dzung et al.

[21]. These researchers sprayed coffee seedlings, grown in the field and in a greenhouse, with solutions of chitosan and its oligomers at a concentration of 20–80 ppm. Following application of chitosan oligomers, the coffee seedlings grown in the field contained more nitrogen, phosphorus, potassium, calcium and magnesium, respectively by 9.49%, 11.76%, 0.98%, 18.75% and 3.77% than the control plants. Similar results regarding significant effect of chitosan on the mineral content in plants were obtained by Nguyen Van et al. [28]. They proved that spraying coffee seedlings with chitosan nanoparticles of high molecular weight, resulted in increased uptake of macronutrients by plants, by 9.8–27.4% for N, 17.3–30.4% for P, and 30–45% for K.

An analysis of micronutrients revealed significant differences in the content of boron and iron, depending on the treatment of *E. autumna-*

Table 2. Effect of fungicides and polysaccharides on macronutrients concentration in the leaves and bulbs of *Eucomis autumnalis* (means for 2012–2013)

Macronutrient content (% DW)	Plant organ	Treatment of bulbs before planting		
		non-treated (control)	dressed with fungicides	coated with polysaccharides
Nitrogen	leaves	2.65 [c]	2.96 [b]	3.22 [a]
	bulbs	0.45 [b]	0.51 [b]	0.69 [a]
Phosphorus	leaves	0.22 [a]	0.19 [a]	0.20 [a]
	bulbs	0.05 [a]	0.07 [a]	0.04 [a]
Potassium	leaves	2.58 [c]	3.28 [b]	4.13 [a]
	bulbs	0.38 [b]	0.49 [a]	0.52 [a]
Calcium	leaves	2.29 [a]	2.55 [a]	2.61 [a]
	bulbs	0.15 [a]	0.19 [a]	0.14 [a]
Magnesium	leaves	0.14 [a]	0.17 [a]	0.18 [a]
	bulbs	0.04 [a]	0.03 [a]	0.04 [a]

DW = dry weight. Date are means of triplicate. Values in the same row followed by the same letter do not differ significantly.

Table 3. Effect of fungicides and polysaccharides on micronutrients concentration in the leaves and bulbs of *Eucomis autumnalis* (means for 2012–2013)

Micronutrient content (mg/kg DW)	Plant organ	Treatment of bulbs before planting		
		non-treated (control)	dressed with fungicides	coated with polysaccharides
Boron	leaves	21.0 [c]	25.0 [b]	39.1 [a]
	bulbs	6.47 [c]	9.10 [b]	13.3 [a]
Copper	leaves	3.02 [a]	3.45 [a]	2.97 [a]
	bulbs	6.30 [a]	5.13 [a]	4.72 [a]
Iron	leaves	70.1 [b]	68.7 [b]	85.5 [a]
	bulbs	34.6 [a]	38.7 [a]	41.6 [a]
Manganese	leaves	38.2 [a]	47.8 [a]	38.0 [a]
	bulbs	5.61 [a]	7.55 [a]	6.48 [a]
Zinc	leaves	36.8 [a]	32.0 [a]	37.0 [a]
	bulbs	21.0 [a]	19.0 [a]	26.0 [a]

DW = dry weight. Date are means of triplicate. Values in the same row followed by the same letter do not differ significantly.

lis bulbs before planting. Bulb treating or coating had no influence on the content of manganese, zinc and copper in the leaves and bulbs or iron in the bulbs (Table 3). The highest content of boron, both in the leaves and bulbs, was found in the plants grown from coated bulbs. The lowest concentration of boron in the leaves and bulbs was found in the control plants. Boron is involved in the formation of cell wall structures, plant growth, and indirectly in carbohydrate metabolism. Monocotyledon plants, including *E. autumnalis*, are characterized by lower boron requirements than dicotyledons [27]. Our experiment showed that the leaves of plants grown from coated bulbs contained significantly more iron than the plants grown from fungicide-treated or control bulbs. The leaves of plants grown from coated bulbs had greater greenness index correlated with chlorophyll content, which was most likely due to an increased iron concentration in the leaves. Iron is involved in the synthesis of chlorophyll and some proteins and serves as an electron carrier in redox reactions [27]. Chatelian et al. [29] investigated the effects of chitooligosaccharides (COS) on growth and accumulation of minerals in beans (*Phaseolus vulgaris* L.) grown in hydroponic conditions and reported diverse micronutrient content in plants treated with chitosan. It was found that chitosan application significantly changed the accumulation of such micronutrients as molybdenum, boron, manganese, iron, copper, sodium, zinc, lead and cadmium by the roots, stems and leaves. Therefore, it can be assumed that chitosan may affect the absorption and accumulation of individual elements by plant tissues.

An analysis of the nutrient content in different parts of the plants indicated that the leaves contained more nitrogen, phosphorus, potassium, calcium, magnesium, boron, zinc, manganese, and iron than the bulbs, which, in turn, contained more copper (Table 2–3). The available literature contains no data on the limiting values defining an optimal range of nutrients for *Eucomis* species and cultivars. The results of the analyzes for the plants may be useful in the development of fertilization schedules for the studied species.

The agricultural industry has recently showed an increasing interest in new, environmentally-friendly technologies aimed at protecting specific biological materials against negative effects of direct contact with the external environment. Coating seeds, cuttings, bulbs and corms in hydrogel seems to be a particularly promising method

[30, 31]. The present study employed a patented method based on the formation of polyelectrolyte complexes [13]. These complexes are formed by an interaction of anionic functional groups of the polyelectrolyte with polyvalent metal cations, or by reaction at the interface of aqueous solutions of polyelectrolytes with functional groups of opposite charges. Raw materials used for coating bulbs in this study included two polysaccharides, oligochitosan and sodium alginate. These compounds are safe for humans, they do not pollute the environment, and help us to reduce the use of harmful chemicals in the agriculture and horticulture.

CONCLUSIONS

Oligochitosan and sodium alginate can be successfully used in practice for the preparation of hydrogel coatings for the bulbs of *E. autumnalis*. Bulb coating had beneficial effect on most morphological traits of the plants, such as height, number and length of leaf, greenness index, number of flowers per inflorescence, fresh weight of the aboveground part and fresh weight of bulbs. Moreover, the leaves and bulbs contained more nitrogen, potassium and boron. Data on the production technology of *E. autumnalis* in the Central European climatic conditions are lacking, and thus the results of this study may be a valuable source of information for producers of ornamental and herbal plants interested in cultivation of this species.

Acknowledgments

The research was supported by the by the Polish National Science Centre (project 7778/B/P01/2011/40).

REFERENCES

1. Cabrera J.C., Wégria G., Onderwater R.C.A., González G., Nápoles M.C., Falcón-Rodríguez A.B., Costales D., Rogers H.J, Diosdado E., González S., Cabrera G., González L, Wattiez R. 2013. Practical use of oligosaccharins in agriculture. Acta Horticulturae 1009, 195–211.

2. Deepmala K., Hemantaranjan A., Bharti S., Nishant Bhanu A. 2014. A future perspective in crop protection: Chitosan and its oligosaccharides. Advances in Plants and Agriculture Research 1(1), 1–8.

3. Hadwiger L.A. 2013. Plant science review: Multiple effects of chitosan on plant systems: Solid science or hype. Plant Science 208, 42–49.

4. Al-Tawaha A.R.M., Al-Ghzawi A.L.A. 2013. Effect of chitosan coating on seed germination and salt tolerance of Lentil (*Lens culinaris* L.). Research on Crops 14 (2), 489–491.

5. Salachna P., Zawadzińska A. 2014. Effect of chitosan on plant growth, flowering and corms yield of potted freesia. Journal of Ecological Engineering 15(3), 97–102.

6. Zahid N., Ali A., Manickam S., Siddiqui Y., Alderson P.G., Maqbool M. 2014. Efficacy of curative applications of submicron chitosan dispersions on anthracnose intensity and vegetative growth of dragon fruit plants. Crop Protection 62, 129–134.

7. Jabeen N., Ahmad R. 2013. The activity of antioxidant enzymes in response to salt stress in safflower (*Carthamus tinctorius* L.) and sunflower (*Helianthus annuus* L.) seedlings raised from seed treated with chitosan. Journal of the Science of Food and Agriculture 93(7), 1699–1705.

8. Salachna P., Wilas J., Zawadzińska P. 2014. The effect of chitosan coating of bulbs on the growth and flowering of *Ornithogalum saundersiae* Baker. Proceedings of the 29th International Horticultural Congress: Sustaining Lives, Livelihoods and Landscapes (Aug. 17-22), Brisbane, Australia.

9. Bixler H.J., Porse H. 2011. A decade of change in the seaweed hydrocolloids industry. Journal of Applied Phycology 23, 321–335.

10. Naeem M., Idrees M., Aftab T., Khan M.M.A., Moinuddin, Varshney L. 2011. Irradiated sodium alginate improves plant growth, physiological activities and active constituents in *Mentha arvensis* L. Journal of Applied Pharmaceutical Science (2) 5, 28–35.

11. Sarfaraz A., Naeem M., Nasir S., Idrees M., Aftab T., Hashmi N., Khan M.A.A., Varshney M., Varshney L. 2011. An evaluation of the effects of irradiated sodium alginate on the growth, physiological activities and essential oil production of fennel (*Foeniculum vulgare* Mill.). Journal of Medicinal Plants Research 5, 15–21.

12. Hussein O.S., Hamideldin N. 2014. Effects of spraying irradiated alginate on *Solanum tuberosum* L. plants: Growth, yield and physiological changes of stored tubers. Journal of Agriculture and Veterinary Science (7)1, 75–79.

13. Bartkowiak A., Startek L., Żurawik P., Salachna P. 2008. Sposób wytwarzania otoczek hydrożelowych na powierzchni organów roślinnych. Patent PL Nr 197101.

14. Luria G., Ziv O., Weiss D. 2011. Effects of temperature, day length and light intensity on Eucomis development and flowering. Acta Horticulturae 886, 167–174.

15. Masondo N.A., Finnie J.F., Van Staden J. 2014. Pharmacological potential and conservation prospect of the genus *Eucomis* (Hyacinthaceae) endemic to southern Africa. Journal of Ethnopharmacology 151, 44–53.

16. Taylor J.L.S., van Staden J. 2001. The effect of age, season and growth conditions on anti-inflammatory activity in *Eucomis autumnalis* (Mill.) Chitt. Plant extracts. Plant Growth Regulation 34(1), 39–47.

17. Startek L., Bartkowiak A., Salachna P., Kamińska M., Mazurkiewicz–Zapałowicz K. 2005. The influence of new method of corm coating on freesia growth, development and health. Acta Horticulturae 673, 611–616.

18. Bartkowiak A. 2001. Binary polyelectrolyte microcapsules based on natural polysaccharides. Wydawnictwo Politechniki Szczecińskiej, Szczecin.

19. Anonymous 1995. Information on special bulbs. International Flower Bulb Center, Hillegom, The Netherlands.

20. Ostrowska A., Gawliński S., Szczubiałka Z. 1991. Metody analizy i oceny właściwości gleb i roślin. Instytut Ochrony Środowiska, Warszawa.

21. Dzung N.A., Khanh V.T.P., Dzung T.T. 2011. Research on impact of chitosan oligomers on biophysical characteristics, growth, development and drought resistance of coffee. Carbohydrate Polymers 84, 751–755.

22. Rahman M.M., Kabir S., Rashid T.U., Nesa B., Nasrin R., Haque P., Khan M.A. 2013. Effect of γ-irradiation on the thermomechanical and morphological properties of chitosan obtained from prawn shell: Evaluation of potential for irradiated chitosan as plant growth stimulator for *Malabar* spinach. Radiation Physics and Chemistry 82, 112–118.

23. Khan Z.A, Khan M.M.A., Aftab T., Idrees M., Naeem M. 2011. Influence of alginate oligosaccharides on growth, yield and alkaloid production of opium poppy (*Papaver somniferum* L.). Frontiers of Agriculture in China 5, 122–127.

24. Idrees M., Naeem M., Aftab T., Hashmi N., Khan M.M.A., Varshney M., Varshney L. 2012. Promotive effect of irradiated sodium alginate on seed germination characteristics of fennel (*Foeniculum vulgare* Mill.). Journal of Physiology and Biochemistry (8) 1, 108–113.

25. Piskornik M., Kurzawińska H., Klimek A. 2001. Wpływ fungicydów użytych do zaprawiania bulw na zdrowotność i plonowanie iksji ogrodowej (*Ixia × hybrida* L.) uprawianej pod osłonami. Zeszyty Naukowe Akademii Rolniczej im. H. Kołłątaja w Krakowie 379, 149–154.

26. Klimek A., Piskornik M. Kurzawińska H. 2001. Wpyw wybranych fungicydów na rośliny jaskra azjatyckiego (*Ranunculus asiaticus* L.). Zeszyty Naukowe Akademii Rolniczej im. H. Kołłątaja w Krakowie 379, 105–109.

27. Kopcewicz J., Lewak S. 2007. Fizjologia roślin. Wydawnictwo Naukowe PWN, Warszawa.

28. Nguyen Van S., Dinh Minh H., Nguyen Anh D. 2013. Study on chitosan nanoparticles on biophysical characteristics and growth of Robusta coffee in green house. Biocatalysis and Agricultural Biotechnology 2(4), 289–294.

29. Chatelain P.G., Pintado M.E., Vasconcelos M.W. 2014. Evaluation of chitooligosaccharide application on mineral accumulation and plant growth in *Phaseolus vulgaris*. Plant Science DOI: 10.1016/j. plantsci.2013.11.009.

30. Salachna P., Zawadzińska A. 2014. Optimization of *Ornithogalum saundersiae* Baker propagation by twin scale cuttings with the use of biopolymers. Journal of Basic and Applied Sciences 10, 514–518.

31. Salachna P., Zawadzińska P., Wilas J. 2014. The use of natural polysaccharides in *Eucomis autumnalis* (Mill.) Chitt. propagation by twin-scale cuttings. Proceedings of the 29th International Horticultural Congress: Sustaining Lives, Livelihoods and Landscapes (Aug. 17-22), Brisbane, Australia.

15

TILLAGE EROSION: THE PRINCIPLES, CONTROLLING FACTORS AND MAIN IMPLICATIONS FOR FUTURE RESEARCH

Agnieszka Wysocka-Czubaszek[1], Robert Czubaszek[1]

[1] Department of Environmental Protection and Management, Białystok University of Technology, Wiejska 45A, 15-351 Białystok, Poland, e-mail: a.wysocka@pb.edu.pl; r.czubaszek@pb.edu.pl

ABSTRACT

Tillage erosion is one of the major contributors to landscape evolution in hummocky agricultural landscapes. This paper summarizes the available data describing tillage erosion caused by hand-held or other simple tillage implements as well as tools used in typical conventional agriculture in Europe and North America. Variations in equipment, tillage speed, depth and direction result in a wide range of soil translocation rates observed all over the world. The variety of tracers both physical and chemical gives a challenge to introduce the reliable model predicting tillage erosion, considering the number and type of tillage operation in the whole tillage sequence.

Keywords: tillage erosion, erosion rates, soil redistribution.

INTRODUCTION

While water and wind erosion are still considered to be the dominant processes degrading soil on agricultural land, there is a growing recognition that another type of erosion is a serious contributor to changes in soils and landscapes. The tillage erosion had got the special attention from researchers in the last two decades. The first attempt of measuring the tillage erosion was made in 1940s, in Poland the investigations on tillage erosion started in the 1950's. The results revealed that soil translocation depends on the ploughing direction and slope gradient. In subsequent years only a few authors conducted studies on tillage erosion. However, this phenomenon did not receive much attention until the 1990's, when more and more papers considered studies on this type of erosion. In Poland tillage erosion rates were measured mainly on loess soils. The relationship between tillage erosion and landscape features as well as the effect of tillage erosion on the soil profiles variability along the slope were investigated [Zgłobicki 2002]. Although large amount of studies have been conducted still some uncertainties exist. This paper describes the principles and effects of tillage erosion; factors controlling soil movement; rates of soil translocation and erosion as a result of using various tillage tools; methods of research; and draw conclusions for further investigations.

DEFINITION, MAIN PRINCIPLES AND EFFECTS

Tillage erosion is, by the definition, the displacement of cultivated layer during tillage. The soil uplift by the tillage tools is always perpendicular to the sloping surface of the land while the soil falls back perpendicular to the horizontal plane due to gravity. The translocation of soil is expressed as its moved mass in a specific direction per meter width. The soil is transported downslope during the tillage operation conducted in the downward direction, while during upward tillage the upslope soil translocation occurred. However due to gravitational forces smaller mass of soil is moved in the upward direction during the upslope cultivation. That is why the net soil distribution on the field is in downslope direction. Topography, especially the slope angle, is the most important control on the redistribution of soil particles by tillage. The steeper the slope gradient the larger is the difference between the vectors of upward and downward movement creating a net movement downslope and that is why tillage erosion depends mainly on tillage direction and slope gradient [Govers et al. 1999]. During te ploughing three phases of motion can be distinguished: (i) drag, when the soil is in contact with tillage tool; (ii) jump, when the soil loses this contact and (iii) rolling, when the clods and particles roll and jump with close contact to soil surface [Torri et al. 2002]. The amount of tillage erosion increases with the number of tillage operations [De Alba et al. 2004], the tillage depth [Van Muysen et al. 2002], and tillage speed [Van Muysen et al. 2000, 2002]. In experimental study with donkey-drawn mouldboard ploughing, in the top slope position, the soil surface level decreased by 0.57 m after 10 operations and by 0.23 after next 10 operations. Further decrease of 0.17 m was observed after next 10 operations [Li et al. 2004]. The increasing tillage depth in mouldboard ploughing by 0.4 m, despite reducing the speed from 1.81 to 1.54 and 1.45 m s^{-1} increased the soil translocation rate from 155 kg m^{-1} per operation to 223 and 281 kg m^{-1} per operation, respectively. The decreasing of the tillage speed by 0.27 m s^{-1} results in reduction of soil translocation rate by 68 kg m^{-1} per operation [Van Muysen et al. 2002]. The initial soil conditions also play an important role in tillage erosion [Van Muysen et al. 2000]. The experiment with mouldboard ploughing on different soil conditions: (i) pre-tilled soil and (ii) grass fallow, resulted in higher soil displacement distance in the tillage conducted on pre-tilled soil than on the compact soil in grass fallow. Greater ploughing depth and mechanical behavior of the pre-tilled soil affected the particle movement. Dense grass roots and high degree of consistency in grass fallow soil resulted in strong soil clods, which were more difficult to move [Van Muysen et al. 1999].

Tillage erosion leads to surface denudation on convex parts of hillslope and soil accumulation on the concave areas [De Alba et al. 2004], which is in opposite with water erosion pattern. Soil erosion modeled by WATEM on the basis of the digital elevation model on the field cultivated for at least 100 years revealed that soil loss caused by water erosion occurred on almost whole slope, with lowest rates near the summit and highest on the backslope position. The accumulation of soil moved with water erosion was on the toeslope and in the depression, where was

the highest. The tillage erosion transported soil mainly from shoulder slope, and accumulation occured on the footslope and toeslope. Opposite to the water erosion the lowest soil loss by tillage erosion was predicted on the lower backslope, because the amount of soil translocated to this part of slope was equal to the amount of soil removed from this slope position. However in complex landscape the soil loss rates on the convexities and accumulation rates in concavities decrease with time [Li et al. 2008].

De Alba et al. [2004] has proposed new theoretical two-dimensional model of soil catena evolution due to soil redistribution by tillage. Soil profile truncation occurs on convexities and in the upper areas of the cultivated hillslopes; while the opposite effect takes place in concavities and the lower areas of the field where the original soil profile becomes buried and deep colluvial soils develop [Heckrath et al. 2005]. At sectors of rectilinear morphology in the hillslope (backslope positions), a null balance of soil translocation takes place, independent of the slope gradient and of the rate of downslope soil translocation. As a result, in those backslope areas, a substitution of soil material in the surface horizon with material coming from upslope areas takes place. This substituted material can produce an inversion of soil horizons in the original soil profile and sometimes the formation of "false truncated soil" [De Alba et al. 2004].

Govers et al. [1996] stated that soil translocation is important geomorphological process and tillage erosion rates may exceed 10 Mg ha^{-1} yr^{-1}, which is equal to water erosion reported in Europe on hilly landscape. This can be confirmed by comparison of the annual sheet and rill erosion rate against tillage erosion in Europe. The actual mean sheet and rill erosion rates in Europe are in the range of 0.1–8.8 Mg ha^{-1} yr^{-1} and mean tillage erosion is between 3.0 and 9.0 Mg ha^{-1} yr^{-1} [Verheijen et al. 2009]. Tillage erosion was found to be a main process redistributing the soil particles in conventionally tilled corn-based production [Lobb et al. 1999] and cereal-based production [Kosmas et al. 2001]. Van Oost et al. [2005] have compared rates of soil erosion by tillage with those by water. By comparing two time periods, they found that there has been a shift from water-dominated to tillage-dominated erosion processes in agricultural areas during the past few decades. This reflects the increase in mechanized agriculture and the authors concluded that where soil is cultivated, tillage erosion may lead to larger losses than overland flow. However, the contributions of water and tillage erosion towards total soil erosion vary across topographically complex landscapes and their patterns are mainly dependent on topographic features. On undulating areas, tillage and water erosion both contribute in similar rate to total soil erosion while on hummocky landscapes, tillage erosion dominates, and the effects of water erosion are minor [Li et al. 2008]. Li and Lindstrom [2001] concluded that water erosion is the main factor responsible for decline in soil quality on the steep slopes in Chinese Loess Plateau, but tillage erosion is an equal contributor in soil quality deterioration on the terraced hill slopes.

Field borders, fences, and vegetated strips that interrupt soil fluxes also contribute to the erosion pattern by leading to the creation of topographic discontinuities or lynchets. The translocation of soil by tillage and water erosion on the terraced hill slope creates lynchets and enriches soils in the lower end of terrace in nitrogen and organic matter [Li, Lindstrom 2001]. The repeated translocation of soil in one direction with tillage tools that preferentially move soil to one side, create berms and a "dead furrow" or channel on opposite sides of the tilled domain [Vieira, Dabney 2011]. Thapa et al. [1999b] attempted to evaluate four tillage systems (i) contour mouldboard ploughing in the open field; (ii) contour soil barriers formed by ridge tillage in the open field; (iii) contour barriers formed by natural grass strips plus mouldboard ploughing; and (iv) contour barriers formed by a combination of ridge tillage and natural grass strips. The results show that both ridge tillage and natural grass barrier strips reduce tillage erosion rates for corn production on steepland soils in the humid tropics. In case of olive fields frequently ploughed by a local, donkey-drawn tillage implement the maximum soil loss values for contour tillage, were almost nine times less than for up and down tillage [Barneveld et al. 2009]. However, the case studies from Yanting, in Sichuan Province, China; Ha Sofonia, in Lesotho; and, Chinamora, in Zimbabwe confirm the importance of tillage erosion and translocation on terraces and contour-strips subjected to cultivation by animal traction. Rates of tillage erosion were comparable or greater than water erosion on the examined fields [Quine et al. 1999a]. On the other hand, land consolidation, typical for European agriculture contributes to acceleration of tillage erosion by the conversion of depositional areas into terrains which generates the soil loss [Chartin et al. 2013].

Tillage erosion has been described as the major cause of physical soil degradation in rolling agricultural landscapes. The long-term effects of soil redistribution by tillage increase the variability of soil properties [Kosmas et al. 2001], transform soil profile morphology and landscapes [De Alba et al. 2004], and lead to a significant decline in soil productivity. Tillage erosion led to truncated soil profiles on the shoulderslopes [De Alba et al. 2004] and developing of nutrient-rich and deep colluvial soil in the concave part of slope. Therefore, the within-field variability of soil properties in arable lands on the slope is controlled by tillage erosion which affected the redistribution of carbon and its field budget and nutrient losses [Heckrath et al. 2005]. Tillage erosion within the field borders is a key driver of net carbon cycle. According to studies based on CORINE, land use and the assumption that soils contain on average 2% of carbon, tillage erosion and deposition results in the burial of c. 7 Tg C y^{-1} [Van Oost et al. 2009]. Although the content of soil organic matter and available nutrients increase in the areas of soil accumulation [Li, Lindstrom 2001, Li et al. 2004] the long term enrichment of lowerings exposed to concentrated water flow may increase the amount of nutrient lost from the field [Heckrath et al. 2005, Van Oost et al. 2009] via water erosion as well as by leaching in more moist environment.

METHODS OF TILLAGE EROSION MEASUREMENT

Tillage translocation, defined as a transport and resultant displacement of soil by tillage [Govers et al. 1999], can be measured with a tracer method, i.e. a volume of soil is labeled and tilled, and then changes in tracer concentrations before and after tillage are used to calculate soil translocation. The tracer method for measuring soil translocation includes physical and chemical ones. Physical tracers are: metal cubes [Van Muysen et al. 1999], flat steel washers [Montgomery et al. 1999], magnetic tracers [Zhang et al. 2009], rock fragments [Nyssen et al. 2000] and gravels [Zhang et al. 2004]. Chemical tracers are radionuclides [Zgłobicki 2002] and chlorides [Lobb et al. 1999].

One of the most common physical tracers are the numbered aluminum cubes, which are placed in a series of holes and their positions are precisely recorded using a theodolite. After the treatment the areas immediately up- and downslope of the origin location are excavated and the position of each tracer is recorded. The use of metal detector to locate the tracers that moves relatively large distance allows a tracer recovery rate higher than 98% [Van Muysen et al. 2002]. Another popular tracers are brightly coloured gravels, or dyed aquarium gravel or stone chips [Turkelboom et al. 1999, Nyssen et al. 2000, Li et al. 2004, Zhang et al. 2004, Tiessen et al. 2007b]. The magnetic tracer is used seldom and it can be derived from the residues of brick and tile kilns, consisting of calcined soil and coal. The plots perpendicular to the tillage direction are established on the study fields. The soil from each plot is excavated and mixed with tracer and then returned to the plot. The magnetic strength of labeled soil in plot and soil prior to the application of the magnetic tracer must be measured to determine background and can be detected with a magnetometer, which commonly is used for the magnetic measurements of soil, rock, mine, brick, tile, cement, semiconductor, etc. The magnetic tracer was introduced to measure the soil translocation in conventional and conservation hoe-tilling in China [Zhang et al. 2009].

There are two methods of calculating translocation using plots filled with physical tracers. The Distribution-Curve Method, which is more common method, allows the calculation of soil translocation directly from the distributions of tracer after tillage. Summation-Curve Method, which is a less frequently used method, calculates the translocation from a summation curve generated from the distribution of tracer after tillage by employing convolution. Lobb et al. [2001] described and compared both methods using hypothetical and experimental data. Both methods provide accurate measures of gross translocation, but the Summation-Curve Method provides a measure of error associated with gross translocation and a more thorough characterization of the dispersion of translocated soil.

Besides using the aggreagate-sized physical tracer tillage erosion can be measured by marking the soil matrix with chemical tracers such as chlorides. The chloride (KCl - greenhouse grade muriate of potash) [Lobb et al. 1999] or sodium chloride solution can be used to measure the till-age erosion. The comparison of aluminium cubes and sodium chloride tracers revealed that there were no significant differences between these two methods [Barneveld et al. 2009].

However, one of the most popular tracers is ^{137}Cs used by many authors who confirmed its reliability and accuracy in measuring of the soil translocation and redistribution on the slope as a result of water and tillage erosion [eg. Pennock 2003, Zgłobicki 2002]. The ^{137}Cs technique provides data which are spatially distributed, shows the net effect of all types of erosion and provides the medium-term average erosion rates, on the basis of just single site visit. It is a manmade radionuclide, which was generated during the atmospheric testing of thermonuclear-weapons conducted in the 1950s and early 1960s and deposited onto the Earth's surface through wet and dry precipitation. After deposition to the Earth's surface, ^{137}Cs is quickly and strongly adsorbed by soil particles which makes it nonexchangeable. Therefore, its redistribution across the landscape is related to the redistribution of soil particles and that is why ^{137}Cs is used as a tracer indicating the physical movement of soil by erosion processes. The ^{137}Cs inventories (total activity in the soil profile per unit area) measured at the study site is compared with an estimate of the total atmospheric input, which is represented by the mean ^{137}Cs inventory obtained at a "reference site". Areas which evidence ^{137}Cs loss are identified as suffering net erosion and net ^{137}Cs gain indicates the deposition.

In order to derive quantitative estimates of erosion rates the calibration is needed which can be done by a mass-balance model or by means of proportional model which uses a simple linear function to convert the loss or gain of ^{137}Cs inventory (compared to a reference level) to a loss or gain of soil mass, respectively [Walling et al. 2002]. The research of Li et al. [2010] on proportional model and three types of mass-balance models proposed by Walling et al. [2002] revealed that all four conversion models are highly sensitive to the input values of the reference ^{137}Cs level, particle size correction factors and tillage depth. Another approach is represented by a model of Van Oost et al. [2003], which integrates a ^{137}Cs mass-balance model with spatially distributed soil erosion models where all processes significantly contributing to the redistribution of soil are independently simulated in a two dimensional spatial context.

One must bear in mind that the ^{137}Cs technique determines the impact of water and tillage erosion, so there still remains the problem of separating the contributions of water erosion and tillage to the pattern of net ^{137}Cs redistribution. The contribution of the individual processes can be identified by comparison of ^{137}Cs-derived soil redistribution rates with the water erosion model predictions. Although the ^{137}Cs inventory is mainly used to evaluate the tillage soil redistribution within single landscape unit the research of Pennock [2003] revealed that ^{137}Cs may be used at regional scale. However, the Chernobyl accident occurred on the 26th of April 1986 resulted in significant fallout of among others ^{137}Cs in Poland and other northern and eastern countries in Europe. In Poland Zgłobicki [2002] used ^{137}Cs tracer as one of the method to investigate the denudation in northwestern part of the Lublin Upland and overcame the problem by indirect quantification of the ^{137}Cs from Chernobyl fallout.

Recently, the measurements using lead-210 ([210]Pb$_{ex}$) has become recognized as an effective tool for documenting the soil translocation in many landscapes. Gaspar et al. [2013] used this tracer to measure soil redistribution caused by erosion and cultivation in mountain Mediterranean landscapes. Also plutonium isotopes ([239]Pu and [240]Pu) originated from atmospheric nuclear weapons tests were used for soil redistribution investigations in a catchment in Australia. The Pu measured with accelerator mass spectrometry (AMS) method allows to use 4-20 g samples and to measure much more samples than ^{137}Cs measured by γ-ray spectroscopy, what allows to perform more detailed investigations [Hoo et al. 2011].

Olson et al. [2002] used the fly ash, the product of high temperature coal combustion, together with magnetic minerals, magnetic susceptibility, and organic C content of a soil to estimate the extent of soil loss as a result of human activities at the cultivated field in Pushkino, Russia. Deposition of fly ash derived from distant railway traffic started around 1851 and increased in 1870 as a result of closer construction of a railway. Tillage and accelerated erosion redistributed the fly ash causing the deposition of sediment rich in fly ash on the lower and upper footslopes of the field. The estimated annual soil loss amounts to an average of 4.7 Mg ha[-1] yr[-1] for the past 60 to 80 years based on loss of fly ash and reduction in magnetic susceptibility [Olson et al. 2002]. Recently Olsen et al. [2013] combined the fly ash technique with ^{137}Cs to determine soil erosion rates for past cropland from 1910 till nowadays. Measuring of soil magnetic susceptibility is another fast and non-destructive method for estimation the amount of soil redistributed by tillage erosion within the landscape [Jordanova et al. 2011].

Several models were proposed to calculate the tillage erosion. Lindstrom et al. [1990] developed the first model of soil translocation by tillage as a statistical relationship between soil displacement and slope gradient. Govers et al. [1999] introduced the transport coefficient k (kg m[-1] per tillage operation) to relate the net unit soil transport rate due to a specific tillage operation to the slope gradient. The Tillage Erosion Prediction (TEP) model developed by Lindstrom et al. [2000], can predict soil redistribution along single slope profiles. Van Oost and Govers [2000] developed the Water and Tillage Erosion Model (WATEM), which simulates 2D patterns of soil redistribution using a diffusion-type equation and assumes that all soil translocation occurs in the direction of steepest slope, irrespective of the pattern of tillage. The SORET model is of the spatial distribution type and can perform 3D simulations of soil redistribution in Digital Terrain Models (DTMs) on the field scale. It can predict soil redistribution arising from different patterns of tillage in a given landscape via computer simulation of a single tillage operation, and is also able to forecast the long-term effects of repeated operations [De Alba 2003]. A tillage translocation model (TillTM) is a two-dimensional model (in the horizontal and vertical dimensions), where there are the topography data and soil constituent concentrations as a function of depth at a series of data points along the tillage direction [Li et al. 2008]. A diffusion-type model Directional Tillage Erosion Model (DirTillEM) was developed to better account for the effect of complex tillage patterns and field boundaries on tillage erosion across an agricultural landscape [Li et al. 2009]. Recently Vieira and Dabney [2011] developed a two-dimensional Tillage Erosion and Landscape Evolution model which allows complex internal boundaries to be defined within the simulation domain i.e. the model allows prediction of the formation of edge-of-field berms by defining alternative boundary conditions.

ERODIBILITY OF TILLAGE OPERATIONS

The tillage erosion was measured for many different tillage operations. In Thailand, Turkelboom et al. [1999] measured the tillage erosion by manual hoeing on steep slopes (32–82%). The experimental data showed that one tillage pass results in soil flux in the range from 390 to 870 kg m^{-1}, depending on the slope angle. The soil loss from the typical field located on the slope of 30–50% was estimated at 8–18 Mg ha^{-1}. Two typical hoe-tilling methods: (i) hoe-tilling in its conventional approach and (ii) protective non-overturning hoeing tillage, were applied on several terraces with different slope angles. The investigations revealed that translocation rates ranged from 46.47 to 113.62 kg m^{-1} per tillage pass, depending on the slop angle. The conservation approach of hoe tillage causes the decrease of soil downslope translocation to a range from 19.45 to 39.62 kg m^{-1} per tillage pass and results in a significant reduction in tillage erosion [Zhang et al. 2009].

Oxen-pulled ard tillage is another simple and popular method of tillage still used in Africa and Asia. The experiment was carried out in Ethiopia, on the terraced slope, where the tillage was parallel to the contour. The soil flux ranged from 4.8 to 38.7 kg m^{-1} and tillage erosion rates were smaller than those observed for mechanized tillage [Nyssen et al. 2000]. The experiment conducted in China with donkey-drawn mouldboard plough along the contours revealed high net accumulation of soil in the lower slope position after 50 operations. According to direct measurement using differential global positioning system the soil surface level at the top of the slope decreased by 1.25 m and increased by 1.33 m at the slope bottom [Li et al. 2004]

Although several authors paid attention to the hand-held or other simple tillage implements, majority of papers contributing to issue of erodibility of tools reported the translocation rates of implements used in typical conventional agriculture in Europe and North America. The displacement distance during mouldboard plough on silt loams were from 0.23 m downslope on the linear-convex backslope to 0.50 m during downslope operation on the convex shoulder [Montgomery et al. 1999]. On the shale-sandstone soils, on 21% slope the translocation of soil during mouldboard ploughing in the downward direction was 0.42 m and 0.16 m during upward tillage pass [Kosmas et al. 2001]. In up and

downward mouldboard tillage the soil translocation in the direction of tillage depends on slope gradient [Van Oost et al. 2000], tillage depth and speed, while in case of contour tillage soil movement is affected by slope gradient together with tillage speed. On the basis of these conclusions, the model was developed which allows to evaluate the effect of tillage depth, speed and/or tillage direction on the soil erosivity of a mouldboard ploughing [Van Muysen et al. 2002]. The mouldboard causes the asymmetric soil movement so the final rate of soil translocation should be determined on the basis of complex interaction between the morphology of the relief and the direction of tillage. The horizontal cutting angle of the mouldboard blades in relation to the forward direction of the tractor plays main role in soil movement intensity in complex landscapes. The direction different than perpendicular to the slope might be the controlling factor in reducing tillage erosion of mouldboard plough. The results of experiment based on tillage performed on sandy loam soils in 3 directions: (i) up- and downwards the slope, (ii) slantwise down and slantwise up, and (iii) contour, revealed that tillage in up and down at 45° to the maximum slope with turning soil upslope was the least erosive. Simulation of tillage erosion in complex topography by mouldboard plough revealed that contour tillage leads to higher average erosion rate what questions the role of this type of tillage in reducing the tillage erosion [De Alba 2003]. Also in potato cropping system the conservation tillage did not reduce the tillage erosion [Tiessen et al. 2007b].

The erosivity of primary tillage operations including mouldboard plough and chisel plough and the erosivity of secondary tillage with offset disc and vibrashank on the field with potato crop in undulating landscape of Canada on loamy soils were also measured. The results showed that both primary and secondary implements were very erosive, the average soil displacement measured for each operation was around 3 m, with maximum equal to 5.6 m for chisel plough and vibrashenk. The mass translocation was the highest for chisel plough, following the mouldboard plough, vibrashenk and offset disc [Tiessen et al. 2007b]. However Lobb et al. [1999] findings were contrary, because in their experiment the highest mass translocation was measured for moudlboard plough (72 kg m^{-1}), following chisel plough (62 kg m^{-1}) and tandem disc (56 kg m^{-1}), and field cultivator (41 kg m^{-1}). In case of chisel plough, tandem disc and field cultivator tillage translocation was slope gradient dependent. The tillage depth and speed affected the rate of erosion as well, but in complex landscape these two parameters are highly variable due to changing topographic and soil conditions [Lobb et al. 1999].

In potato production the tertiary tillage operations such as planting, hilling and harvesting result in significant soil translocation and can be equally as erosive as primary and secondary tillage operations. The harvester and sequence of planting, hilling and harvesting displace the soil up to 6.0 m and 23.6 m respectively, which is much greater distance than those resulted from primary and secondary operations [Tiessen et al. 2007a]. The tandem disc and field cultivator were also found as the erosive implements [Lobb et al. 1999]. Chisel tillage is another very erosive operation. In Belgian Loam Belt chisel tillage caused denudation rate more than 1 mm per tillage operation. This type of operation is usually combined with mouldboard ploughing, also one of the most erosive operations resulting in total annual tillage erosion rate equal to 3 mm per year [Van Muysen et al. 2000]. Chisel tillage was the most erosive operation in potato cultivation with mass movement per one operation equal to 64.4 kg m^{-1} [Tiessen et al. 2007b]. This type of tillage operation translocate the fine earths over larger distance than coarse material.

Although the recognition of erosivity of one operation is very important, under the normal conditions, farmers use several operations, required various equipment, during the year for crop cultivation. Transport coefficient for the whole tillage experiment using an implement sequence of a rotary harrow and seeder was 123 kg m^{-1} per tillage operation and suggests that these operations contribute significantly to soil displacement and tillage erosion [Van Muysen, Govers 2002]. The soil movement resulted from a typical tillage sequence, including multiple mouldboard, chisel and harrow passes was studied by Van Muysen et al. [2006] on Luvisols, Cambisols and Regosols, which have developed in loess deposits in Belgian Loess Belt. The soil displacement rate for tillage sequence was 2342 kg m^{-1} per tillage sequence and 167 kg m^{-1} per tillage operation. The results also revealed that total erosivity of different tillage operation cannot be calculated by summing up the erosivity of single operations, because total erosivity of tillage sequence is highly dependent on the tillage direction of every pass [Van Muysen et al. 2006], which can be difficult or even impossible to obtain. The study on four tillage implements: air-seeder, spring-tooth-harrow, light-cultivator and deep-tiller used as a typical conventional tillage sequence for cereal-based production in Canadian Prairies revealed that erosivity of air-seeder and spring-tooth-harrow were much lower than that of light-cultivator and deep-tiller, but their effect on total erosion has to be taken into account especially when those implements are used just after other tillage operations. However, the erosivity of full sequence in this tillage system was considerably lower than those with a mouldboard plough [Li et al. 2007].

THE RATES OF TILLAGE EROSION

The rates of tillage erosion have been reported all over the world. In Denmark average tillage erosion on glacial till on the typical hillslope of terminal moraine amounts to 27 Mg ha^{-1} yr^{-1} on the shoulderslopes, while deposition of 12 Mg ha^{-1} yr^{-1} was measured on foot- and toeslopes [Heckrath et al. 2005]. In Canada on glacial till in the hummocky landscape, convex slopeshoulders had the highest mean soil loss rates of 33 Mg ha^{-1} yr^{-1}, with the mean deposition rate on concave footslope was equal to 10 Mg ha^{-1} yr^{-1} [Pennock 2003]. The estimated soil erosion rates for pasture was 21 Mg ha^{-1} yr^{-1} and 38 Mg ha^{-1} yr^{-1} for crop rotation with potatoes in area of Prince Edward Island, Canada. The highest losses occurring on the slope shoulder suggest that tillage erosion is the major contributor in overall erosion [Kachanoski, Carter 1999]. In humid climate the mean annual soil loss from the contour mouldboard ploughing in the open field amounted to 63 Mg ha^{-1} yr^{-1}, while the soil loss was reduced by 30% during contour mouldboard ploughing within contour natural grass barrier strips, the reduction for contour ridge tillage in the open field was 45% and for contour and for natural grass barrier strips plus ridge tillage was 53% [Thapa et al. 1999b]. Conservation hoeing tillage reduced the tillage erosion rates from 78 to 28 Mg ha^{-1} yr^{-1} in the hilly areas in China [Zhang et al. 2009]. In China the tillage erosion was estimated from 50 to 150 Mg ha^{-1} yr^{-1} [Zhang et al. 2004]. The estimated mean tillage erosion rates on ribbon terraces was equal to 55 Mg ha^{-1} yr^{-1}, while on the shoulder terraces decreased to 14 Mg ha^{-1} yr^{-1} [Quine et al. 1999a].

CONCLUSIONS

Tillage erosion is major contributor in within-field variability of soil properties with important implications for nutrient losses and decline of crop productivity. The widespread use of tillage practices and high redistribution rates associated with process indicate that tillage erosion should be considered in soil landscape studies and when developing environmentally sustainable farming practices. The implications for further investigations are as follows:

1. According to the authors' knowledge there is no investigation comparing the rate of tillage erosion according to the age of equipment. In Poland many farmers have recently bought the latest equipment, so this may be an accelerating factor for tillage erosion.

2. The total erosivity of full sequence in various tillage systems was paid little attention except a few papers considering this issue. The proper methodology and interactions of various tillage operations in one crop rotation need further investigations.

3. The influence of tillage erosion on changes in old-glacial landscape, especially in Poland requires more research. Understanding of erosion processes on the gentle sandy slopes and their quantification may contribute to better principles in modern agriculture, which may control the nutrient accumulation in lowerings and valley bottoms.

4. The variety of tracers both physicals and chemicals gives a challenge to introduce a reliable model predicting not only tillage erosion but also water erosion, considering the number and type of tillage operation in the whole tillage sequence. The new tracers, especially isotopic radionuclides, need more investigations to provide reliable models for erosion rate calculation.

Acknowledgments

Financial support for this research was provided within the project S/WBiIŚ/1/14.

REFERENCES

1. Barneveld R.J., Bruggeman A., Sterk G., Turkelboom F. 2009. Comparison of two methods for quantification of tillage erosion rates in olive orchards of north-west Syria. Soil Till. Res. 103, 105–112.

2. Chartin C., Evrard O., Salvador-Blanes S., Hinschberger F., Van Oost K., Lefèvre I., Daroussin J., Macaire J.-J. 2013. Quantifying and modelling the impact of land consolidation and field borders on soil redistribution in agricultural landscapes (1954–2009). Catena 110, 184–195.

3. De Alba S. 2003. Simulating long-term soil redistribution generated by different patterns of mouldboard ploughing in landscapes of complex topography. Soil Till. Res. 71, 7–86.

4. De Alba S., Lindstrom M., Schumacher T.E., Malo D.D. 2004. Soil landscape evolution due to soil redistribution by tillage: a new conceptual model of soil catena evolution in agricultural landscapes. Catena 58, 77–100.

5. Gaspar L., Navas A., Machin J., Walling D.E. 2013. Using $^{210}Pb_{ex}$ measurements to quantify soil redistribution along two complex toposequences in Mediterranean agroecosystems, northern Spain. Soil Till. Res. 130, 81–90.

6. Govers G., Lobb D.A., Quine T.A. 1999. Tillage erosion and translocation: emergence of a new paradigm in soil erosion research. Soil Till. Res. 51, 167–174.

7. Govers G., Quine T.A., Desmet P.J.J. Walling D.E. 1996. The relative contribution of soil tillage and overland flow erosion to soil redistribution on agricultural land. Earth Surf. Process. Landforms 21, 929–946.

8. Heckrath G., Djurhuus J., Quine T.A., Van Oost K., Govers G., Zhang Y. 2005. Tillage Erosion and Its Effect on Soil Properties and Crop Yield in Denmark. J. Environ. Qual. 34, 312–324.

9. Hoo W.T., Fifield L.K., Tims S.G., Fujioka T., Mueller N. 2011. Using fallout plutonium as a probe for erosion assessment. J. Environ. Radioactivity 102, 937–942.

10. Jordanova D., Jordanova N., Atanasova A., Tsacheva T., Petrov P. 2011. Soil tillage erosion estimated by using magnetism of soils—a case study from Bulgaria. Environ Monit Assess. 183, 381–394.

11. Kachanoski R.G., Carter M.R. 1999. Landscape position and soil redistribution under three soil types and land use practices in Prince Edward Island. Soil Till. Res. 51, 211–217.

12. Kosmas C., Gerontidis St., Marathianou M., Detsis B., Zafiriou Th., Van Muysen W., Govers G., Quine T.A., Van Oost K. 2001. The effect of tillage displaced soil on soil properties and wheat biomass. J. Soil Till. Res. 58, 31–44.

13. Li S., Lobb D.A., Lindstrom M.J. 2007. Tillage translocation and tillage erosion in cereal-based production in Manitoba, Canada. Soil Till. Res. 94, 164–182.

14. Li S., Lobb D.A., Lindstrom M.J., Farenhorst A. 2008. Patterns of water and tillage erosion on topographically complex landscapes in the North American Great Plains. J. Soil Water Conserv. 63 (1), 37–46.

15. Li S., Lobb D.A., Tiessen K.H.D. 2009. Modeling tillage-induced morphological features in cultivated landscapes. Soil & Tillage Research 103, 33–45.

16. Li S., Lobb D.A., Tiessen K.H.D., McConkey B.G. 2010. Selecting and Applying Cesium-137 Conversion Models to Estimate Soil Erosion Rates in Cultivated Fields. J. Environ. Qual. 39, 204–219.

17. Li Y., Lindstrom M.J. 2001. Evaluating Soil Quality–Soil Redistribution Relationship on Terraces and Steep Hillslope. Soil Sci. Soc. Am. J. 65, 1500–1508.

18. Li Y., Tian G., Lindstrom M.J., Bork H.R., 2004. Variation of Surface Soil Quality Parameters by Intensive Donkey-Drawn Tillage on Steep Slope. Soil Sci. Soc. Am. J. 68, 907–913.

19. Lindstrom M.J., Nelson W.W., Schumacher T.E., Lemme G.D. 1990. Soil movement by tillage as affected by slope. J. Soil Till. Res. 17, 255–264.

20. Lindstrom M.J., Schumacher J.A., Schumacher T.E., 2000. A Tillage Erosion Prediction model to calculate soil translocation rates from tillage. J. Soil Water Conserv. 55, 105–108.

21. Lobb D.A., Kachanoski R.G., Miller M.H. 1999. Tillage translocation and tillage erosion in the complex upland landscapes of southwestern Ontario, Canada. Soil Till. Res. 51, 189–209.

22. Lobb D.A., Quine T.A., Govers G., Heckrath G., 2001. Comparison of methods used to calculate tillage translocation using plot-tracers. J. Soil Water Conserv. 56, 321–328.

23. Montgomery J.A., McCool D.K., Busacca A.J., Frazier B.E. 1999. Quantifying tillage translocation and deposition rates due to moldboard plowing in the Palouse region of the Pacific Northwest, USA. Soil Till. Res. 51, 175–187.

24. Nyssen J., Poesen J., Haile M., Moeyersons J., Deckers J. 2000. Tillage erosion on slopes with soil conservation structures in the Ethiopian highlands. Soil Till. Res. 57, 115–127.

25. Olson K. R., Gennadiyev A.N., Jones R.L., Chernyanskii S. 2002. Erosion Patterns on Cultivated and Reforested Hillslopes in Moscow Region, Russia. Soil Sci. Soc. Am. J. 66, 193–201.

26. Olson K.R, Gennadiyev A.N., Zhidkin A.P, Markelov M.V., Golosov V.N., Lang J.M. 2013. Use of magnetic tracer and radio-cesium methods to determine past cropland soil erosion amounts and rates. Catena 104, 103–110.

27. Pennock D.J. 2003. Terrain attributes, landform segmentation, and soil redistribution. Soil Till. Res. 69, 15–26.

28. Quine T.A., Walling D.E., Chakela Q.K., Mandiringana O.T., Zhang X. 1999a. Rates and patterns of tillage and water erosion on terraces and contour strips: evidence from caesium-137 measurements. Catena 36, 115–142.

29. Thapa B.B., Cassel D.K., Garrity D.P. 1999b. Ridge tillage and contour natural grass barrier strips reduce tillage erosion. Soil Till. Res. 51, 341–356.

30. Tiessen K.H.D., Lobb D.A., Mehuys G.R., Rees H.W. 2007a. Tillage translocation and tillage erosivity by planting, hilling and harvesting operations common to potato production in Atlantic Canada. Soil Till. Res. 97, 123–139.

31. Tiessen K.H.D., Mehuys G.R., Lobb D.A., Rees H.W. 2007b. Tillage erosion within potato production systems in Atlantic Canada I. Measurement of tillage translocation by implements used in seedbed preparation. Soil Till. Res. 95, 308–319.

32. Torri D., Borselli L. 2002. Clod movement and tillage tool characteristics for modelling tillage erosion. J. Soil Water Conserv. 57, 24–28.

33. Turkelboom F., Poesen J., Ohler I., Van Keer K., Ongprasert S., Vlassak K. 1999. Reassessment of tillage erosion rates by manual tillage on steep slope in northern Thailand. Soil Till. Res. 51, 245–259.

34. Van Muysen W., Govers G. 2002. Soil displacement and tillage erosion during secondary tillage operations: the case of rotary harrow and seeding equipment. Soil Till. Res. 65, 185–191.

35. Van Muysen W., Govers G., Bergkamp G., Roxo M., Poesen J. 1999. Measurement and modelling of the effects of initial soil conditions and slope gradient on soil translocation by tillage. Soil Till. Res. 51, 303–316.

36. Van Muysen W., Govers G., Van Oost K. 2002. Identification of important factors in the process of tillage erosion: the case of mouldboard tillage. Soil Till. Res. 65, 77–93.

37. Van Muysen W., Govers G., Van Oost K., Van Rompaey A. 2000. The effect of tillage depth, tillage speed and soil condition on chisel tillage erosivity. J. Soil Water Conserv. 55, 354–363.

38. Van Muysen W., Van Oost K., Govers G. 2006. Soil translocation resulting from multiple passes of tillage under normal field operating conditions. Soil Till. Res. 87, 218–230.

39. Van Oost K., Cerdan O., Quine T.A. 2009. Accelerated sediment fluxes by water and tillage erosion on European agricultural land. Earth Surf. Process. Landforms 34, 1625–1634.

40. Van Oost K., Govers G. 2000. WATEM, online manual. Catholic University of Leuven. http://www.kuleuven.ac.be/facdep/geo/fgk/leg/pages/downloads/WaTEM/.

41. WatemHome.htm.

42. Van Oost K., Govers G., Van Muysen W., Quine T.A. 2000. Modeling Translocation and Dispersion of Soil Constituents by Tillage on Sloping Land. Soil Sci. Soc. Am. J. 64, 1733–1739.

43. Van Oost K., Govers G., Van Muyssen W. 2003. A process-based conversion model for Caesium-137 derived erosion rates on agricultural land: an integrated spatial approach. Earth Surf. Process. Landforms 28, 187–207.

44. Van Oost K., Van Muysen W., Govers G., Deckers J., Quine T.A. 2005. From water to tillage erosion dominated landform evolution. Geomorphology 72, 193–203.

45. Verheijen F.G.A., Jones R.J.A., Rickson R.J., Smith C.J. 2009. Tolerable versus actual soil erosion rates in Europe. Earth-Science Reviews 94, 23–38.

46. Vieira D.A.N., Dabney S.M. 2011. Modeling edge effects of tillage erosion. Soil Till. Res. 111, 197–207.

47. Walling D.E., He Q., Appleby P.C. 2002. Conversion models for use in soil-erosion, soil-redistribution, and sedimentation investigations.. *In:* Zapata F. (ed.) Handbook for the assessment of soil erosion and sedimentation using environmental radioactivity. Kluwer Academic Publ., Dordrecht, the Netherlands. 111–164.

48. Zgłobicki W. 2002. Dynamika współczesnych procesów denudacyjnych w północno-zachodniej części Wyżyny Lubelskiej. Wyd. UMCS, Lublin.

49. Zhang J.H., Lobb D.A., Li Y., Liu G.C. 2004. Assessment for tillage translocation and tillage erosion by hoeing on the steep land in hilly areas of Sichuan, China. Soil Till. Res. 75, 99–107.

50. Zhang J.H., Su Z.A., Nie X.J. 2009. An investigation of soil translocation and erosion by conservation hoeing tillage on steep lands using a magnetic tracer. Soil Till. Res. 105, 177–183.

16

FUNCTION OF WATER IN THE LANDSCAPE OF THE VILLAGES IN THE PAST AND IN PRESENT, ON EXAMPLE OF VILLAGES IN LOWER SILESIA

Irena Niedźwiecka-Filipiak[1], Liliana Serafin[1]

[1] Institute of Landscape Architecture, Wrocław Universiy of Environmental and Life Sciences, pl. Grunwaldzki 24a, 50-363 Wrocław, Poland, e-mail: irena.niedzwiecka-filipiak@up.wroc.pl

ABSTRACT

Since ancient times, water was associated with colonization, being one of the main factors determining the localization of both cities and villages. Rivers, streams, water reservoirs were also the element contributing to the attractiveness of the rural landscape. Initially, the function of surface waters in the rural areas was limited to utility and connected with farm production. With time, the surface waters started being used for energy production and for industrial purposes. Proper management of surface waters also contributes to increase retention and reduce the risk of flooding. With time, streams and ponds became being used in mansion parks, which have been the endeavor enriching the composition. Today, rivers and water reservoirs in the villages no longer play such a significant utility and industrial role. Their function changed into mainly decorative and recreational. However, in many places the potential of using the areas adjoining rivers and ponds is not used by the village residents, which result in backfilling small water reservoirs in the villages and closing the visibility of flowing streams.

Keywords: surface waters, rural landscape, rivers, ponds.

INTRODUCTION

Rural landscape was shaped by many factors. One of them were the anthropogenic processes that are a blend of cultural, social and economic factors [Marcucci 2000]. Significant element of the rural landscape are the surface waters. They are the one of the most important components of the natural environment, having the impact on the attractiveness of the landscape [Iwicki 1997]. Water for centuries had a profound influence on the location and development of settlements. The water was the element, which was the basis for the existence of rural residents. Initially settlements were located in close proximity or near the natural forms of surface waters. Thus, in the rural landscape there are ponds, rivers and streams, which frequently were performing utility, farm and disposal functions [Borcz, Pogodziński 1994]. In rural areas, reservoirs and watercourses, although performing utility functions, influenced on a large diversity of landscapes [Wagner 2005].

With the development of civilization, the man used the water, increasingly transforming the landscape by creating new forms of water reservoirs and watercourses [Borcz, Potyrała 1993]. During this time, the rank of water has increased from basic, connected with the everyday life of rural residents, to more advance use of water as an energy source, which was the basis for creating a variety of industries. The other new function of water was a decorative one, within the development of the palace and park complexes. This application appeared most frequently in the case of palace parks and gardens, which often were located deliberately near rivers. In the composition both natural elements like oxbow lakes, springs, streams and other watercourses, as well as artificial water features (ponds, canals, fountains) were used [Bernat 2003]. In many villages the whole systems of water management were created, allowing retention of the excess water, as well as the possibility of obtaining it in times of shortage. Today we often underestimate the importance of water in shaping the rural landscape. Many of the old water reservoirs inside the built-up areas of the villages are backfilled without analyzing their historical role and the watercourses are covered and hidden. This causes not only the consequence of depletion of the rural landscape in terms of scenic view, but also periodic flooding of buildings, which is often a big surprise for residents.

The question must be asked whether nowadays the role of surface waters in shaping the rural landscape is appreciated and whether these resources are used by the rural society in the right way. Natural capital that brings many benefits for society are varied and multifunctional landscapes [de Groot 2006]. One of the most important elements of these landscapes are surface waters. Therefore, so important objective of water management in agriculture and rural areas is protection of surface waters against improper or excessive exploitation. Now a serious threat against the achievement of these objectives is intense urbanization and fragmentation of rural areas. As a result of these processes, the existing natural, water supply and drainage systems are threatened by destruction [Kaca 2009].

MATERIAL AND METHODS

Surface waters in the villages in Wisznia Mała Commune

Initially, the surface waters in the villages were used mostly for utility purposes. During this time, together with buildings and whole village development of each village, surface waters fitted to the natural environment. Work and life of the rural society in a natural way was linked with nature and its cycles. Both, the watercourses and water reservoirs had influence on rural landscape. The build-up areas in the villages were localized and shaped depending on the location of the village relative to the flowing stream. The strongest impact had the streams flowing inside the village, along the buildings, due to the need of taking into account the flood plain, without the possibility of situating any buildings there. It was necessary to build watercourse crossings to individual farms located on the other side of the stream. Sometimes two roads on both sides of the river with several major crossings connecting the two parts of the village were built. Along the rivers, high vegetation was planted, mostly fruit trees. An equally important role in the village was played by water reservoirs whose function and importance varied

with changes of the rural areas. Ponds located in the center of the village, one or more, over time became in many cases fire protection reservoirs with concreted edges and gridded, metal fence. Previously available for all village residents, at this time completely cut off from the possibility of close contact with them. This process has also unfavorable visual reflection by isolation of the space and raw concrete edges of the banks.

In order to test these issues the villages from the Wisznia Mała commune from Lower Silesia Region were selected. In these villages the changes in terms of surface waters, that occurred up over time, were examined. Analyses were carried out on the basis of the comparison of the messtischblatt maps from the early twentieth century with contemporary maps and on the basis of field surveys. Also the studies conducted in 2002 [Niedźwiecka-Filipiak 2002]were used to determine whether last years had positive impact on the approach of the village residents and local authorities to the subject. In 2002 it was demonstrated that the role of surface waters had still great importance, despite the loss of a number of water reservoirs in recent years. The aim of the analysis was to determine what is the quantitative situation of water reservoirs now and also if the actions of the village residents and local authorities have recently had a positive impact on this situation. Another issue is whether the landscape potential of surface waters in the villages is used through proper management of their surroundings.

The commune of Wisznia Mała is located in the immediate vicinity of Wroclaw, bordering it from the north side. It is therefore situated in the area of strong suburbanisation processes, but it is also a potential place of recreation for Wroclaw's inhabitants [Niedźwiecka-Filipiak, 2013]. Figure 1 shows on a graphical scheme the changes that have taken place in this area in comparison to the year 1942. In most of the villages the built-up areas has expanded and some of them have merged with each other (Rogoż with Kryniczna or Ligota Piękna with Wisznia Mała). The villages Raków and Cienin also underwent significant changes. These villages have been almost totally razed to the ground (Cienin) and the greater part of the land is utilized as a military training ground. Currently in part of the former Raków built-up area the Toya golf course has been built, together with a housing estate.

The studied area is located in the river basin of Widawa, which runs through the south-western part of the area. Table 1 shows that at present only one village in the whole Wisznia Mała commune has no surface waters. It is the village Piotrkowiczki. However, in the first half of the twentieth century, also in this village there were two small ponds within residential areas, which have not survived to the present day. In three villages there are small watercourses Ława and Mienia

Figure 1. The development of the commune of Wisznia Mała between 1942 and 2014 year

Table 1. Summary of surface waters in the villages from the commune of Wisznia Mała

No.	Name of the village	Watercourses, streams and rivers	Stagnant waters ponds (numer)		Total
			within residential areas at the beginning of the twentieth century[1]/ 2002[2]/2014[3]	in the immediate vicinity of the built-up areas at beginning of the twentieth century 2002/2014	
1	Kryniczno	0	4/3/1	4/0/1	8/3/2
2	Krzyżanowice	*Widawa* – nearby	1/0/0	0/0/0	1/0/0
3	Ligota Piękna	0	1/1/1	2/2/3	3/3/4
4	Machnice	*Ława* – nearby	4/2/4	0/0/0	4/2/4
5	Malin	0	1/1/1	2/2/2	3/3/3
6	Mienice	*Mienia* – along	1/0/0	0/0/0	1/0/0
7	Ozorowice	*Ława* – along	1/1/1	0/0/1	1/1/2
8	Pierwoszów	*Ława* – crosswise	3/1/0	0/0/1	3/1/1
9	Piotrkowiczki	0	2/0/0	0/0/0	2/0/0
10	Psary	*Widawa* – nearby	2/2/2	2/2/2	4/4/3
11	Rogoż	0	4/3/3	0/1/1	4/4/4
12	Strzeszów	*Ława* – along	4/2/2	3/0/0	7/2/2
13	Szewce	0	3/1/1	0/0/0	3/1/1
14	Szymanów	*Widawa* – nearby	1/0/0	2/3/3	3/3/3
15	Wisznia Mała	0	8/4/4	1/1/1	9/5/5
16	Wysoki Kościół	0	1/0/1	0/1/1	1/1/2
	TOTAL		41/21/21	16/12/16	57/33/37

Comments: 1 – messtischblatt [amzp.pl]; 2 – [Niedźwiecka–Filipiak 2002]; 3 – based on field research.

flowing along the built-up areas (Mienice, Ozorowice and Strzeszów), and in one example also flowing crosswise to the village (Pierwoszów). Krzyżanowice, Psary and Szymanów are located in the vicinity of the river Widawa that flows beside the built-up areas.

During the analysis of the amount of water reservoirs in the studied localities it was concluded that within approx. 100 years their number in urban areas have declined sharply, it is now a half smaller. Liquidated water reservoirs inside the old manor complexes (eg. Mienice), in the center of the village (eg. Ozorowice), the mill ponds (eg. Strzeszów) and other, particularly the small ones. Over the last 10 years the total number of all ponds inside the village has not changed, but the differences are in the various localities. Unfortunately, water reservoirs, especially those small are still being backfilled. Such examples are villages Kryniczno and Pierwoszów. In Kryniczno, following the map from year 1942, there were four water reservoirs inside the village (Figure 3). One of them was located inside the manor farm buildings, another was situated nearby and the other two had probably economic and retention functions. Even in 2002, in the village three water reservoirs were located there. The one situated near the mansion have been backfilled, but the another one located nearby was still preserved. However, it was overgrown by vegetation and neglected. The third pond, which was situated near the church, in the meantime was turned into a concrete fire protection reservoir, fenced with a mesh and became inaccessible due to the dangerous steep banks. The last pond was located within the private property, during this time its size had been reduced but the pond is now well-maintained with the addition of the bridge, however, not very visible from the road. Over the last 10 years, another two ponds were filled. The only one which remained was the one located within the private area. The new reservoir located in the vicinity of built-up areas of the village is overgrown and neglected (Figure 4).

In terms of water reservoirs situated in the immediate vicinity of the built-up areas the quantitative changes are much smaller. Their number decreased till 2002, but in the last 10 years has risen to the status of the early twentieth century (Figure 2). The biggest changes concern the development of the two recreational areas, in which new water reservoirs were created. The first is the restaurant Miłocin with water reservoir in Pier-

woszów. The other is a golf course, in which there are currently six major water reservoirs and a few smaller ones, along with the Toya Golf housing estate. These sites are an example of good use of water reservoirs and creating them attractive not only for the residents of the commune, but also for Wroclaw's inhabitants and in the case of golf courses also the region.

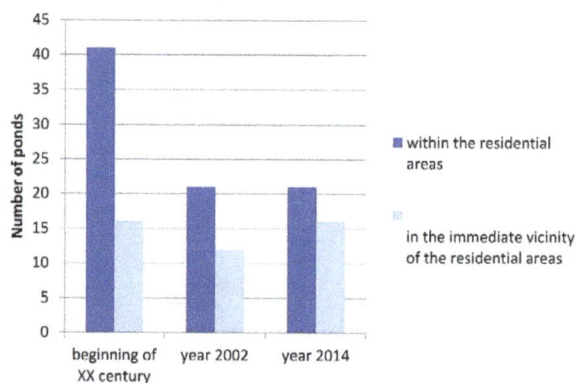

Figure 2. Number of ponds with division on ponds within the residential areas and ponds which are in the immediate vicinity of the residential areas

However, it should be noted that in any villages from the studied area the potential of watercourses flowing through the villages was not used to improve the quality of space around. In the village of Mienice a small part of the river surroundings within the private property have been developed in a model way (Figure 5), while within the area around the common building located by the river, the values of water haven't been taken into account. The building is fenced, cut off from the watercourse, both functionally and visually (Fig.6). Although the public space could be enlarged by designing a descent to the water and including the river into the project, as was it done in the private property. It should be added that this is not a bad will of the inhabitants, more the lack of awareness, because the village is well cared for, and its inhabitants are active.

RESULTS

Contemporary functions of water reservoirs. Good practices from Poland and from abroad

The example of the right approach to the potential of existing resources and the use of surface waters in shaping the rural landscape is the village Rumbau located in Rhineland-Palatinate in Germany. The village currently has 480 residents, the

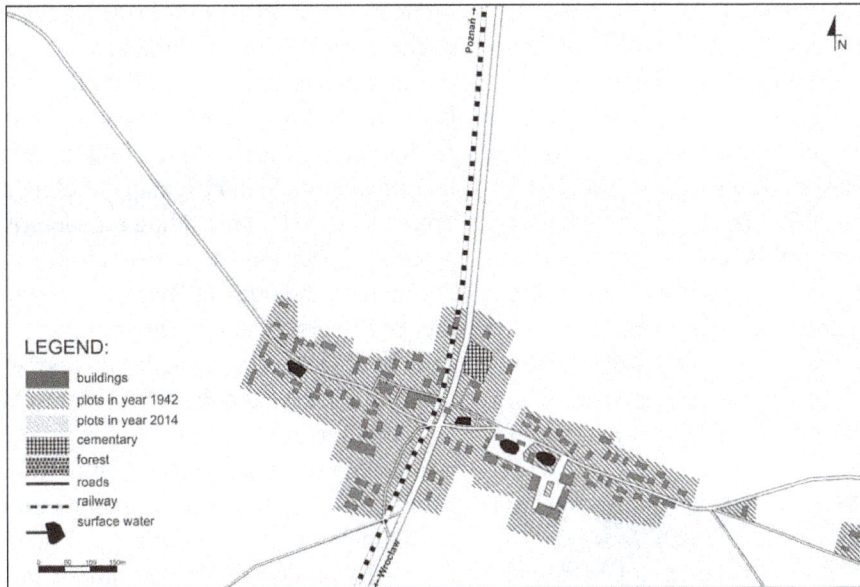

Figure 3. Kryniczno in the year 1942, there are 4 ponds visible

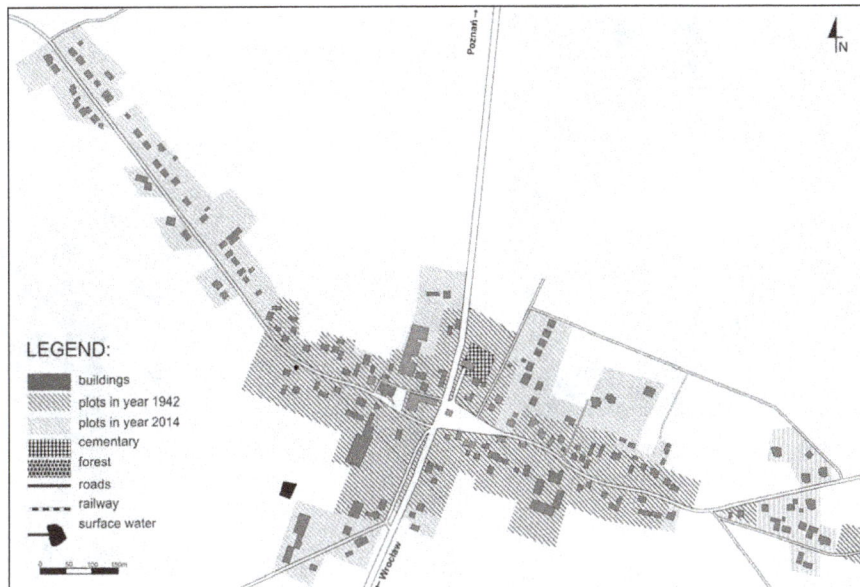

Figure 4. Kryniczno in the present time, with one small pond

Figure 5. Mienice. The development of the neighbourhood of the river on a private property

Figure 6. Mienice. On the first plan the undeveloped surroundings of the river, which is isolated from the common building by a fence

area of the village covers about 1,500 hectares, of which 1,000 hectares are forests. The village has a pond and touristic trail called "Rumbach and its springs". Each of the seven springs and wells is properly exposed and beard with the information board. Examples are shown on Figure 7 and 8. Inside the village there is also a small water reservoir. The project of this reservoir and its surroundings takes into account children's playing area, recreation for the elderly, as well as places where all residents can make barbecue or play active sports. It is now under construction. The idea of Sebastian Kneipp, which involves treat-

ment by immersion in cold water was used also in the project. Most treatment is based on walking in water or putting hands into it. The idea is being used in Germany for example in nurseries for hardening the health of children. In the end of the nineteenth century Kneip developed the comprehensive system including regenerative-healing program based on the beneficial effect of water. Inspired by this idea in Rumbau, next to the pond the engineers planned small oblong pool with comfortable handrails enabling easy entrance to it for all interested persons, including the elderly.

Another example of good practice is an ongoing project of land development of area by a small river in the village of Ottersheim in Germany, Rhineland-Palatinate (Figure 11). This project is worth showing because without the residents of the village the realization of it would not be possible. Earlier the area adjoining the river was consisted of the private plots. Because this was the area of floodplains, there were gardens and or-

Figure 7. Fragment of the plan of development of the interior of the village with use of the pond, below Kneipp therapy pool

Figure 9. Spring number 3. The water flows freely in the gutter along the street

Figure 8. Sketch of the stream flowing through the village along with showing the location of seven springs

Figure 10. The so-called Mayor Spring

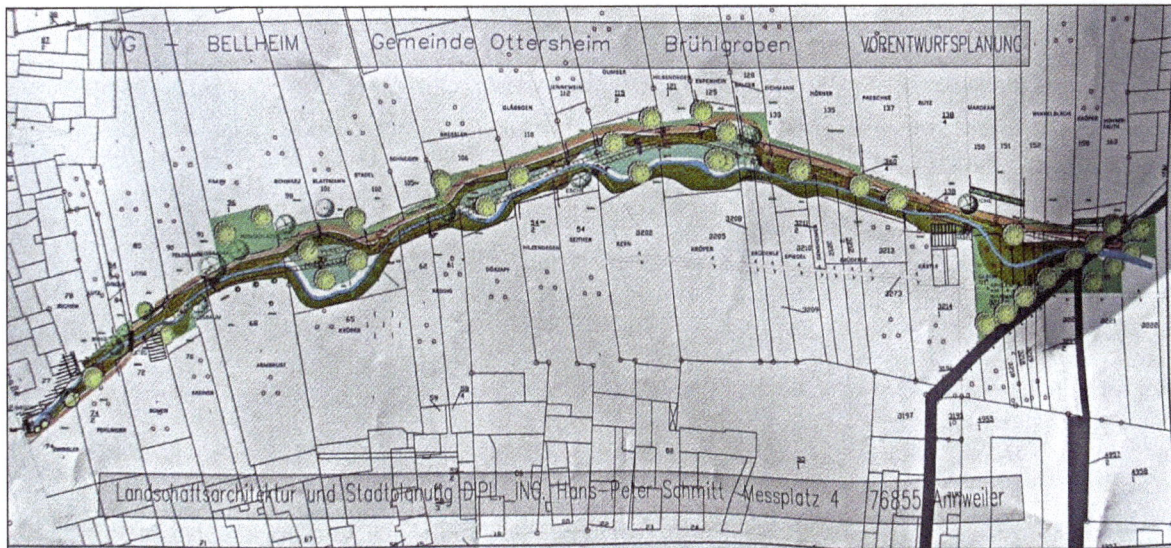

Figure 11. Ottersheim, the development plan of the area by the river-realization is planned till year 2018

chards. In the village there was little public space being favorable for integration of the inhabitants. It was therefore decided by the commune to buy the fragments of private plots adjoining the river at a symbolic price (with the consent of the inhabitants). That enabled the creation of the promenade with scenic places to rest and new public space favorable for the integration of the inhabitants.

Different example of the village, where the surface waters are partially used, is the village of Bagieniec located in the commune of Świdnica. The surface waters in the village were used as early as at the beginning of the twentieth century (Figure 12). Then the inhabitants of the nearby town of Świdnica were comming to Bagieniec, searching for the ability of resting in peace among the ponds and green areas. Today, the complex of preserved ponds is used again for fishing and as a recreational area (Figure 13).

CONCLUSION

Former economic function of surface waters in the villages has changed and now they have lost their former rank. As shown on the example of the analyzed commune, the biggest changes concern small water reservoirs within residential areas, from which many were liquidated during the last 100 years. Although having the potential of performing the representative functions, being a place for recreation and the decoration of the village, only a small proportion of the small water reservoirs, which remained, is properly managed. Residents do not see their potential, rather only trouble and breeding area for onerous mosquitoes. Similarly, the potential of rivers, streams and watercourses flowing through the villages is not used. With appropriate land development they could also conduct representative and recre-

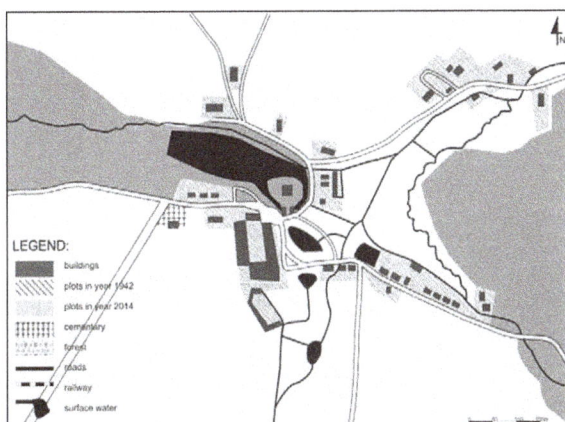

Figurte 12. The village Bagieniec in 1942

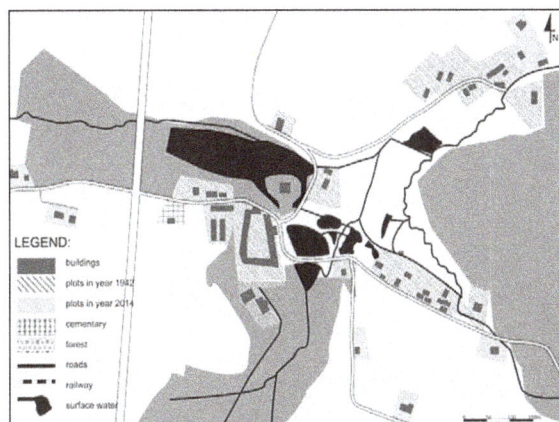

Figure 13. The village Bagieniec now

ational function. The best used examples of water reservoirs are larger ponds located outside the villages. They are used as for fishing and swimming purposes. These water reservoirs are the base around which the areas with recreational and touristic function are built.

REFERENCES

1. Bernat S. 2003. Woda w Krajobrazie parków wiejskich doliny Bugu. Woda w przestrzeni przyrodniczej i kulturowej. Prace komisji krajobrazu kulturowego t II. Wydawnictwo Komisja Krajobrazu Kulturowego PTG, Sosnowiec, p. 38.

2. Borcz Z., Pogodziński Z. 1994. Woda w krajobrazie wiejskim - zagrożenia i ochrona. Zeszyty Naukowe AR we Wrocławiu, Monografie, p. 236.

3. Borcz Z., Potyrała J. 1993. Egzystencja i przyszłość wiejskich stawów. Zeszyty Naukowe AR we Wrocławiu, 231, 335–343.

4. De Groot R. 2006. Function-analysis and valuation as a tool to assess land use conflicts in planning for sustainable, multi-functional landscapes. Landscape and Urban Planning 75, 175–186.

5. Iwicki S. 1997. Znaczenie zasobów wodnych w rozwoju turystyki na obszarach wiejskich. Woda jako czynnik warunkujący wielofunkcyjny i zrównoważony rozwój wsi i rolnictwa. Wydawnictwo IMUZ, Falenty, p. 209.

6. Kaca E. 2009. Gospodarka wodna wsi i rolnictwa. I Kongres Nauk Rolniczych, Nauka w Praktyce: przyszłość sektora rolno-spożywczego i obszarów wiejskich, Puławy, 93–104.

7. Marcucci D.J. 2000. Landscape history as a planning tool. Landscape and Urban Planning 49, 67–81.

8. Niedźwiecka-Filipiak I. 2002. Wpływ wód powierzchniowych na układ przestrzenny wsi. Inżynieria Rolnicza 3 (36), 383–392.

9. Niedźwiecka-Filipiak I., Zielińska L. 2013. Zieleń w krajobrazie wsi podmiejskich Wrocławia. Planowanie krajobrazu – wybrane zagadnienia, Wydawnictwo PAN, Uniwersytet Przyrodniczy w Lublinie.

10. Przyjazne naturze kształtowanie rzek i potoków. 2006, tłumaczenie z Manual of River Restoration Techniques, Wydawnictwo Polska Zielona Sieć, Kraków.

11. Wagner A. 2005. Znaczenie zbiorników wodnych w rozwoju ekoturystyki i agroturystyki w wybranych rejonach wiejskich w okolicach Krakowa. Przegląd Naukowy: Inżynieria i Kształtowanie Środowiska 2(32), 140–146.

17

COST OF MUNICIPAL WATER TREATMENT PLANT IN THE BIGGEST POLISH TOWN IN PODLASKIE PROVINCE FOR THE YEARS 2010–2012

Agnieszka Kisło[1], Iwona Skoczko[1]

[1] Department of Technology in Engineering and Environmental Protection, Faculty of Civil and Environmental Engineering, University of Technology in Bialystok, 45A Wiejska Str., 15-351 Bialystok, Poland, e-mail: agnieszka.kislo@wp.pl

ABSTRACT

In this paper the operation costs of the municipal water treatment plant in the biggest Polish town in Podlaskie province was analyzed. Capacity of this WTF is 600 m³/h. Water treatment processes are primarily focused on removal of iron, mangnese and turbidity and disinfection by UV rays. Water is taken by 19 wells and then it is oxygenated. From aerators water is addressed to ten filters, which filter water at a speed of 8.5 m/h. The analysis of the operation costs of the municipal water treatment plant was carried out by a method of testing and interpretation of the materials provided by the Water and Sewerage Company in a big town in the Podlaskie Province. It was established that, groundwater treatment plant operation costs, carried out in 2010–2012, showed the highest share of depreciation and remuneration costs.

Keywords: underground waters, water treatment plant, operation costs analysis.

INTRODUCTION

Groundwater is water that lies beneath the Earth's surface at different depths, resulting from a variety of geological processes. Its total volume is approx. 60 000 thousand km³, which is approx. 4.12% of the total volume of the Earth's hydrosphere resources [Chełmicki 2002].

The determinant of the usefulness of natural waters for a particular purpose is its physico-chemical and bacteriological composition. Physico-chemical and bacteriological composition of groundwater is variable and depends on many factors, which include, inter alia, the contact time with the layers of rock, season of the year, the amount and quality of rainwater, and land cover [Nawrocki, Biłozór 2000].

Groundwater, due to its large resources and high quality, is a very important source of drinking water. Great economic importance and commonly occurring groundwater threat forces to maintain a quality control by organizing a groundwater monitoring system. Water to be safe for health must meet the parameters speci-fied in the Regulation of the Minister of Health of 20.04.2010 "The quality of water intended for human consumption" (Dz. U. No. 72, item. 466 dated 29.04.2010) [Regulation of the Minister of Health of 20.04.2010].

Today, the price of water contains a number of indirect costs, i.e.: the costs of its recognition, treatment and delivery to the recipient. The increase of water prices may be due to continuous modernization and introduction of new treatment technologies. This significantly improves the quality of water supplied to our homes. It should be also noted that the cost of "producing" water affect other factors, such as the cost of building water intake, water treatment technology, water transmission networks, as well as operating costs of these systems [Dąbrowski, Mountain2004; Dziembowski 1983].

Water whose parameters do not meet the legal requirements needs treatment. Nowadays level of technology and technical knowledge allow to treat even the most polluted waters. However, the process is economically inefficient if high pollution treatment costs grow disproportionately with the degree of water pollution. Economic considerations support the concept of water apprehension of the best quality. The treatment of most groundwater is limited to the reduction of the content of iron and manganese. Such a treatment is relatively simple technically [Heidrich 1985].

METHODOLOGY

In order to analyze the costs of the Municipal Water Treatment Plant (WTP) a method of testing and interpreting the materials provided by the Water and Sewerage Company in the big town in the Podlaskie Province was used. The data covered the period 2010–2012 and presented an overall picture of the financial operations of the Municipal Water Treatment Plant.

DISCUSSION

The most important operating costs of groundwater treatment plants include [Nawrocki, Biłozór 2000; Zadrożna 2011]:

- energy costs,
- salary related costs,
- costs of materials and repairs,
- the cost of installation of water meters,
- general and administrative costs,
- other costs.

In order to recognize the distribution of the various costs, the groundwater treatment plant (with a capacity of 600 m³/h) was analyzed. The main problems of water quality are related to an extensive content of iron, manganese and turbidity. Therefore, the water treatment processes are primarily focused on removal of iron, mangnese and turbidity [The material provided by the Water and Sewage in Suwałki].

Water is taken by 19 wells. Its treatment technology at the WTP is as follows: from the well the water is addressed to two pressure reaction chambers. There the water is oxygenated using compressed air supplied by three oil-free compressors. The objective of aeration is to introduce oxygen into the water to allow the partial oxidation of iron and manganese from the form of Fe^{2+} and Mn^{2+} to the form of Fe^{3+} and Mn^{4+} and their primary precipitation. Water aeration allows also its de-acidification. Crowding undissolved gas is run by automatic valves. From aerators the water is addressed to ten filters, with filtration rate of

8.5 m/h. The filters are filled with a fluid layered "CULSORB M" and work fully automatically thanks to programmable controllers, dampers and diaphragms. Behind the filters, water is collected in four reservoir's tanks of clean water with a total capacity of 9600 m³. Clean water tanks provide water supply to compensate for uneven hourly partitions. From the tanks, equipped with dual air filtration system, water flows by gravity into the pumps' hall. Next it meets continuous UV disinfection lamp and goes to recipients using horizontal pumps.

Water treatment plant provides treated water to almost 100% of the town's population (70 thous.). The total length of the water supply network in the studied city (connections) is 214.5 km. Yearly WTP produces about 3 million m³ of water (approx. 8 thousand. m³ per day) and sales approx. 2.5 million m³. Water (2 million m³) is supplied mainly to households. Approx. 60 thousand. m³ of water is spent for production [Documentation hydrogeological... 1993].

Total cost of Municipal Water Treatment Plant include the following types of costs:

- depreciation,
- fuel costs,
- costs of oil,
- costs of consumables,
- costs of materials for repairs,
- costs of electricity consumption,
- costs of thermal energy,
- the costs of transport services,
- the costs of conservation and modernization,
- the costs of other services,
- salaries,
- social security,
- employees' benefits,
- taxes and fees,
- the costs of business trips,
- the cost of banking services,
- the costs of representation and advertising,
- other costs,
- settlement costs.

Figure 1 shows that, the largest share in total cost of ownership of groundwater treatment plant are depreciation costs (45.40–47.10%) and expenses (19.7%). The lowest part is the cost of fuel (0.01–0.04%). Analyzing the cost statement for the period 2010–2012 for the Water Treatment Plant, the authors have noted a gradual increase in operating costs over the years. It can be assumed that this relationship is caused by the increase of

2010

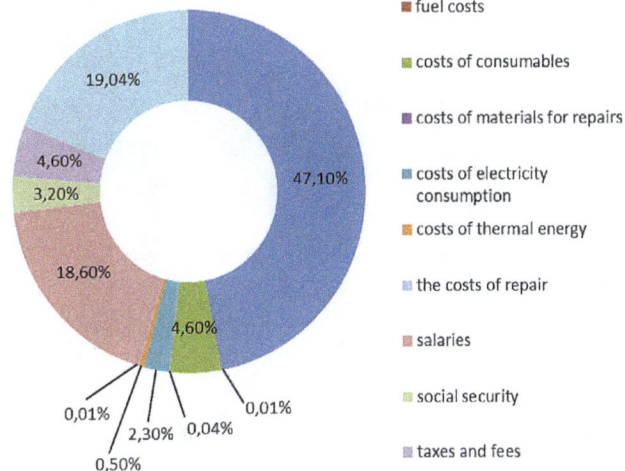

- depreciation
- fuel costs
- costs of consumables
- costs of materials for repairs
- costs of electricity consumption
- costs of thermal energy
- the costs of repair
- salaries
- social security
- taxes and fees
- other costs

2011

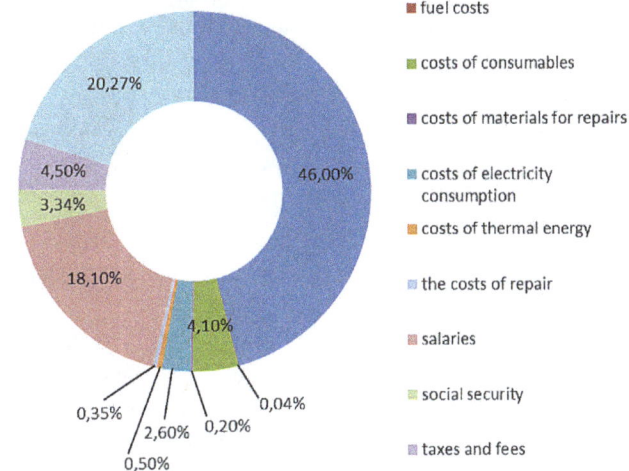

- depreciation
- fuel costs
- costs of consumables
- costs of materials for repairs
- costs of electricity consumption
- costs of thermal energy
- the costs of repair
- salaries
- social security
- taxes and fees
- other costs

2012

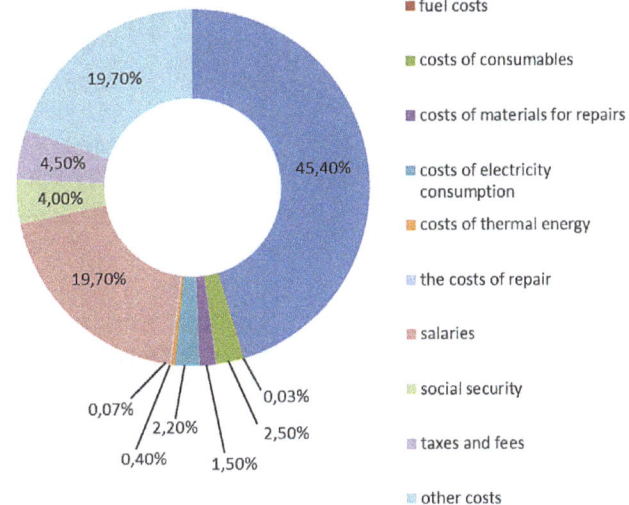

- depreciation
- fuel costs
- costs of consumables
- costs of materials for repairs
- costs of electricity consumption
- costs of thermal energy
- the costs of repair
- salaries
- social security
- taxes and fees
- other costs

Figure 1. Percentage of individual costs' share in relation to the total outlays of the Water Treatment Plant in 2010–2012

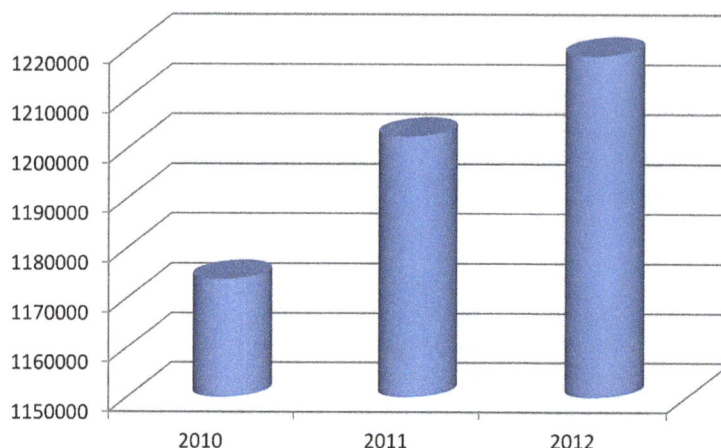

Figure 2. Graph of the changes in the total annual outlays of the Water Treatment Plant
in the big city in the Podlaskie Province in 2010–2012

market prices and the cost of maintenance of the facilities. Over the three analyzed years (2010–2012) the WTP management decided to make only one quite expensive conservation (Figure 2). Therefore, it can be concluded that the groundwater treatment plant is characterized by high operational reliability.

CONCLUSIONS

On the basis of the analyzed materials the authors have found that the processes used in Municipal Water Treatment Plant allowed for reduction of iron, manganese and turbidity from the treated water to the required level. The largest share of total operating costs of groundwater treatment station are depreciation costs and the costs of wages. The studied WTP is characterized by low fuel costs. The increase of the market price of energy and materials used for water treatment caused an increase in total cost of ownership of underground water treatment station in Suwałki in 2010–2012 by about 5%.

REFERENCES

1. Chełmicki W. 2002. Water: resources, degradation, protection. PWN, Warsaw.

2. Dąbrowski S., Mountain J. 2004. Methodology for determining the reserves of fresh groundwater intakes. Ministry of the Environment, Warsaw.

3. Dziembowski Z 1983. Municipal enterprise economics. PWE, Warsaw.

4. Heidrich Z. 1985. Technical and economic aspects of water treatment. Environment, ed. PZITS, Wroclaw.

5. Nawrocki J., Biłozór S. 2000. Water treatment. Chemical and biological processes. PWN, Warsaw-Poznan.

6. Weber L. 2010. Exploitation of underground water treatment station. Part I. Types and operation of aeration systems underground water pressure. Water Technology, 1(3).

7. Weber L. 2010. Exploitation of underground water treatment station. Part IV. Checking the condition of filter beds. Water Technology, 4(6).

8. Zadrożna S. 2011. Evaluation of groundwater treatment technologies for the chosen example. Thesis Engineering. University of Bialystok, Bialystok.

9. Documentation hydrogeological groundwater resources area Suwałki. Department of Service Design "Eco-Geo" M. Tatarata in Suwałki. Suwałki 1993.

10. Regulation of the Minister of Health of 20.04.2010 on the quality of water intended for human consumption (Dz. U. No. 72, item. 466 dated 29.04.2010).

11. The material provided by the Water and Sewage in Suwałki. Website www.pwik.suwalki.pl

18

PHYSICAL PROPERTIES, PERMEABILITY AND RETENTIVENESS OF SILT LOAM AND ITS COMPOSITES WITH SAND FOR CONSTRUCTING CARRYING LAYER OF A FOOTBALL FIELD

Tomasz Kowalik[1], Włodzimierz Rajda[1]

[1] Department of Land Reclamation and Environmental Development, University of Agriculture in Krakow, Mickiewicza 24-28, 30-050 Kraków, Poland, e-mail: rmkowali@cyf-kr.edu.pl, rmrajda@cyf-kr.edu.pl

ABSTRACT

Physical and water properties of silt loam from the area of planned football field were tested and compared with analogous properties of several composites made in laboratory conditions from the collected material with a dominant sand share. The research was conducted in a view of silt loam and its composites usefulness for constructing a carrying layer of football fields. Water permeability of silt loam and composites, as well as retention abilities were tested. The created composites met the water permeability requirements specified by DIN 18035 standard for constructing carrying layer of football fields. On the other hand, silt loam without sand admixture did not meet the requirements, but revealed a high retention capacity and water availability to plants. Among the composites the best retention capacity characterised the mixtures with the biggest content of silt loam, but the best water availability was registered in composites with medium content of silt loam from the football field area. The obtained results may be useful for more precise determination of the standards for grain size distribution of the composites used for constructing the carrying layer of a football field.

Keywords: silt loam, composites, water permeability, retention of soil, water availability.

INTRODUCTION

Natural turf covered football fields, apart from their shape and geometric dimensions [Żegocińska-Tyżuk 1988] should be characterized by an appropriate abundance in nutrients and proper sorption capacity, but also by high water permeability and retentiveness [Policht_Latawiec 2008, Gołąb and Gondek 2013, Milivojević et al. 2011]. Due to the use of wrong materials and (or) inappropriate construction technologies, the football pitches usually do not meet water permeability requirements [Rajda et al. 2011], and therefore, their functionality is limited. Then grass does not have proper aesthetic values [James et al. 2007a, James et al. 2007b]. On sand grounds during rainless periods, the turf requires frequent sprinkling, otherwise it dries. On the other hand, on compact grounds, football pitch, which usually has a con-siderable retention capacity, is hardly permeable [Pereira et al. 2007] and at irrigation or excessive rainfall, water stagnates on the field surface and causes falling out of grass. It both cases it causes difficulties for football players; the results of conservation measures are poorer and the maintenance costs of football field grow [James et al. 2007a, James et al. 2007b].

According to Żegocińska-Tyżuk [1988], properly constructed football pitch on compact ground should be made up of two several-centimetre thick layers placed on an appropriate foundation (Figure 1). The carrying layer should be constructed of a mixture (composite) of the indigenous ground from the humus layer and appropriate sand admixture, so that the fraction would be dominant in the composite. A drainage layer built of sand only should be placed under the carrying layer [James et al. 2007b]. Such composition is recommended by DIN 18035 standard [Deutche norm]. The carrying layer composite, at the same time constituting grass root layer should be compacted in order to obtain adequate elasticity and dynamic strength.

Compacted carrying layer at maximum capillary capacity should have water permeability no less than 0.3 mm·min⁻¹. It ensures good conditions for excessive rainfall seepage to the drainage layer and drains. The carrying layer should be also characterized by good retention properties, determining the frequency and doses of irrigation, on which grass vegetation and conditions of football filed exploitation depend [Oleszczuk end Truba 2013, James et al. 2007a, James et al. 2007b].

Physical and water properties of indigenous ground from the area of a football field planned in Muchacz were analyzed in the paper and compared with analogous properties of laboratory made ground and sand composites.

Suitability of the tested materials for constructing of the football filed carrying layer were estimated from the perspective of water permeability, but also selected physical properties and retention capacities of the ground and composites were analyzed. The obtained results may be useful for precise formulation of the standards of granulometric composition of composites, but also for the construction of football field planned in Muchacz, in wadowicki county.

METHODS

Indigenous ground

Samples used for determining the granulometric composition were collected from 3 shallow pits in the area of planned football field (Figure 2), ground samples from the 5–30 cm humus layer and from 31–45 cm sub-arable layer. For determining physical and water properties, 18 samples with undisturbed structure were collected by means of 100 cm³ rings (3 pits, 2 horizons, 3 replications).

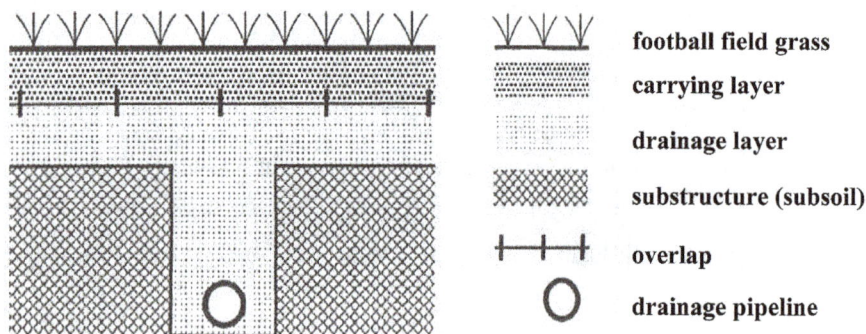

Figure 1. Cross-section of grass covered football field on compact ground (according to DIN 18035)

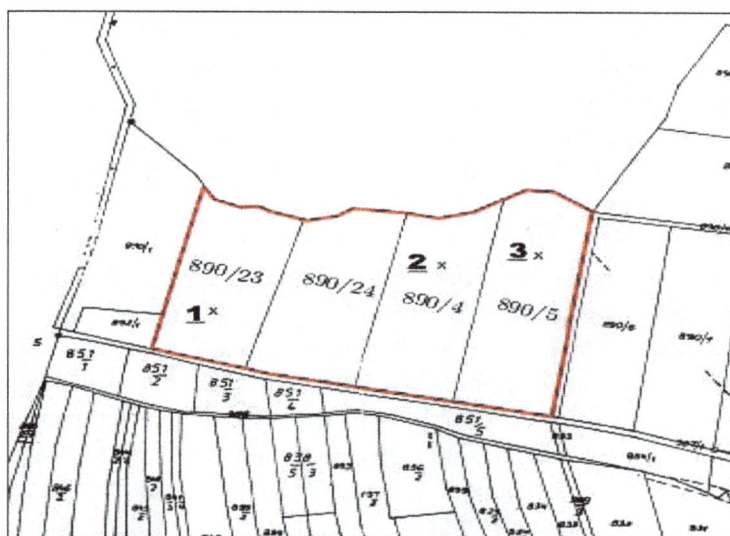

-------- outer boundaries of plots destined for the football field; 1,2,3 – sampling points

Figure 2. Distribution of pits in the area of planned football field

Grain size distribution was assessed using Casagrande's sedimentation method in Prószyński's modification, specific density by pycnometer method and bulk density by gravimetric method. Permeability coefficients ($K_{10}{}^g$) were determined by means of laboratory Ejikelkamp permeability meter [Rajda et al. 2011b], whereas water potential (pF^g), on the basis of which material retention capacity was assessed, in 2-chamber pressure extractor. Assessment of permeability coefficients was conducted depending on the ground permeability at constant or variable water pressure, however, their values were established taking into consideration water viscosity acc. to Ostromecki for water temperature t = 10 °C.

Composites

6 composites were formed from the indigenous ground, sand and peat substrate (Photo 1) (Photo 2, Table 1) of which a total of 18 samples were collected in three replications to the rings of 100cm³ capacity. Granular composition and physical properties of the composites , except for sand and gravel fractions assessed by sieve method, were analysed in the same way as in case of the indigenous ground. Composite samples were ini-

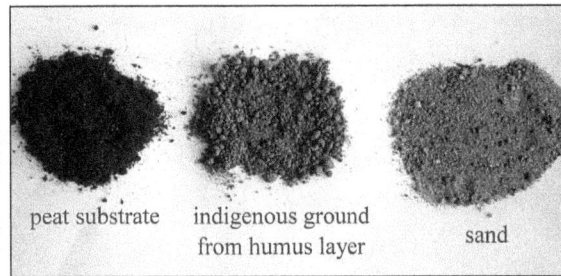

Photo 1. Components of the composites

Photo 2. Composites

Table 1. Weighed portions and planned proportional share of the components in composites

Composite	Symbol	Planned share in composite [g] at initial moisture	Planned share in composite [g] absolutely dry matter	[%]	Organic matter in component [%]	Organic matter in component [g]	Organic matter in composite [%]
Ia	p*	1999	1998	90	0.06	10	
	g*	215	178	8	2.50	4	1.98
	s.o.*	69	44	2	74.8	30	
Total Ia		2283	2220	100	–	44	
Ib	p*	1999	1998	90	0.06	10	
	g*	187	155	7	2.50	4	2.97
	s.o.*	104	66	3	74.8	49	
Total Ib		2290	2219	100	–	63	
IIa	p*	1790	1780	80	0.06	10	
	g*	480	400	18	2.50	10	1.98
	s.o.*	69	44	2	74.8	30	
Total IIa		2339	2224	100	–	50	
IIb	p*	1790	1780	80	0.06	10	
	g*	460	380	17	2.50	9	3.0
	s.o.*	104	66	3	74.8	49	
Total IIb		2354	2226	100	–	68	
IIIa	p*	1560	1550	70	0.06	9	
	g*	750	620	28	2.50	16	2.0
	s.o.*	31	44	2	74.8	20	
Total IIIa		2341	2214	100	–	45	
IIIb	p*	1560	1550	70	0.06	9	
	g*	720	600	27	2.50	15	3.3
	s.o.*	104	66	3	74.8	49	
Total IIIb		2384	2216	100	–	73	

* p – sand, g – indigenous ground, s.o. – organic matter.

tially compacted under pressure of $p_1 = 12$ N·cm^{-2} and their bulk density p_1, porosity n_1, permeability coefficient K_1^k and water potential $-pF_1$ were assessed. Analogous properties (p_2, n_2, K_2^k and pF_2^k) were determined after compacting under the pressure of $p_2 = 78$ N·cm^{-2}.

RESULTS

Properties of mineral components

Indigenous ground with its lower proportion in the composites was classified to silt loams with strongly diversified grain size distribution ($U = 17.1$) In the variation area permitted by DIN 18035 standard it definitely differed from the recommended grain size distribution (Figure 3). In the sub-arable layer (31–34 cm) as compared with the humus layer, granular composition of the ground was more uniform and characterized by a higher share of sand fractions and lower content of organic matter (Table 2).

The sand used for the composites was classified by PN-R-04033 [19] standard to coarse grained deposits (Table 2, Figure 3) with uniform graining ($U = 2.1$). In relation to the mixtures recommended by DIN 18035 standard for construction of the carrying layer, it contained even about 20% less fractions with diameters smaller than 0.8mm (Figure 3), whose share should be between 10 and 30%. Moreover, it was almost devoid of fractions which are smaller than 0.2 mm and contained a small amount of organic matter.

Table 2. Grain size distribution and organic matter of sand and indigenous ground (means from 3 replications)

Fraction symbol		Equivalent grain diameters [mm]	Proportional grain content in a component	
			Sand	Silt loam
Clay (i)		<0.002		$\frac{15^*}{15}$
Silt (π)	d	0.002–0.005		$\frac{12^*}{10}$
	s	0.005–0.02	1	$\frac{27^*}{27}$
	r	0.02–0.05		$\frac{31^*}{29}$
Sand (p)	d	0.05–0.10		$\frac{8^*}{8}$
		0.10–0.25	2	
	s	0.25–0.50	22	$\frac{7^*}{11}$
		0.50–1.0	52	
	r	1.0–2.0	15	$\frac{0}{0}$
Gravel (ż)		2.0–5.0	8	$\frac{0}{0}$
Kind of deposit			Pg	$\frac{Gπ^*}{Gπ}$
Effective grain size		d_{10}	0.38	$\frac{0.0014}{0.0023}$ **
		d_{60}	0.80	$\frac{0.024^*}{0.026}$
		d_{90}	1.80	$\frac{0.070^*}{0.130}$
Index and degree of non-uniformity $U = d_{60}{:}d_{10}$			2.1	$\frac{17.1^*}{11.3}$
			uniform grained	very non-uniform grained
Organic matter			0.58	$\frac{2.88^*}{1.85}$

* Layer: $\frac{\text{50-30 cm}}{\text{31-45 cm}}$

** Extrapolated from grain-size distribution curves (Figure 3).

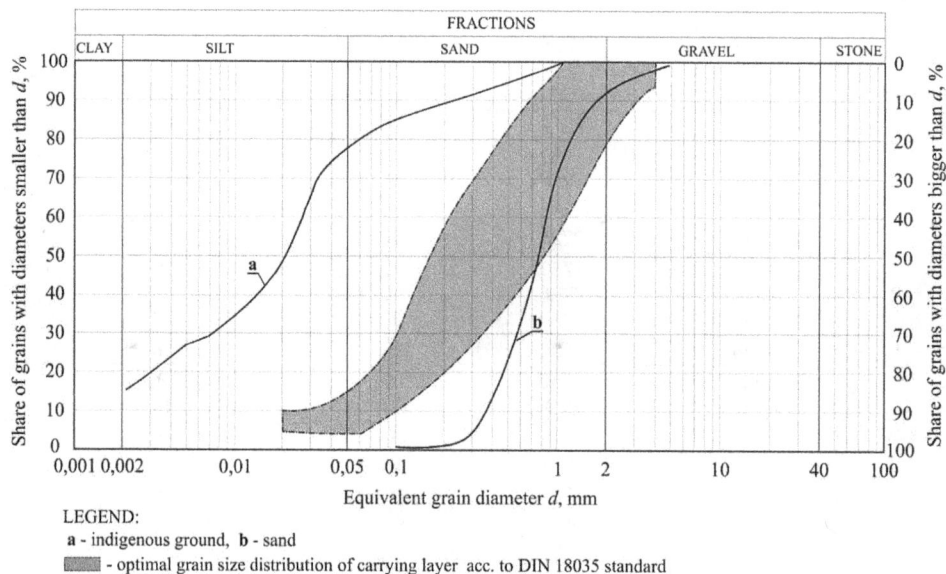

LEGEND:
a - indigenous ground, **b** - sand

▓ - optimal grain size distribution of carrying layer acc. to DIN 18035 standard

Figure 3. Grain-size distribution curves of components against the normal grain-size distribution

Table 3. Some physical and indigenous ground properties (means for 3 pits)

Layer [cm\	Specific density ρ_s [g×cm⁻³]	Bulk density ρ_o [g×cm⁻³]	Porosity n [%]	Porosity index e [–]
Humus 5–30	2.64	1.42	46.2	0.86
Sub-arable 31–45	2.67	1.61	39.8	0.66

Mineral components definitely differed by their equivalent diameters d_{10}, d_{60} and d_{90} and the content of organic parts (Table 2).

Due to arable use, bulk density of ground from the humus layer was about 12% (0.19 g·cm⁻³) smaller than bulk density of ground from the sub-arable layer. As a result, total porosity in the ground humus layer was higher in comparison with the sub-arable layer; in this case the absolute difference was 6.4%$_{vol.}$ (Table 3).

Properties of composites

Grain size distribution of composites mostly differed from the assumptions for the benefit of sand and gravel fractions (see Table 1 and 4). In composites Ia and Ib sand and gravel fractions constituted, respectively 90% and 94% in comparison with the assumed 90%. In composites II and IIb they made up respectively 86% and 89%, in relation to assumed 80%, whereas in composites III a and IIIb – 75% and 79% against 70% (see: Table 1 and Table 4). Composite I was classified to uniform grained deposits, the other two to very non-uniform grained. Equivalent diameters of the composites were also different, particularly d_{10}, whereas grain non-uniformity indices differed to a lesser degree (Table 4).

Composite II proved the best adjusted to the normal grain size distribution interval. In comparison with the permissible grain size distribution interval, composite I contained between several and over 10% less of 0.2–0.8 mm fractions in relation to the lower limit of this interval, whereas composite III had over 10% more of fractions smaller than 0.05 mm in relation to the interval upper limit (Figure 4).

Bulk density of equally compacted composites changed slightly with increase in the content of fine particles and 1% higher content of organic matter (variant b of the composites) (Table 5). At compaction pressure $p_1 = 12$ N·cm⁻², bulk density of the composites was higher than the density of indigenous ground from the humus layer, where after compacting under $p_2 = 78$ N·cm⁻² pressure it was also higher than ground density from the sub-arable layer (see Table 5 and 3). Total poros-

Table 4. Grain size distribution and organic matter in composites (means for 3 replications)

Fraction symbol		Equivalent grain diameters [mm]	Proportional content in composite	
			Sand	Silt loam
Clay(i)		<0.002		15* / 15
Silt (π)	d	0.002–0.005		12* / 10
	s	0.005–0.02	1	27* / 27
	r	0.02–0.05		31* / 29
Sand (p)	d	0.05–0.10		8* / 8
		0.10–0.25	2	7* / 11
	s	0.25–0.50	22	
		0.50–1.0	52	
	r	1.0–2.0	15	0 / 0
Gravel (ż)		2.0–5.0	8	0 / 0
Kind of deposit		Pg		Gπ* / Gπ
Effective grain size	d_{10}		0.38	0.0014 ** / 0.0023
	d_{60}		0.80	0.024* / 0.026
	d_{90}		1.80	0.070* / 0.130
Index and degree of graining non-uniformity $U = d_{60} : d_{10}$ and kind of deposit			2.1	17.1* / 11.3
			uniform grained	very non uniform grained
Organic matter			0.58	2.88* / 1.85

* d – fine. s – medium. r – coarse.

ity formed accordingly. After compacting under pressure of $p_2 = 78$ N·cm⁻² it diminished definitely in comparison with the porosity obtained after compacting under the pressure of $p_1 = 12$ N·cm⁻², but due to grain size distribution, both values were clearly lower than the indigenous ground porosity. Increase in organic matter content (variant *b* of the composites) was visible as a slight increasing tendency of porosity (Table 5).

Considering the organic matter composites Ia and Ib met the assumed research requirements, whereas composites IIa and IIb and IIIa and IIIb contained respectively 0.6% and 0.8% less than the assumed value (see: Table 3 and Table 4).

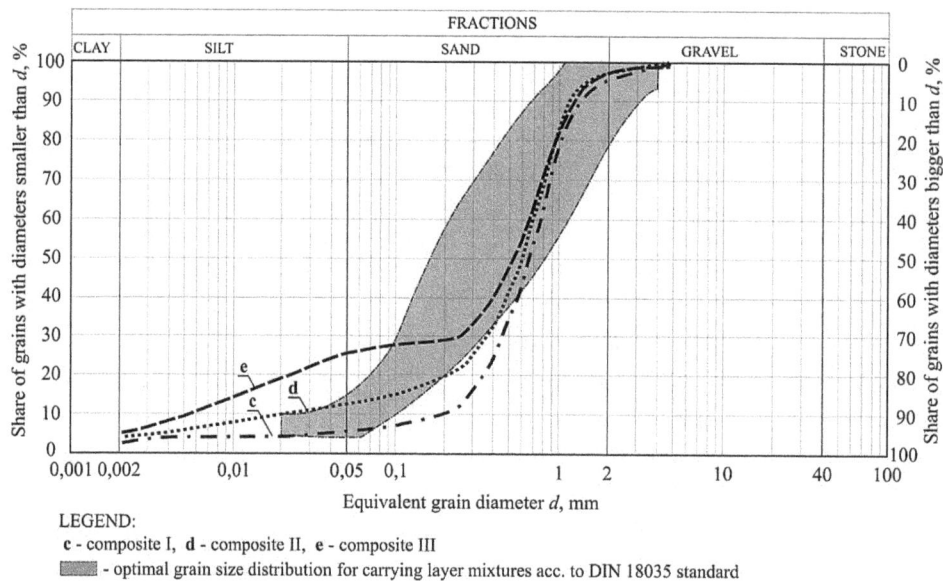

LEGEND:
c - composite I, d - composite II, e - composite III
▓ - optimal grain size distribution for carrying layer mixtures acc. to DIN 18035 standard

Figure 4. Grain size distribution curves against normal grain size distribution interval

Table 5. Physical properties (p_1, n_1, e_1) and (p_2, n_2, e_2) of composites compacted under pressure of respectively: p_1= 12 N·cm^{-2} and p_2= 78 N·cm^{-2}

Properties	Symbol	Composites						Mean I–III	
		I		II		III			
		a	b	a	b	a	b	a	b
Specific density [g×cm^{-3}]	ρ_s	2.60	2.60	2.56	2.60	2.61	2.57	2.59	2.59
Bulk density [g×cm^{-3}]	ρ_1	1.52	1.48	1.47	1.46	1.51	1.52	1.50	1.49
	ρ_2	1.64	1.69	1.62	1.63	1.62	1.58	1.63	1.63
Porosity [%]	n_1	41.6	42.8	42.6	43.9	42.2	40.9	42.1	42.5
	n_2	36.9	36.9	36.7	37.3	37.9	38.5	37.2	37.6
Porosity index [–]	e_1	0.71	0.75	0.74	0.78	0.73	0.83	0.73	0.79
	e_2	0.58	0.58	0.58	0.63	0.61	0.63	0.59	0.59

Water permeability and retention capacity of silt loam and composites

Permeability coefficient of silt loam $(K_{10}{}^g)$ revealed a high changeability. In the humus layer of respective pits it ranged from 0.15 to 0.26 mm ·min^{-1} with mean for 3 pits and 3 replications 0.23 mm ·min^{-1} (Table 6). In the sub-arable layer mean $K_{10}{}^g$ values fell within the range from 0.000 to 0.005 mm ·min^{-1} and classified the material to impermeable deposits.

Diminishing sand proportion in the composites and increase in silt and clay fractions from the admixture of silt loam caused a decrease in permeability coefficient $(K_{10}{}^k)$ – especially after compaction under the pressure of p_2 = 78 N·cm^{-2} leading to a significant decrease in porosity (Table 7 and 5). At the biggest share of silt loam and considerable grain non-uniformity (composite III), after compacting under the pressure of p_1 = 12 N·cm^{-2},

permeability coefficient $K_{10}{}^k$ decreased by about 2.5-fold on average for variants a and b, in comparison with composites I and II (Table 7), whereas at the same grain size distribution relations, after compaction under the pressure of p_2 = 78 N·cm^{-2} it changed almost 15-fold in comparison with the mean for composites Ia and Ib and about 4.5-fold as compared with composites IIa and IIb (Table 7).

The increase in organic matter content, on average from about 2.5% (composite variant *a*) to over 3.5% (variant *b*), at compaction under the pressure of p_1 = 12 N·cm^{-2}, was visible for composites II and III as a slight decrease in permeability coefficient, whereas at compaction under the pressure of p_2 = 78 N·cm^{-2} no evident effect of organic matter was registered (Table 7).

At high total porosity silt loam was characterized by a considerable retention capacity and water availability to plants (Table 8, Figure 5). Differences in moisture for pF 2.5 and pF 4.2 de-

Table 6. Permeability coefficient $K_{10}{}^g$ of silt loam [mm·min^{-1}] (means for 3 replications)

Layer [cm]	Pit No.			Mean for layer
	1	2	3	
5–30	0.15	0.26	0.29	0.23
31–45	0.005	0.003	0.000	0.003

Table 7. Permeability coefficients $K_{10}{}^k$ [mm·min^{-1}] of composites compacted under the pressure of $p_1 = 12$ N·cm^{-2} and $p_2 = 78$ N·cm^{-2} (means for 3 replications)

For compaction under the pressure	$K_{10}{}^k$ of composite					
	I		II		III	
	a	b	a	b	a	b
$p_1 = 12$ N·cm^{-2}	9.2	15.6	15.1	9.8	6.2	3.6
$p_2 = 78$ N·cm^{-2}	5.0	6.8	1.9	1.8	0.4	0.4

termining the content of widely available water in humus layer constituted over 30% of bulk. Easily accessible water (moisture differences at pF 2.5 to pF 3.7) made up 17%, whereas unavailable water (at pF 4.2) was below 10% (Figure 5, Table 8). Less advantageous retention properties of silt loam were in the sub-arable layer.

Composites were characterized by a much worse retention capacity. Subjected to the pressure of $p_1 = 12$ N·cm^{-2}, at identical values of water potential pF, revealed a lower water content (per $\%_{cap.}$) than composites compacted under the pressure of $p_2 = 78$ N·cm^{-2} (p_2) (Figure 5, Table 9), however, their grain size distribution had a greater effect on water potential than compaction. Percentage increases in moisture, within the range of pH = 1.6 to pF = 4.2, differed slightly for the respective compaction levels, but for composite III moisture at analogous pF values was about twice higher than the moisture of composite I (Figure 5). The outcome of these relations were small percent differences in water supply, and therefore small, only several millimeter supply in the 15-centimeter composite layer (Table 9).

Table 8. Characteristic moisture states and water supply in silt loam (means for 3 replications)

Layer [cm]	Percentage water content at:						Water supply [mm] for 15 cm layer at:		
	pF 2.5	pF 3.7	pF 4.2	pF 2.5–4.2	pF 2.5–3.7	pF 3.7–4.2	pF 2.5–4.2	pF 2.5–3.7	pF 3.7–4.2
5–30	39.8	12.8	9.7	30.1	17.0	3.1	45.2	25.5	4.7
31–45	35.6	19.0	16.3	19.3	16.6	2.7	29.0	24.9	4.1

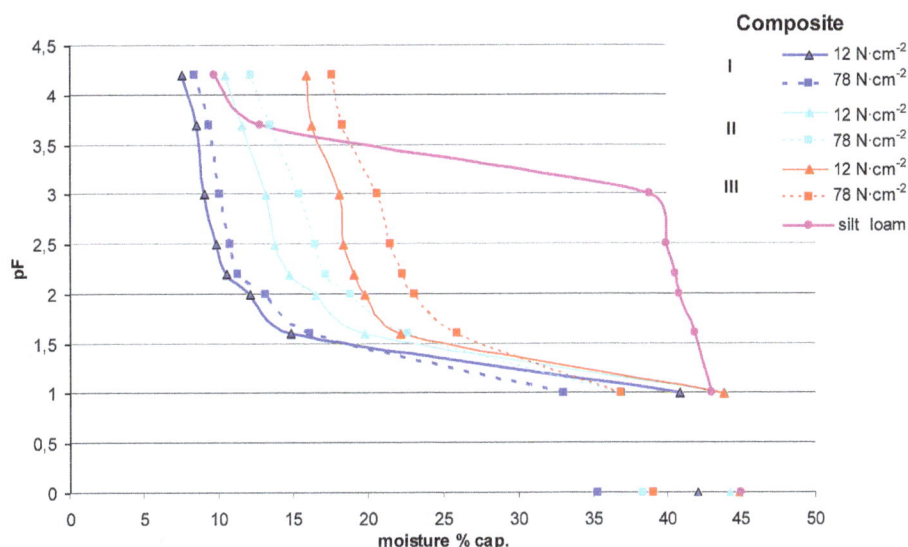

Figure 5. Water potential (pF) of silt loam from humus layer and composites (means for 3 replications)

Table 9. Characteristic moisture states of composites compacted under the pressure of $p_1 = 12$ N·cm^{-2} and $p_2 = 78$ N·cm^{-2} and water supply in 15 cm layer (mean for 3 replications)

Water potential		Composite					
		I		II		III	
		a	b	a	b	a	b
for $p_1 = 12$ N·cm^{-2}							
Moisture in % cap. for:	pF 2.5	9.8	9.7	14.0	13.7	17.7	19.0
	pF 3.7	8.1	8.9	11.3	11.9	16.0	16.5
	pF 4.2	7.2	7.8	10.1	10.7	15.7	16.1
Percentage water content *	OD; pF 2.5–4.2	2.6	1.9	3.9	3.0	2.0	2.9
	ŁD; pF 2.5–3.7	1.7	0.8	2.7	1.8	1.7	2.5
	TD; pF 3.7–4.2	0.9	1.1	1.2	1.2	0.7	0.4
Water supply [mm] *	OD	3.9	2.9	5.9	4.5	3.0	4.4
	ŁD	2.6	1.2	4.1	2.7	2.6	3.8
	TD	1.4	1.7	1.8	1.8	1.1	0.6
for $p_2 = 78$ N·cm^{-2}							
Moisture in % cap. for:	pF 2.5	10.2	11.4	16.4	16.5	20.5	22.5
	pF 3.7	9.0	9.8	13.2	13.8	17.8	18.7
	pF 4.2	8.2	8.7	11.9	12.5	17.2	18.0
Percentage water content *	OD; pF (2.5–4.2)	2.0	2.7	4.5	4.0	3.3	4.5
	ŁD; pF (2.5–3.7)	1.2	1.6	3.2	2.7	2.7	3.8
	TD; pF (3.7–4.2)	0.8	1.1	1.3	1.3	0.6	0.7
Water supply [mm] *	OD	3.0	4.1	6.8	6.0	5.0	6.8
	ŁD	1.8	2.4	4.8	4.1	4.1	5.7
	TD	1.2	1.7	2.0	2.0	0.9	1.1

* Widely available (OD), easily accessible (ŁD), hardly available (TD).

The same moisture values and water supplies for 15 cm thick carrying layer of the WISŁA sport club football field were about twice higher than the value obtained for more thickly compacted composites II and III containing less of fine particles than in the compared carrying layer of the football field. On the other hand, moisture values and water supplies in composite I with the lowest content of fine particles (cc. 5%) were approximately equal to the adequate values for the drainage layer on the same football field formed only from the sand Rajda and Kanownik 2006].

DISCUSSION

Material composed of light loam with 62% content of sand incorporated in 2002 into the 0–40 cm layer during renovation of the WISŁA S.A. sport stadium pitch was characterized by a bulk density of 1.48 g·cm^{-3} and porosity of 40.1%. Analogous parameters of silt loam containing 68% of sand, incorporated into the 40–80 cm layer after compaction were 1.70 g·cm^{-3} and 30.7%. Measured infiltration rate from the field surface was then on the level of 0.040 mm·min^{-1} [Rajda et al. 2011a], i.e. about ten times less than the lowest value recommended by the DIN 18035 standard [Deutche norm]. In result, after heavier rainfall water stagnated on the football field surface. Following the next alteration in 2004, the owners carried out renovation works in compliance with recommendations of the above-mentioned standard: the carrying layer of a composite with 91–95% of sand fraction and underlain by drainage layer had bulk density of 1.40 g·cm^{-3}, total porosity 46.7% and permeability of 4.6 mm·min^{-1}, while permeability coefficient K$_{10}$ measured in the laboratory in 5 replications, was 13.1 mm·min^{-1} (3.0–24.0 mm ·min^{-1}) [Policht-Latawiec 2008, Rajda et al. 2011b]. i.e. on the level corresponding to the analyzed less compacted composites I and II with porosity of 42–43%, containing respectively 96 and 98% of sand.

CONCLUSIONS

On the basis of conducted experiments it was stated that:

1. Composites with various silt loam and sand and organic matter proportions allow to shape water permeability, therefore, the time of water draining from the football pitch surface may be formed according to needs.

2. Apart from grain size distribution, water permeability and retention capacity of the composites were affected by compaction, however greater influence of grain size distribution was marked at stronger compaction.

3. The tested composites fulfilled the requirements of DIN 18035 standard for water permeability of the carrying layer of the football field; even composite IIIb with the highest share of silt loam and compacted under the pressure of 78 N·cm^{-2} also met the requirements.

4. Increase in the share of fine particle fraction considerably affected water permeability of the composites, but to a lesser extent their water capacity.

5. Increasing the content of organic matter from 2.5% to over 3.5% did not influence water permeability or water capacity of the composites.

6. Composites IIIa and IIIb revealed the highest retention capacity, whereas composites IIa and IIb were characterised by the best water availability to plants.

7. At sprinkling of 15-centimeter thick carrying layer formed of composites IIa and IIb, a single sprinkling dose, considering the drainage layer, should not exceed 10mm ($10 \, dm^3$ per $1 \, m^2$).

REFERENCES

1. Deutsche Norm, DIN 18035, cz. 4. Sportplätze. Rasenflächen, Teil 4, 1991.

2. Głąb T., Gondek K. 2013. The Influence of Soil Compaction on Chemical Properties of Mollic Fluvisol Soil under Lucerne (Medicago sativa L.). Polish Journal of Environmental Studies, 22 (1), 107–113.

3. James I. T., Blackburn D.W.K., Godwin R. J. 2007a. Mole drainage as an alternative to sand slitting in natural turf sports surfaces on clays. Soil Use and Management, 23, 28–35.

4. James I.T., Hann M.J., Godwin R.J. 2007b. Design and operational considerations for the use of mole ploughing in the drainage of sports pitches. Biosystems Engineering 97, 99–107.

5. Milivojević J., Nikezic D., Krstic D., Jelic M., Dalović I. 2011. Influence of Physical-Chemical Characteristics of Soil on Zinc Distribution and Availability for Plants in Vertisols of Serbia. Polish Journal of Environmental Studies, 20(4), 993–1000.

6. Oleszczuk R., Truba M. 2013. The analysis of some physical properties of drained peat-moorsh soil Layers. Annals of Warsaw University of Life Sciences – SGGW, Land Reclamation No 45 (1), 41–48.

7. Pereira J.O., Défossez P., Richard G. 2007. Soil susceptibility to compaction by wheeling as a function of some properties of a silty soil as affected by the tillage system. European Journal of Soil Science 58 (1), 34–44.

8. Policht-Latawiec A. 2008. Badanie wodoprzepuszczalności kompozytów gleby pyłowo-ilastej, piasku i substratu torfowego. Acta Scientiarum Polonorum, Formatio Circumiectus, 7 (4), 21–30.

9. Rajda W., Kanownik W. 2006. Retencja użyteczna warstwy nośnej i drenażowej płyty boiska do piłki nożnej. Acta Sci. Pol., Architektura 5 (2), 65–74.

10. Rajda W., Żarnowiec w., Stachura T. 2011a. Właściwości fizyczne i przesiąkliwość płyty boiska TS Wisła Kraków S.A. po renowacji. Przegląd Naukowy Inżynierii i Kształtowania Środowiska 20(2), 153–159.

11. Rajda W., Stachura T., Żarnowiec W. 2011b. Własności fizyczne i przesiąkliwość kompozytu warstwy nośnej i piasku warstwy drenażowej płyty boiska Wisły Kraków S.A. po przebudowie w 2004 r. Acta Sci. Pol., Architektura 10 (1), 31–41.

12. Żegocińska–Tyżuk B. 1988. Terenowe urządzenia sportowo rekreacyjne. Wyd. Politechniki Krakowskiej, Kraków.

19

VARIATION IN LIQUID WASTE COMPOSITION SUPPLYING SELECTED COLLECTION POINT

Elżbieta Halina Grygorczuk-Petersons[1], Józefa Wiater[1]

[1] Department of Technology in Engineering and Environmental Protection, Bialystok University of Technology, Wiejska 45, 15-351 Bialystok, Poland, e-mail: e.petersons@pb.edu.pl; j.wiater@pb.edu.pl

ABSTRACT

The problem of liquid wastes is still current not only in Poland, but also in the world. The primary source of liquid wastes are single-family and multi-family houses. To a lesser extent, public service or production facilities, mainly in urban areas, are equipped with no-outlet reservoirs. Low concentrations in liquid wastes and their high density are often some difficulty to work with not only by sewage treatment plants, but also collection points. Therefore, the knowledge of the composition of liquid wastes supplied to the collection points is important. The paper presents the results of research as well as variability of concentrations of selected parameters of liquid wastes supplied to the collection point in Bialystok, that accepts both municipal and industrial sewage. Statistical processing of the obtained results and those derived from The Waterworks Bialystok showed the presence of high variability of total suspended matter and electrolytic conductivity as well as organic impurities expressed as BOD_5 and COD.

Keywords: collection points, liquid wastes, composition variability, permissible value

INTRODUCTION

Even at the present time, there is a specific solution of the sewage system in the cities, i.e. the no-outflow sewage system also called the "no-network system". In most cases, it serves as a temporary solution to the sewage system, at a time when investments are ahead of the construction of the sewerage system or if there is no economic justification for its construction [Kisiel, Bień 2005; Błażejewski, Nawrot 2009]. One of the elements of such system is collection points. The obligation of their construction, maintenance, and operation (individual or shared with other municipalities) is governed by the act requiring municipalities to maintain cleanliness and order in their locations [Act 1996]. Preferably, the collection point should be located at a smallest distance from the accumulation of liquid wastes, and the best solution is considered to locate it in the sewage treatment plant. The advantage of the collection point located in the sewage system at a distance from sewage treatment plant is that introduced impurities are mixed with sewage with-

in the network, thus they are diluted and homogenized [Tomczuk 2011b].

Liquid wastes transported to the collection point are mixtures of various organic and inorganic matter dispersed in water, occurring in all states of matter. The degree of fragmentation of substances contained is very uneven. These may be large solid particles, fine particles of suspended solids, and emulsified colloidal and dissolved substances. The pH of the liquid waste can be very different: from the acid through neutral to alkaline, depending on their chemical composition [Maksymowicz, Opęchowski 2006].

Concentrations of selected parameters of liquid wastes, municipal and municipal-like ones, are usually far above the concentration in the wastewater influent through the sewage network, because spontaneously occurring biological processes leading to the degradation of organic matter and, consequently, to deposition of portion of solids at the bottom of the reservoir, in which they are temporarily stored [Grygorczuk-Petersons, Tałałaj 2007]. This portion of impurities, in respect of their composition, is similar to bottom sediments of a very high hydration degree [Bartoszewski et al. 2011].

Liquid wastes with large loads of biodegradable organic compounds emit odor of hydrogen sulfide, have black-gray color, and after filtration, are much less transparent than «fresh» wastewaters [Maksymowicz, Opęchowski 2006].

At the same time, in such wastes under the influence of anaerobic processes occurring in no-outflow reservoirs (rotting processes), an increasing in the concentration of selected parameters occurs. From sanitary point of view, this kind of wastes is very dangerous, because they contain large amounts of rotting or fermenting organic substances, pathogenic bacteria and viruses, as well as all types of human and animal-origin parasites.

Liquid wastes from service facilities, public buildings, or industry, may also contain infectious organisms, hazardous and toxic substances of organic and inorganic character.

It should be noted that liquid wastes accumulated in no-outflow reservoirs can be supplied to the collection point if this does not endanger the health of the station's service, construction, and proper operation of equipment, or the sewage treatment plant. For this reason, knowledge of the composition of liquid waste delivered to the collection point is extremely important.

Therefore, the work was undertaken, the purpose of which was to determine the variability of the physicochemical composition of liquid wastes supplied to the collection point in Bialystok, that accepts both municipal and industrial sewage, mainly from the food industry.

MATERIALS AND METHODS

Characteristics of Liquid Wastes Collection point in Bialystok

Liquid wastes generated in Bialystok and the surrounding areas, should go to The Liquid Wastes Collection point, that was taken over from the City Government of Bialystok in 1992 by Bialystok Waterworks Ltd. The Collection point was a well sized 3.20×3.20 m, to which the liquid wastes were poured directly from vacuum trucks. They were then mixed with municipal-farming sewage supplied by a collector with a diameter of 800 mm, from part of the district Jaroszówka, Wyżyny estates, and the town Wasilkowo. As mixed wastes, they flew (also through a collector with a diameter of 800 mm) to the wastewater treatment plant.

A new-generation container of the collection point with corresponding equipment (flow-meter Danfoss, device for sampling and measurement of wastewater pH, computer with appropriate software) was installed in the collection point in March 2000.

In 2010–2013, the liquid wastes from average of 23 vacuum trucks were discharged into the collection point during a day. The average monthly amount of wastes supplied to the collection point was approximately 2500 m^3 at the permitted quantity of 3500 m^3/month [Memo 2011].

The collection point accepts liquid wastes generated by households and industrial plants (dominated by the food industry), the provided documentation of the source of their origin and when they are not a threat to the safety and health of persons servicing the collection point and wastewater facilities. These wastes come mainly from Bialystok, but a small part of them is transported from nearby communities (Wasilków, Zaścianki, Juchnowiec Kościelny, Choroszcz, Sokółka, Turośń Kościelna, Dobrzyniewo Duże).

The amount of introduced liquid wastes is determined on the basis of indications of the measuring device owned by the Company.

Waterworks Bialystok lead the control of liquid wastes delivered to the collection point. This is done by control sampling during emptying the tank of the vacuum truck. Samples of liquid wastes from vacuum trucks are taken randomly. Sampling is done manually on the gravity pipeline. In case of exceeding the permitted contamination indices in liquid wastes as defined in Annex 2 and 3 to the agreement on the supplying of liquid wastes to the collection point [Contract 2011], the company pays an extra fee and bears the cost of the analyses. When the permissible limits of contaminants are not exceeded, the supplier bears only the cost of the liquid waste supplying to the collection point.

Methods

The study used the results of own research and those obtained from the Waterworks Bialystok. Samples of liquid wastes were collected from The Collection point in Bialystok in 2012. The samples were manually collected according to the norm [PN-ISO 5667-10: 1997] on the gravity pipeline located in the station area during emptying the vacuum trucks in the presence of the station's employee.

Each time, 10 samples of liquid wastes from successively emptied vacuum trucks were collected, which were then transported to the laboratory of Technical University in Bialystok to determine selected indicators of contamination.

In order to reduce any errors in the research work, each sample was described by specifying the date, time, declared sources of liquid wastes and its address, the capacity of the vacuum truck, and the provider of liquid wastes.

The analyses included determining the following parameters performed in accordance with Polish norm: potentiometric pH with a pH-meter Hach Session 4, electrolytic conductivity by means of conductivity method, total suspended solids by gravimetric method, organic substances measured by using the biochemical oxygen demand (BOD_5 by dilution method), and chemical oxygen demand by dichromate method (COD_{Cr}). The results of the liquid wastes parameters were compared with acceptable indicators determined in the agreement on the introduction of liquid wastes into the collection point in Bialystok being in force in 2012 [Contract 2011].

The results obtained during the research (80 samples) are presented in tabular, graphical, and descriptive forms. Table 1 shows the statistics characterizing tested waste accumulation indicators calculated using STATISTICA v. 10.0 software: arithmetic mean, median, standard deviation, minimum, maximum, skewness (A).

In order to determine the compliance of the studied traits (determination results of pollution indicators) with normal distribution, the null hypothesis H0 was tested: the distribution of the test indicator is the normal distribution, against the alternative hypothesis H1: the distribution is not a normal one. To verify the hypothesis, the $\chi2$ test was applied, assuming the null hypothesis H0 at the significance level of $p > 0.05$. Since the tested characteristics did not show any conformity with

normal distribution, their compatibility with other types of distributions were analyzed, similarly as in the previous case.

Graphics shows "box with a mustache" type plots (Figure 1) and histograms (Figure 2) plotted using STATISTICA v.10.0 software, which show statistical differences between the tested parameters.

RESULTS AND DISCUSSION

The average value of electrolytic conductivity of liquid wastes delivered to the collection point (Table 1) was within the range of permissible limit values set by the Waterworks Bialystok and contained in the contract [Contract 2011] to their supply, while mean values of BOD_5 and COD_{Cr} as well as total suspended solids significantly exceeded those levels. The average BOD_5 value determined for 2012 amounted to 2489 g/m³ (Table 1), i.e. it exceeded the permissible level by almost 2.5-fold; the mean value of the suspension exceeded "the norm" more than 8-fold and the average chemical oxygen demand 4.5-fold.

Analyzing values of the minimum indicators of pollutants in liquid wastes, it should be noted that in all cases they reached values below the limit (Table 1), while the maximum values exceeded them many times as similar as average values. The maximum electrolytic conductivity was 24.6 mS/cm, BOD_5 – 25 000 g O_2/m³, maximum COD_{Cr} up to 803 306 g O_2/m³ (141 times higher than acceptable limit), and the maximum content of total suspended solids 90 896 g/m³ (61 times higher than permissible level).

Such a large excess, in particular in the case of suspensions, is evidence that the septic tanks may be leaking hence liquid infiltration into the ground occurs, which results in concentrating the liquid wastes. The second reason may be to provide to the collection point highly concentrated, industrial liquid wastes and leachates from the

Table 1. Statistics characterizing supplied liquid wastes

Parameters	Reaction pH	Electrolytic conductivity [mS/cm]	BOD_5 [g O_2/m³]	COD_{Cr} [g O_2/m³]	Total suspended solids [g/m³]
Threshold value [Contract 2011]	6.5 – 9.5	0.2 – 18	1 000	5 700	1 500
Mean	–	6.80	2 489	28 850	12 056
Minimum	2.3	0.1	100	503	275
Maximum	11.6	24.6	25 000	803 306	90 896
Median	7.4	3.75	1 500	4 090	2 495
Standard deviation	–	6.33	3 691	125 540	23 056
Skewness (A)	-1.04	1.26	4.61	6.09	2.73

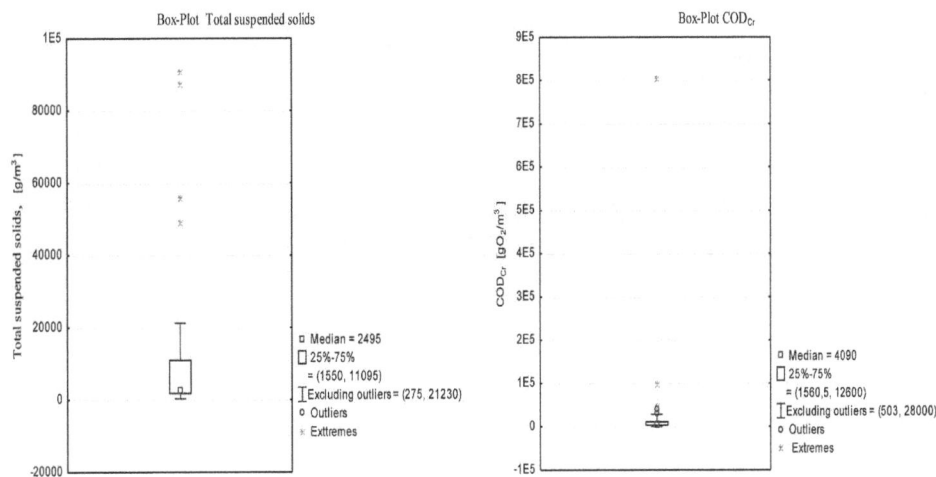

Figure 1. Range of oscillations of total suspended solids and COD_{Cr} in liquid wastes

Waste Treatment Plant in Hryniewicze characterized by, among others, high electrolytic conductivity value and high levels of pollutants expressed as COD [Leszczyński 2011 and 2013].

Statistical processing of results (Table 1) showed the existence of large disparities between pollutant concentrations expressed as the contents of BOD_5 and COD_{Cr}, total suspended solids, and electrolytic conductivity, due to the very large difference between the median values and mean values of the analyzed parameters.

Based on the minimum and maximum values (Table 1, Figure 1), very large differences between obtained values of the individual parameters of liquid wastes were recorded. Standard deviations show their considerable dispersion. Also, graphs of "boxes with a mustache" type exhibit large fluctuations in the analyzed indicators of impurities in the liquid wastes supplied to the collection point (Figure 1).

It was found that span values of the analyzed parameters were remarkable. In all cases, the occurrence of outliers and extreme for pH, electrolytic conductivity, BOD_5, COD_{Cr} and total suspended solids, was observed (Figure 1).

It was found that the resulting properties (results of pollution indices determination) in the liquid wastes were not to conform with the normal distribution, therefore, the compatibility of the distribution of analyzed features with other distributions were verified. Following distributions were taken into account: exponential, gamma, lognormal, and $\chi2$. The assessment of the compliance with those distributions was carried out on the basis of the $\chi2$ test.

The hypothesis on the compliance with the exponential distribution was assumed for BOD_5

(Figure 2c). Other characteristics (pH, conductivity, COD, and suspended solids), due to the low probability of the test (less than 0.05) showed no statistically significant compliance with any of the analyzed distribution.

On the basis of the presented histograms, it can be also specified a range of the most frequently achieved results and the significant range of the indicator occurrence (Figure 2). Therefore, it was concluded that liquid wastes were characterized by:
- in 79.1% pH from the range of 6–8 pH (Figure 2a),
- in 68.3% COD_{Cr} from the range of 503 – 10 000 gO_2/m^3,
- in 61% electrolytic conductivity from the range of 0.1 – 6 mS/cm, including in 36.6% values from 2 to 4 mS/cm (Figure 2b);
- in 86.6% BOD_5 from the range of 100 – 4000 g O_2/m^3, including in 54.6% values from 100 to 2000 g O_2/m^3 (Figure 2c),
- in 72.9% total suspended solids from the range of 275 – 20 000g/m³ (Figure 2d).

In the case of electrical conductivity, total suspended solids, BOD_5, and $COD_{(Cr)}$ histograms, the existence of a strong the right-hand asymmetry of the distribution was showed, which means that for these indicators of pollution, there are individual high and very high values (Figure 2, Table 1), while the vast majority of results ranged below the average. Only in the case of pH, a strong left-hand asymmetry (A = –1.04) was observed indicating that most of the analyzed samples of liquid wastes had a pH above average (Table 1, Figure 2a).

A comparison of indicators of liquid waste delivered and municipal wastewater shown that the minimum values of parameters of liquid wastes

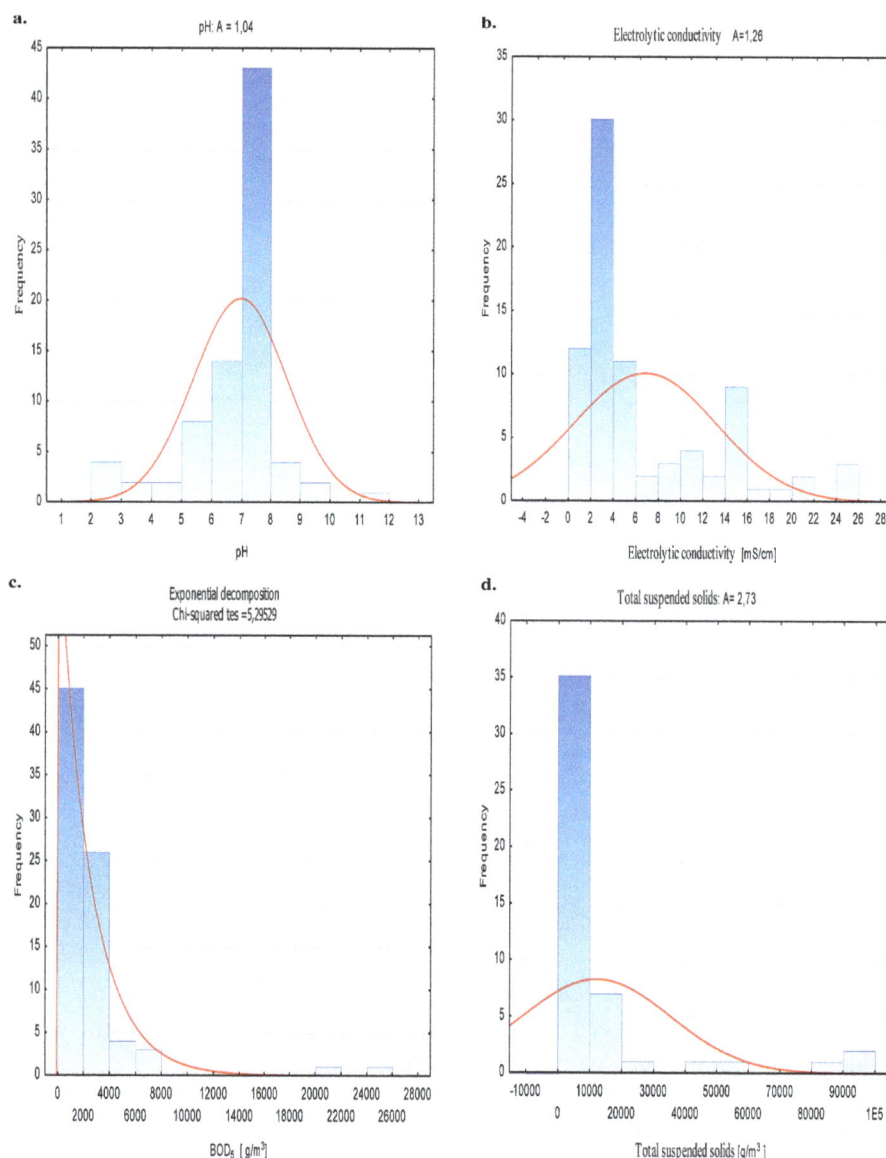

Figure 2. Histograms of selected parameters

supplied to the collection point in Bialystok had much smaller values than the municipal sewage quoted by various authors: Maksymowicz and Opęchowski [2001] as well as Simpson [2008] (Table 2). At the same time, the maximum values exceeded them repeatedly. For example, according to own research, BOD_5 for liquid wastes accepted a value from 100 to 25 000 g O_2/m^3, while BOD_5 for municipal wastewater according to research by Maksymowicz and Opęchowski from 150 to 500 g O_2/m^3, and the research by Simson - from 346 to 527 g O_2/m^3 (Table 2).

The analysis of liquid waste pollution indicators showed the existence of large span between values of the analyzed parameters (Table 2). The variation range of selected parameters was higher than that in the studies by Maksymowicz and Opęchowski [2001] (Table 2), and the mean val-

ues several times higher than values obtained by Tomczuk [2011a] (Table 1 and 2).

Largely dispersed values of the analyzed contamination indicators of liquid wastes delivered to the collection point in Bialystok confirm, among others, the diversity of liquid waste sources (single-family and multi-family residences, waste treatment plant in Hryniewicze, service facilities, industrial plants), no sealing of some no-outflow reservoirs, and low frequency of their transport.

CONCLUSIONS

1. Liquid wastes delivered to the collection point in Bialystok in 2012 were characterized by a high variability of composition. This resulted mainly from the diversity of the liquid waste

Table 2. Characteristics of sewage and liquid wastes

Index	Volues				
	Municipal sewage		Liquid wastes		
	Simson 2008	Maksymowicz. Opęchowski 2001	Maksymowicz. Opęchowski 2001	Tomczuk 2011a (mean)	Own and Waterworks
Total suspended solids [g/m³]	417 – 676	1 200 – 3 200	2 500 – 50 000	116	275 – 90 896
BOD₅ [gO₂/m³]	346 – 527	150 – 500	1 000 – 7 000	1080	100 – 25 000
COD_Cr [gO₂/m³]	796 – 1 239	250– 1000	2 500 – 15 000	1790	503 – 803 306
Electrolytic conductivity [mS/cm]	–	–	–	2.9	0.1 – 24.6
Reaction pH	6.8 – 7.9	–	–	7.7	2.3 – 11.6

types originating both from municipal housing, and public facilities, service establishments, or industry.

2. The values of the limit parameters set by the Waterworks Bialystok, which are indicated in the agreements between the collection point administrator and waste suppliers are notoriously exceeded.

3. The results of the study consist of determined, real parameters of liquid wastes delivered to the collection point in Bialystok, which demonstrate the need for their more frequent checks. Such tests are performed too rarely, which means that Suppliers of the wastes in a situation of exceeded permissible levels specified in the contract [Contract 2011] do not bear the actual costs associated with the introduction of impurities determined by the exceeding levels of pollutants contained therein.

REFERENCES

1. Bartoszewski K. at al. 2011. Poradnik eksploatatora oczyszczalni ścieków. Wyd. PZIiTS, Oddz. Wlkp. Poznań. [In Polish].

2. Błażejewski R., Nawrot T. 2009. Jak uszczelnić system gromadzenia i dowożenia nieczystości ciekłych?. GWiTS, 9. [In Polish].

3. Grygorczuk-Petersons E., Tałałaj I. 2007. Kształtowanie gospodarki odpadami w gminie. Podlaska Agencja Zarządzania Energią. Białystok. [In Polish].

4. Heidrich Z. 2005. Gospodarka wodno-ściekowa. Centrum Informacyjne Lasów Państwowych. [In Polish].

5. Kisiel J., Bień J. 2005. Wspołdzialanie retencyjnego zbiornika stacji zlewnej z oczyszczalnią ścieków. [In:] Zintegrowane, inteligentne systemy wykorzystywania energii odnawialnej. Mat. Konf. Częstochowa-Podlesice. [In Polish].

6. Leszczyński J. 2011. Podczyszczanie odcieków ze składowiska odpadów stałych metodą koagulacji. Inżynieria Ekologiczna, 25, 242–250. [In Polish].

7. Leszczyński J. 2013. Zastosowanie procesu Fentona do podczyszczania odcieków składowiskowych. Instal, 6(341), 52–54. [In Polish].

8. Maksymowicz B., Opęchowski S. 2006. Zasady gospodarowania nieczystościami ciekłymi. Poradnik. Łódź. [In Polish].

9. Maksymowicz B., Opęchowski S. 2001. Zasady sporządzania przez gminy programów postępowania z nieczystościami ciekłymi z terenów nieskanalizowanych z uwzględnieniem dyrektyw Unii Europejskiej. Poradnik praktyczny dla gmin. OBREM, Łódź. [In Polish].

10. Memo 2011. Notatka służbowa Wodociągów Białostockich Sp. z.o.o.. Białystok. Unpublished typescript. [In Polish].

11. PN-ISO 5667-10: 1997. Jakość wody. Pobranie próbek. Wytyczne pobierania próbek ścieków. [In Polish].

12. Simson G. 2008. Pierwsze doświadczenia – test technologiczny z zastosowaniem preparatu BRENNTAPLUS VP1, jako zewnętrzne źródło węgla organicznego do intensyfikacji procesu denitryfikacji w Białostockiej Oczyszczalni. Forum Ekspoatatora, 6, 21–23. [In Polish].

13. Tomczuk B. 2011a. Ścieki dowożone czy nieczystości ciekłe? Forum Eksploatatora, 2(53), 82–83. [In Polish].

14. Tomczuk B. 2011b. Stacje zlewne nieczystości ciekłych – przegląd dostępnych rozwiązań. Forum Eksploatatora, 3(54), 70–71. [In Polish].

15. Contract 2011. Umowa o wprowadzanie nieczystości ciekłych do stacji zlewnej. 2011. Białystok. Maszyn. niepubl [In Polish].

16. Act 1996: Ustawa z dnia 13 września 1996 r. o utrzymaniu czystości i porządku w gminach (tekst jednolity z 2012 r. Dz.U. 2011 poz. 391 z poźn. zm.). [In Polish].

20

WASTEWATER MANAGEMENT IN FOOD PROCESSING ENTERPRISES – A CASE STUDY OF THE CIECHANÓW DAIRY COOPERATIVE

Marek Gugała[1], Krystyna Zarzecka[1], Anna Sikorska[1]

[1] Siedlce University of Natural Sciences and Humanities, B. Prusa 14, 08-110 Siedlce, Poland, e-mail: gugala@uph.edu.pl; kzarzecka@uph.edu.pl

ABSTRACT

The paper reports on wastewater management in food processing enterprises, using the Ciechanów Dairy Cooperative as an example. The efficiency of preliminary treatment of wastewater entering the municipal sewerage system was evaluated using basic pollution indices for wastewater pre-treated in the process of chemical neutralisation as well as results of physical and chemical analyses of raw wastewater from fromage fraise and cottage cheese production lines. The number of exceedances of permissible values of pollution indices in pre-treated wastewater was determined based on values set in the water legal permits issued by the relevant authority. Pre-treated wastewater entering the municipal sewerage system met the standards for permissible wastewater pollution indicators excluding BOD_5 (biochemical oxygen demand) and COD (chemical oxygen demand). Physical and chemical analysis was performed of total raw wastewater and wastewater discharged from fromage fraise and cottage cheese production lines. Pollution indicators (COD, total nitrogen, orthophosphates, total suspensions, fats, pH) had high values but they remained within the ranges typical for dairy food processing.

Keywords: dairy processing wastewater, dairy processing, wastewater management, pollution indicators.

INTRODUCTION

Dairy industry is one of major branches of food processing industry in Poland. Milk production has increased by 58.9% over 2000–2010. The yearly milk production now is almost 12 milliard litres. Assuming the indicator of wastewater volume per unit of production is 3.5 m³/ 1 m³ of milk processed, it can be estimated that in Poland the daily discharge of dairy wastewater is approximately 92,000 m³ [16]. The wastewater is characterised by much higher pollution indicators and varied volumes discharged compared with municipal wastewater. As a result, it is much more difficult to exploit such dairy processing facilities [1]. Varied wastewater discharges by milk processing plants constitute an additional difficulty. Efficient treatment of dairy wastewater is of paramount importance from the standpoint of water environment protection.

MATERIALS AND METHODS

The work is based on data obtained from the Ciechanów Dairy Cooperative. The paper presents values of basic pollution indicators in wastewater treated using neutralisation. Efficiency of preliminary treatment of dairy wastewater was evaluated. Physical and chemical analyses were conducted to obtain values of basic pollution indicators in total (raw) wastewater discharged from fromage fraise and cottage cheese production lines of the Ciechanów Dairy Cooperative.

Physical and chemical analyses were conducted in 2007 at the Department of Environmental Engineering of the University of Warmia and Mazury in Olsztyn. Raw (total) wastewater as well as the wastewater discharged from fromage fraise and cottage cheese production lines were analysed. Wastewater was sampled in the morning and in the afternoon at the premises of the dairy processing plant. The following components were determined: total nitrogen, orthophosphates, total suspensions, fats, organic compounds and pH. An indirect method, based on determination of the amount of oxygen used to oxidise organic compounds, was used to determine these compounds. Chemical Oxygen Demand (COD) was determined by means of the potassium dichromate method with sulphuric acid [11]. Wet mineralisation of wastewater samples was followed by determination of total nitrogen using the distillation method. The samples were filtered and used to determine spectrophotometrically the orthophosphates with the Phosver reagent (ascorbic acid) at 890 nm. The total suspended solids were determined with the gravimetric method, whereas fats were analysed using SOXHLET extraction – a water bath was used to evaporate acidified wastewater samples, ether was removed from the extract, and the remains were dried and used to gravimetrically determine fats [11].

DESCRIPTION OF WASTEWATER DISCHARGED BY MILK PROCESSING INDUSTRY

Wastewater is one of major environmental concerns of milk processing industry. In terms of volume, dairy processing plants discharge mainly wastewater produced during washing and cleaning processes. The wastewater contains 3-4% processed milk, on average. Milk losses take place when processing begins, after periodic cleaning, when old products are replaced by new ones and when milk is spilled or leaks [3]. Dairy wastewater contains easily biodegradable organic substances. Due to rapid fermentation, resulting in a marked pH decrease (to pH=4.5) and promoting rapid oxygen consumption, the wastewater has to be pre-treated before it is sent to drain [16, 18]. Due to the fact that dairy processing wastewater has got much higher values of pollution indicators, compared with domestic or municipal wastewater, it is necessary to pre-treat the wastewater before it is discharged to sewers and a sewage processing plant [1, 12]. Wastewater produced by dairy industry is a mixture of organic and inorganic sewage [8]. It also contains condensates and sewage remaining after milk cooling. The wastewater is produced at various stages and during various processes of the milk processing cycle [14]. Problems associated with

pre-treatment of dairy industry wastewater arise from its specificity (it contains e.g. dissolved and crystallised fat, carbohydrates (lactose), both colloidal and clotted protein as well as cleaning-related substances) [1, 13].

In addition to the aforementioned proteins, carbohydrates and fats, there is a high concentration of organic nitrogen and the following ions: NH_4^+, NO_2^- and NO_3^-. Also, organic and inorganic phosphorus is found and the following elements are detected: Na, K, Ca, Mg, Fe, Co, Ni and Mn. High sodium content indicates that large quantities of alkaline cleaning chemicals are used [10].

The problems with pre-treatment of dairy processing wastewater arise not only from high values of pollution parameters but also from varying daily wastewater discharges, production profile, raw materials, technological level, cleaning and disinfection processes, quantity of water used and high exploitation costs [1, 13]. Discharges of dairy wastewater and its concentration of pollutants differ over time [5]. They are affected by the production profile of the plant which is usually a sequence of periodic processes. An example of this is production of processed milk for consumption, which is a sequence of the following technological stages: milk collection, centrifugation, pasteurisation, normalisation of fat content in milk, cleaning and filling of containers and cleaning of machines and rooms after the process has been finished. Wastewater with varied pH levels and concentration of pollutants is produced at each of these stages.

RESULTS AND DISCUSSION

The most important dairy industry-related pollutants in wastewater are as follows: varying pH, high BOD_5, high fat content and high total suspended solids [3].

Raw (total) wastewater discharged by the Ciechanów Dairy Cooperative had high COD (4,895 mg O_2/dm^3, on average). It was much higher in the afternoon than in the morning. Raw wastewater sampled in discharges from fromage fraise and cottage cheese production lines had much higher COD (Table 1) which fell within the range reported by Chudzik [4].

COD values in typical wastewater discharged by dairy processing plants in Poland range from 1,800 to 9,000 mg O_2/dm^3 [4]. According to Struk-Sokołowska [16], a typical dairy processing plant discharges from 450 to 600 m^3/d wastewater with an average COD value of 2,077 mg O_2/dm^3. Other authors have reported COD values of 3,700 mg O_2/dm^3. The COD values obtained in the study reported here were much higher than the findings mentioned by Bartkiewicz [2]. The wastewater discharged by milk processing plants in Poland normally contains an average of 1,500 mg total suspended solids. The total wastewater discharged by the Ciechanów Dairy Cooperative had the concentration of total suspended soils which was twice as high in the afternoon (1,100 mg/dm^3). The quantity of total suspended solids in the wastewater analysed was quite high: 6,818 mg/dm^3 for the cottage cheese production line and 6,992 mg/dm^3 for the fromage fraise production line. The solids were much lower than the values reported by Grala [9].

The highest pollution levels are determined in wastewater generated during cheese production as it contains 30 g/kg solids [9]. Other authors [6, 8, 15] claim that the total suspended solids in dairy wastewater may reach the level of 2,000 mg/dm^3. Fat content in the total wastewater sampled in the morning and in the evening was 130 and 1,300 mg/dm^3, respectively. The values fell within the range reported by FAPA and others [6, 8, 15]. Fat content in normal wastewater generated by milk processing plants in Poland is 500 mg per one litre of milk processed [7]. Wastewater sampled from the fromage fraise production line discharges had a much higher concentration of fats than the total wastewater (Table 1). The orthophosphate content in dairy wastewater reported in Polish literature falls within the range of 11.9–45.9 mg PO_4^{3-}/dm^3 [17]. In the present work, the content was very low because organic phosphorus was the dominant form in the wastewater.

The average concentration of orthophosphates in the total wastewater was 1.13 mg/dm^3, including 1.58 mg PO_4^{3-}/dm^3 for the cottage cheese production line and 0.35 mg PO_4^{3-}/dm^3 for fromage fraise. The respective concentrations of total nitrogen were 1,258.6 mg/dm^3 (which is a high value) and 325.4 mg/dm^3 (Table 1). The average concentration of total nitrogen in raw wastewater was 96.5 mg/dm^3 and was much lower than the values reported by Anielak [1].

pH values in dairy wastewater range from 4.5 to 9.2 [2]. The samples of raw wastewater taken in the morning and in the afternoon were alkaline whereas the pH for wastewater associated with both the production lines was very acidic and ranged from 4.8 to 5.1 (Table 1).

Total wastewater, including industrial sewage, domestic sewage and rainfall, generated by the Ciechanów Dairy Cooperative was transferred, using one sewer, to TOS10 device which is a sedimentary flotation separator of mud, oils and fats. The device produces average COD_5 and pH values and removes solid particles, fats as well as oils. The permissible values of pollution indicators are set in the water legal permit which is issued by the Water and Sanitation Company in Ciechanów. The wastewater leaving the separator has to be characterised by certain pollution parameters. Only then can it enter the municipal sewers and the wastewater treatment facility in Ciechanów.

Table 2 presents minimum, average and maximum values of total wastewater pollution parameters following neutralisation as well as the standards set in the water legal permit.

The values of pollution parameters in wastewater pre-treated by means of neutralisation indicated that BOD_5 was the lowest in winter (230

Table 1. Characteristics chemical physics stricte sewers – general and line of curd cheeses and homogenized cheeses

Indicators of pollution	Unit	Stricte sewers			
		Sewers general		Line of curd cheeses	Line of homogenized cheeses
		7[30]	11[30]	8[30]	9[00]
ChZT*	mg O_2/dm^3	3900.0	5890.0	51200.0	68280.0
General suspensions	mg/dm^3	7000.0	1500.0	6818.0	6992.0
General nitrogen	mg N_{og}/dm^3	79.1	113.96	1258.6	325.4
Orthophosphates	mg PO_4^{3-}/dm^3	1.51	0.75	1.58	0.35
Reaction	pH	8.07	8.27	5.1	4.8
Canailles	mg/dm^3	130.0	1500.0	3360.0	10844.0

* ChZT – Chemical demand for oxygen.

Table 2. Parameters of General sewers after the neutralization accompanied from the Ciechanów Dairy Cooperative for urban canalization

Indicators of pollution	Unit	Minimum value	Maximum value	Average value	Norms of sewers accompanied to canalization
BZT$_5$ *	mg O$_2$/dm^3	230.0	2000.0	827.7	750.0
ChZT**	mg O$_2$/dm^3	374.0	2874.0	1388.4	1200.0
BZT$_5$/ChZT	–	0.61	0.69	0.59	–
General suspensions	mg/dm^3	94.0	280.0	154.5	450.0
General nitrogen	mg N/dm^3	13.0	75.0	40.0	100.0
General phosphorus	mg P/dm^3	1.8	17.5	8.7	15.0
Extracting substances oneself with petroleum ether	mg/dm^3	6.0	66.0	27.8	100.0
Reaction	pH	7.01	8.43	7.52	6.5 – 9.5

*   BZT5 – Biochemical demand for oxygen.
** ChZT – Chemical demand for oxygen.

mg O$_2$/dm^3), whereas COD fell within the range of 332 to 374 mg O$_2$/dm^3. Average permissible BOD$_5$ and COD values were exceeded in March as they increased to 1,550 and 2,874 mg O$_2$/dm^3, respectively. In April and May, the respective ranges for BOD$_5$ and COD were 580–660 mg O$_2$/dm^3 and 998–1,112 mg O$_2$/dm^3. In late June and early July, the values of both parameters slightly exceeded the permissible averages; in August they were the highest (2,000 and 2,720 mg O$_2$/dm^3, respectively) but dropped in September (to the level of 400 and 1,100 mgO$_2$/dm^3, respectively). In December, both BOD$_5$ and COD increased again and amounted to 1,000 and 1,450 mg O$_2$/dm^3, respectively (Figure 1).

The lowest concentration of total suspended solids (94 mg/dm^3) was obtained in February but in March it was the highest (280 mg/dm^3). In April solids decreased and amounted to 95 mg/dm^3, in late June and early July they increased to the level of 197 mg/dm^3. The increase contin-

ued in August till the value of 276 mg/dm^3 was reached (Figure 2).

Total nitrogen content of neutralised wastewater ranged from 13 to 75 mg N/dm^3, the lowest being in February and the highest in August. The values of pollution parameters for total suspended soils and total nitrogen demonstrated that wastewater entering the municipal sewage system met the standards set in the water legal permit (Figure 2).

The highest ether extract concentration was recorded in January and the lowest during late June-early July period (66 and 6 mg/dm^3, respectively) (Figure 3). The lowest total phosphorus concentration (1.8 mg/dm^3) was determined in January; it increased in March (911.5 mg/dm^3) and was the highest in April (17.5 mg/dm^3), which meant that the average permissible value for the parameter was slightly exceeded. After a decrease to the level of 10.6 mg/dm^3 in May, total phosphorus increased again in August (14.6 mg/

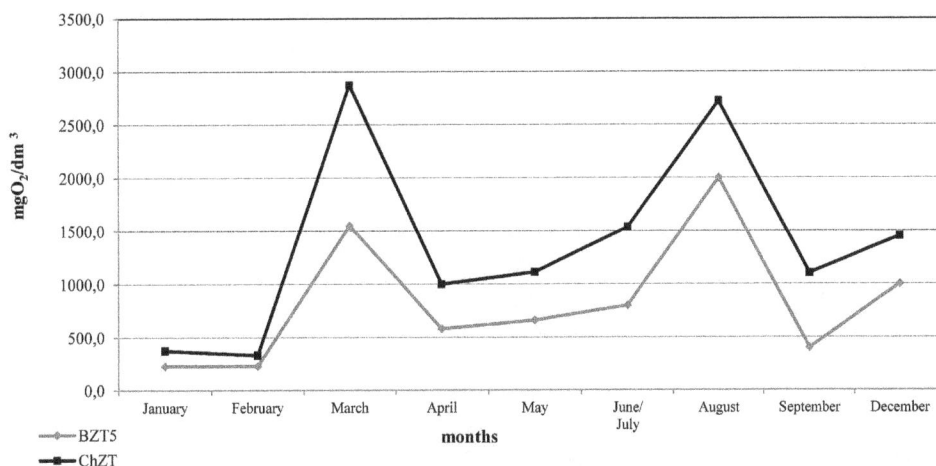

Figure 1. Characteristics physics-chemical of general sewers after neutralization (BZT5, ChZT)

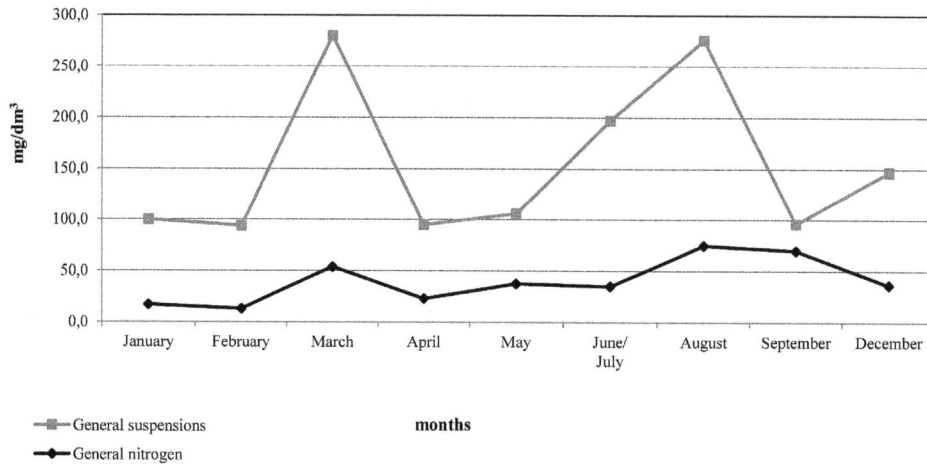

Figure 2. Characteristics physics-chemical of general sewers after neutralization
(General suspensions, General nitrogen)

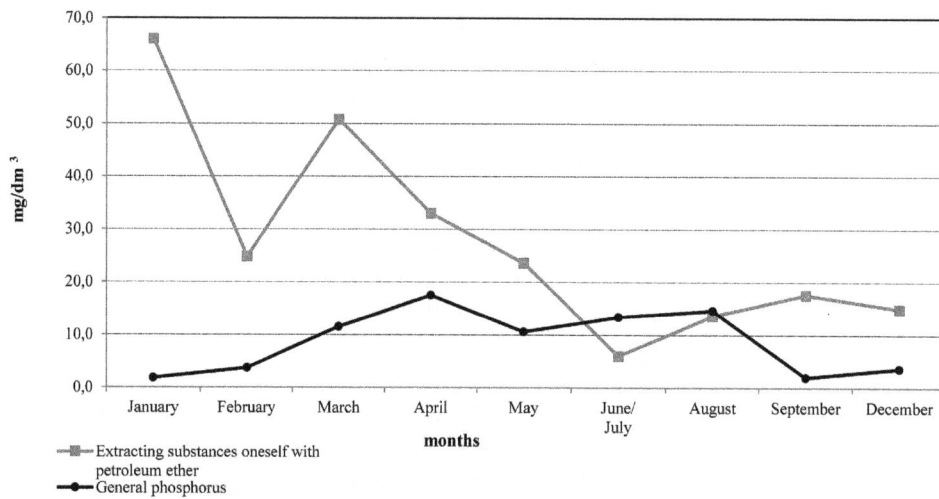

Figure 3. Characteristics physics-chemical of general sewers after neutralization
(Extracting substances oneself with petroleum ether, General phosphorus)

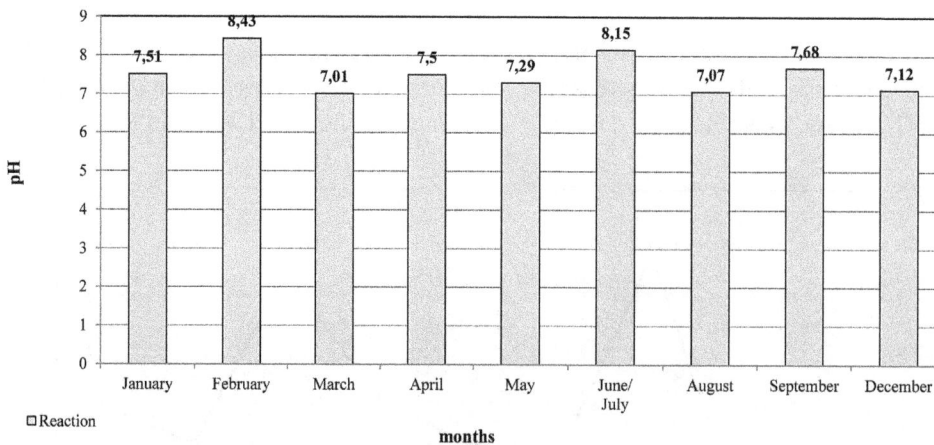

Figure 4. Characteristics physics-chemical of general sewers after neutralization (Reaction)

dm^3) and then dropped in September (1.96 mg P_{og}/dm^3) (Figure 3).

Wastewater pH ranged from 7.01 to 8.15, the lowest being in March and August (7.01 and 7.07, respectively) and the highest in February and during the late June-early July period. The pH values in the remaining months ranged from 7.29 to 7.68 (Figure 4).

CONCLUSIONS

Both raw wastewater and sewage from the fromage fraise and cottage cheese lines entering the separator had high concentrations of pollutants. Despite this they fell within the range of typical values determined in dairy wastewater. Chemical neutralisation reduced COD, total solids and total nitrogen by 75.8, 85 and 58.5%, respectively.

Dairy wastewater discharged by the Ciechanów Dairy Cooperative met most of the standards set for wastewater entering the municipal sewerage system. COD and BOD_5 values for pretreated wastewater only periodically exceeded the permissible values set in the water legal permit. The remaining parameters remained below the acceptable levels.

REFERENCES

1. Anielak A.M. 2008. Gospodarka wodno-ściekowa przemysłu mleczarskiego. Agro Przemysł, 2, 57–59.

2. Bartkiewicz B., Umiejewska K. 2010. Oczyszczanie ścieków przemysłowych. PWN, Warszawa.

3. BAT – Best Available Techniques. 2005. Najlepsze dostępne techniki (BAT) wytyczne dla branży mleczarskiej, praca wykonana przez WS ATKINS - POLSKA Sp. z o. o. na zamówienie Ministerstwa Środowiska.

4. Chudzik B. 1997. Charakterystyka ścieków przemysłowych. Ekoinżynieria, 2, 19–21.

5. Danków R, Cais-Sokolińska D, Pikul J. 2009. Wykorzystanie popłuczyn mleczarskich w celu odzyskania masy białkowej. Nauka Przyroda Technol. 3(4), 116.

6. Dyatlova G., Pevnev S. and Fedorovskaya T. 2008. Treatment of milk plants wastewater. Water Supply and Sanitary Technique, 2, 12–15.

7. FAPA 1998, Ochrona środowiska w przemyśle mleczarskim. Publikacja sfinansowana ze środków Unii Europejskiej programu PHARE będących w dyspozycji Fundacji Programów Pomocy dla Rolnictwa (FAPA). Projekt P9312/04 - 02, Warszawa.

8. Gorban N. 2007. Purification facilities for the meat and dairy industries. Book of Abstracts Scientific and Practical Conference "Water & Environment" V International Water Forum "Aqua Ukraine – 2007", 109–110.

9. Grala A, Zieliński M, Dudek M, Dębowski M. 2010. Efektywność oczyszczania ścieków mleczarskich w reaktorze beztlenowym o przepływie pionowym. Inżynieria Ekologiczna 22, 97–105.

10. Guillen-Jimenez E., Alvarez-Mateos P., Romero-Guzman F., Pereda-Martin J. 2000. Bio-mineralization of organic matter as affected by pH. The evolution of ammonium and phosphates. Water Research, 34, 1215–1224.

11. Hermanowicz W., Dojlido J., 2010. Fizyczno-chemiczne badanie wody i ścieków, Arkady Publishing House.

12. Kajurek M., Dąbrowski W. 2003. Przeróbka i zagospodarowanie osadów ściekowych z oczyszczalni ścieków mleczarskich na przykładzie S.M. Mlekovita. II Międzynarodowa Konferencja Nowe Spojrzenie na osady ściekowe, odnawialne źródła energii, Częstochowa.

13. Krzemińska D., Neczaj E., Grosser A. 2011. Application of Advanced Oxidation Processes (AOP's) for the Industrial Wastewater Treatment. Acta Biochimica Polonica, IV Congress of Polish Biotechnology and IV EUROBIOTECH "Four Colours of Biotechnology" Central European Congress of Life Sciences, Vol. 58, Suppl. 4, Kraków.

14. Kushwaha J.P., Srivastava V.Ch., Mall I.D. 2011. An overview of various technologies for the treatment of dairy wastewaters. Critical Reviews in Food Science and Nutrition, 51, 442–452.

15. Sabliy L., Konontsev S. 2003. Wastewater treatment biotechnology for the plants of milk industry. Visnyk of UDUVGP, 2 (21), 142–150.

16. Struk-Sokołowska J. 2011. Wpływ ścieków mleczarskich na frakcje ChZT ścieków komunalnych. Inżynieria Ekologiczna, 24, 130–144.

17. Świerczyńska A., Bohdziewicz J., 2013. Współoczyszczanie odcieków ze ściekami mleczarskimi w sekwencyjnym bioreaktorze membranowym. Journal of Civil Engineering, Environment and Architecture, Vol. XXX, Issue 60 (3/13), 82.

18. Wierzbicki T.L., Dąbrowski W., Magrel L., 2005. Oczyszczanie ścieków, unieszkodliwianie i przeróbka osadów ściekowych pochodzących z zakładów przetwórstwa mleczarskiego. Research Project No 7 TO7G 029 11. Białystok University of Technology.

21

SIMULATIONS OF THE INFLUENCE OF CHANGES IN WASTE COMPOSITION ON THEIR ENERGETIC PROPERTIES

Halina Marczak[1]

[1] Mechanical Engineering Faculty, Lublin Technology University, Nadbystrzycka 36, 20-618 Lublin, Poland, e-mail: h.marczak@pollub.pl

ABSTRACT

The objective to obtain the recommended (at least 50% by weight) level of recycled and reprocessed raw fractions of municipal waste in the perspective of 2020 might in turn contribute to the deterioration of the fuel properties of waste stream that is intended for incineration. In order to avoid oversizing heat recovery plants their construction should be based on well-defined properties of the waste fuel. The author carried out simulation calculations of the impact of changes in the composition of municipal waste on their energetic properties. The calculations of the calorific value of the waste fraction (so called combustible fraction) with the potential to be used as an alternative fuel were done.

Keywords: waste management, municipal waste, recycling, calorific value of waste.

INTRODUCTION

The regulations implemented in the area of municipal waste management are aimed at increasing their economic use, and in the case of disposing waste in a landfill, at minimizing their impact on the environment.

It is a legal requirement that municipalities take action to increase the level of recycling and reprocessing of raw fraction of municipal waste by the end of 2020: paper, plastic, glass and metals by a minimum of 50% by weight [Resolution… Dz. U. 2013]. In fulfilling this obligation a special role is attributed to selective collection of recyclable waste fractions, while mechanical and biological processing plants (MBP) of mixed waste may only be a supplementary solution.

The laws also oblige for use in waste management solutions that reduce to an acceptable minimum fraction of organic biodegradable municipal waste disposed of by landfill. To achieve the quantitative requirements in this area may be helpful in backyard composting method of organic waste and the existing and built installations for composting or fermentation of biodegradable waste installations MBP preparing mixed municipal waste to the processing and biochemical plants thermal treatment of mixed municipal waste.

Designing the installations of thermal treatment of waste, such parameters as the amount and characteristics of waste fuel are taken into account. In particular, the characteristics of waste fuel determine the authothermicity of thermal treatment process of waste and the cost of functioning the installations. Waste fuel properties determine the justifiability of subjecting them to a process of thermal treatment. Reducing the share of plastic and paper waste stream may affect the deterioration of calorific value of waste.

The change of morphological composition of combustible fraction of mixed municipal waste may also affect the calorific value of the alternative fuel produced from waste. For example, a reduction of plastic fraction in the waste stream will contribute to lowering the fuel properties of the waste.

By definition, legal regulations are designed to have a positive impact on the management of waste, however, a legal requirement to obtain the recommended level of recycling and reprocessing of selected waste fractions by 2020, may have an adverse impact on the profitability of some waste treatment processes. This applies in particular to the existing plants and the construction of waste thermal treatment, as complying with the afore-mentioned legal requirement may worsen fuel properties of the waste stream intended for in-cineration.

The work contained calculations to illustrate the effect of reducing the share of paper and plas-tic waste stream on the properties of energy waste intended for incineration. The calculation results indicate that it may be necessary to take action to increase the calorific value of the waste directed to the combustion process. Waste fuel properties can be corrected by adding fuels to them, e.g. fu-els formed, but keep in mind the adverse impact of such actions on the financial results of compa-nies waste incineration.

GENERAL ASSUMPTIONS OF WASTE MANAGEMENT IN LUBELSKIE VOIVODESHIP

Currently, the system of municipal waste management in Lubelskie Voivodeship includes selective waste collection, waste treatment in order to prepare them for reuse or safe storage and disposal of unprocessed waste in a landfill. Waste processing is carried out in the so-called segregation installations, so called dry waste frac-tion in installations for mechanical treatment of mixed waste in composting anaerobic facilities for waste treatment. In some installations, me-chanical treatment of mixed municipal waste is focused on the production of alternative fuel with simultaneous production of the organic fraction directed to composting. To increase the level of waste recovery, works aimed at the expansion of existing and construction of new installations for this type of waste are carried out. Moreover, in the municipality of Lublin there are plans to build an incineration plant and the construction of MBP [Szyszkowski et al. 2012], which according to the assumptions, will play an important role, along-side selective collection, in fulfilling the obliga-tion to obtain level of recycling and preparation for reusing paper, cardboard, plastics, glass and metals by at least 50% by weight by the end of 2020. To fulfil this obligation, it is equally nec-essary to implement efficient sorting installa-tions for separately collected waste fractions to ensure the acquisition of high quality, mainly for recycling and to run the installations which will

implement the recycling processes. It is believed that the MBP installations for mixed waste will be of limited importance in achieving the required level of recycling and preparation for reusing such material fractions, such as paper, glass, plas-tics and metals. The materials separated in the re-cycling systems are inferior in quality, compared to the collected selectively. Consequently, there can be problems in their disposal or recycling due to higher costs as a result of having to undergo an additional treatment such as cleaning. In turn, MBP plants may be important in the production of alternative fuel from waste, currently used in the cement industry. The technological process of the production of alternative fuels from waste should ensure that you obtain the fuel quality pa-rameters which meet customers' requirements. An important parameter is the calorific value of waste, which should be about 20 MJ/kg for ap-plications in the cement industry.

The increase in the selective collection of pa-per and plastic factions "at source" will reduce the amount of other types of waste, and this in turn can help to reduce the calorific value of the waste [Skowron 2006]. In order to avoid oversiz-ing heat recovery plants, they should be based on well-defined properties of the waste.

Limiting the share of combustible components, plastics in particular, may reduce the calorific value of the fuel. In this regard, paper and cardboard are less significant, as in comparison to plastics they are less calories and more hygroscopic.

When selecting waste treatment techniques and technologies, in addition to the composition and properties of the waste one should consider, the purpose of the processing, the requirements of entrepreneurs involved in recovering the materi-als obtained in the treatment processes, the eco-nomics of the solutions and legal requirements [Marczak 2013].

Based on the data on the amount of municipal waste in Lubelskie Voivodeship, the calculations of the impact of changes in the composition of waste on their energetic properties was made.

EXPERIMENTAL PART

Characteristics of the material

The estimated weight of waste generated in 2012 in Lubelskie Voivodeship amounted to 546920.7 Mg (Table 1). This value is calculated using the rates of waste generation, amounting to

Table 1. Municipal waste generated in Lubelskie Voivodeship in 2012 (own calculations in the basis of [Szyszkowski et al. 2012])

Waste fraction	Waste generation index [kg/M/year]			Amount of waste [Mg]				
	Cities above 50 000	Cities below 50 000	Rural areas	Cities above 50 000	Cities below 50 000	Rural areas	Total	
							[Mg]	[%]
Paper and cardboard	64	32	9	34 374	15 117	10 642	60 133	11
Glass	33	34	18	17 780	15 896	21 128	54 804	10
Metal	9	5	4	4742	2338	5104	12 183	2.2
Plastics	51	37	19	27 432	17 299	21 969	66 699	12.2
Multi-material waste	8	13	8	4403	6234	8726	19 363	3.5
Kitchen and garden waste	95	121	60	51 137	56 416	69 500	177 053	32.4
Mineral waste	10	10	11	5588	4519	13 302	23 409	4.3
Fraction < 10 mm	14	23	31	7281	10 597	35 489	53 368	9.8
Textiles	8	14	4	4064	6390	4546	14 999	2.7
Wood	1	1	1	508	468	1397	2372	0.4
Hazardous waste	3	2	2	1355	1091	1742	4188	0.8
Other categories	11	15	9	5926	7169	10 530	23 625	4.3
Bulky waste	9	9	2	4742	4052	2716	11 509	2.1
Green waste	22	18	3	11 706	8260	3250	23 215	4.2
Total	337	333	181	181 036	155 844	210 040	546 921	100
Number of inhabitants (according to Central Statistical Office - Lublin)	537 200	468 000	1 160 444					

337 kg per capita per year (kg/M/year) for the areas of large cities (more than 50 thousand inhabitants), 333 kg/M/year for small cities (up to 50 thousand residents) and 181 kg/M/year for rural areas. The values of total waste generation and different material fractions were adopted from the document „Plan gospodarki odpadami dla województwa lubelskiego 2017" [Szyszkowski et al. 2012].

The estimated share of paper and cardboard in the stream of the generated waste mass was approx. 11%, while the plastic was approx. 12.2% (Figure 1). Municipal waste collected (including the gathered waste) in Lubelskie Voivodeship in 2012, accounted for approximately 346600

Mg [CSO 2013]. Table 2 shows the data on the amount of municipal waste collected in 2012 [CSO 2013].

Based on the data contained in Tables 1 and 2, the amount of waste taken into account in further analysis was estimated. It was determined by subtracting the weight of separately collected waste from the mass of waste generated in 2012, the quantities of waste analysed are presented in Table 3. For ease of calculations the waste was divided into four groups. Group I includes paper, cardboard, glass, metals, plastics, mineral waste, textiles, wood, multi-material waste, hazardous waste, other waste categories. Group II is kitchen wastes. Group III is a waste of grain size <10 mm,

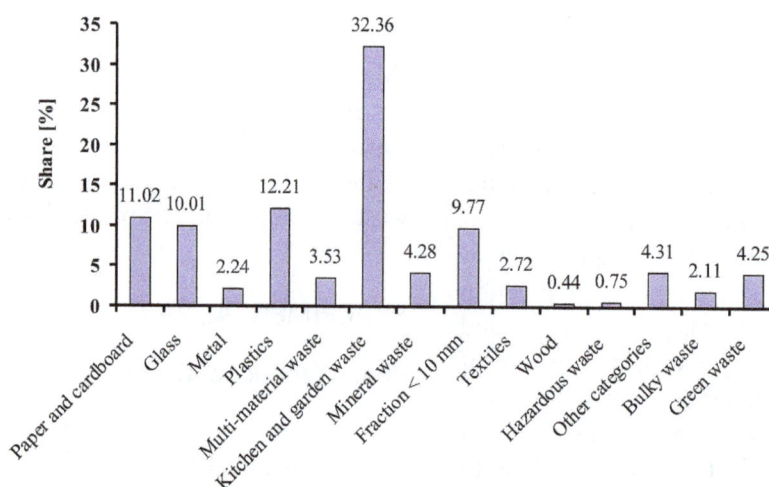

Figure 1. Estimated share of waste fractions in the stream of the generated waste mass in Lubelskie Voivodeship in 2012

Table 2. Amount of waste in Lubelskie Voivodeship in 2012

No	Item	Collected waste [Mg]	Share [%]
1.	Municipal waste (segregated and selectively collected):		
1.1	paper and cardboard	6063.60	1.75
1.2	glass	11693	3.37
1.3	textiles	1947.20	0.56
1.4	hazardous	53.8	0.02
1.5	plastics	5331.50	1.54
1.6	metal	714.90	0.21
1.7	electronic and electric appliances	1395.2	0.40
2.	Waste from parks and gardens:		
2.1	biodegradable	7357.7	2.12
3.	Other municipal waste:		
3.1	mixed	310 545.87	89.60
3.2	bulky waste	1474.5	0.43
4.	Total collected waste	346 577.27	100

while the fourth group is bulky waste and green waste. The masses of individual plastics fractions were determined following Jaglarz [2014] that in the municipal waste stream the share of PE is 27%, PP 6%, and the proportion of PS and PET are 10% and 57%, respectively.

Table 3 shows the moisture content, combustible and non-combustible components in specified fractions. The values of these parameters for waste classified as group IV (including bulky waste and green waste) were not included. This group of waste was not considered in the calculation of the calorific value, since it was assumed that green waste is directed to a biochemical conversion (composting process), while the bulky waste is treated in order to recover recyclable materials which are then passed on to authorized recipients.

Multi-material waste include, among others, packaging and non-packaging composites. For these wastes approximated values of moisture

Table 3. Amounts, composition and characteristics of analysed waste

Group	Waste fraction	Mass waste [Mg]	Share [%]	Moisture* [%]	Combustible components* [%]	Non-combustible components* [%]
I	Paper and cardboard	54 069.88	10.58	25	60	15
	Glass	43 110.54	8.44	0	0	100
	Metal	11 468.19	2.24	0	0	100
	Plastics, in it	61 367.72	12.01			
	PE..................................	16 569.29	3.24	0.13**	99.3**	0.57**
	PP..................................	3682.06	0.72	0.09**	98.78**	1.13**
	PS..................................	6136.77	1.20	0	98.62	1.38
	PET................................	34 979.60	6.85	0	96.64	3.36
	Mineral waste	23 409.22	4.58	0	0	100
	Textiles	13 052.06	2.55	0	88.93	11.07
	Wood	2371.86	0.46	20	79.2	0.8
	Multi-material waste	17 967.51	3.52	25	60	15
	Hazardous waste	4133.74	0.81	–	–	–
	Other categories	23 624.86	4.62	–	–	–
II	Kitchen and garden waste	177 053.10	34.66	71.59	21.84	6.57
III	Fraction < 10 mm	53 367.92	10.45	6.47	76.93	16.6
IV	Bulky waste	10 034.97	1.97	–	–	–
	Biodegradable greek waste	15 857.79	3.10	–	–	–
	Total group I, II, III	484 996.60	94.93	–	–	–
	Total group I, II, III, IV	510 889.36	100	–	–	–

* – According to [Jaglarz 2014].

** – According to [Czop 2013].

Table 4. Moisture content, combustible and non-combustible components for selected groups of waste

Group	Share [%]	Moisture [%]	Combustible components [%]	Non-combustible components [%]
I	52.49	7.27*	45.82*	46.90*
II	36.51	71.59	21.84	6.57
III	11.00	6.47	76.93	16.6
Total group	100	30.66*	40.49*	28.85*

* – Weighted average.

content, combustible and non-combustible components were adopted, as for paper and cardboard. Due to lack of data, Table 3 does not indicate the characteristics of hazardous waste and fractions: other categories of waste.

Table 4 shows the characteristics of the three groups of waste which are included in the calculations of the calorific value. Moisture content of combustible and non-combustible components of the waste in Group I were calculated as weighted averages of the values of these parameters given in Table 3.

Calculations of calorific values of the stream of mixed municipal waste

There is a relationship between calorific value Q_i and the heat of combustion Q_s:

$$Q_i = Q_s - r(W^r + 8.94H^r), \ kJ/kg \qquad (1)$$

where: r – heat of water evaporation; $r = 2442$ kJ/kg
W^r, H^r – accordingly the content (mass) of moisture and hydrogen in the waste in the operational state, expressed as a decimal fraction (8.94 is the conversion rate of hydrogen to water).

Heat of combustion was determined with Dulong's formula [Chudzinski et al. 1974]:

$$Q_s = 32800 \ C^r + 120040 \left(H^r - \frac{O^r}{8}\right), \ kJ/kg \quad (2)$$

where: C^r, H^r, O^r – respectively the content (mass fraction) of carbon, hydrogen, oxygen in the waste in the operational state, expressed as a decimal fraction.

To convert the contents of individual elements in the waste from the dry to the operating status the following relationship was used:

$$P^r = P^s \left(100 - W^r\right)/100 \qquad (3)$$

where: P^r – the content of carbon, hydrogen, oxygen in the waste in the operational state, %
P^s – the content of carbon, hydrogen, oxygen in the waste in the operational state, %
W^r – moisture content in waste, %.

The content of carbon, hydrogen, oxygen in the dry mass of the waste was estimated using the following formula:

$$P^s = \sum_{i=1}^{n} P_i^s u_i \qquad (4)$$

where: P_i^s – the content of carbon, hydrogen, oxygen and in different material fractions i,
u_i – shares of individual material fractions in the total mass of waste,
n – the number of material fractions.

Based on the chemical formulas of the fractions, the content of carbon, hydrogen and oxygen were estimated (Table 5). For textiles a simplifying assumption was taken that they are composed of polyester, and the fraction of <10 mm was described in the chemical formula $(C_6H_{10}O_5)_n$. Table 6 shows the shares of carbon, hydrogen and oxygen throughout the waste in the dry state and the operational state calculated according to the equation (3) and (4) taking into account the data given in Table 5. The calculated heat of combustion is:

$$Q_s = 32800 \cdot 0.245 + 120040 \left(0.030 - \frac{0.201}{8}\right) =$$
$$= 8621.2 \ kJ/kg$$

The calorific value of the waste with calculated moisture 30.66% (Table 4) is:

$$Q_i = 8147.7 - 2442(0.3066 + 8.94 \cdot 0.03) =$$
$$= 7217.5 \ kJ/kg$$

The amount of obtainable heat energy E_t from waste can be determined from the relationship:

$$E_t = M\eta u Q_i \qquad (5)$$

where: M – waste mass,
η – efficiency of electricity production in cogeneration,
u – share of thermal energy in total energy production in cogeneration.

The projected amount of thermal energy that may be acquired from the waste stream with a mass of 484997 Mg/year (Table 3) and the calorific value of 7217.5 kJ/kg and assuming $\eta = 85\%$ and $u = 75\%$ is 2231.5 TJ/year.

Table 5. Content of carbon (C^s), hydrogen (H^s) and oxygen (O^s) in the fractions of waste in the dry state

Waste fraction	Chemical formula	C_i^s	H_i^s	O_i^s
Paper and cardboard	$(C_6H_{10}O_5)_n$	0.444	0.062	0.494
PE	$(C_2H_4)_n$	0.857	0.143	0
PP	$(C_3H_6)_n$	0.857	0.143	0
PS	$(C_8H_8)_n$	0.923	0.077	0
PET	$(C_{10}H_8O_4)_n$	0.625	0.042	0.333
Textiles	$(C_2O_3R)_n$	0.333	0	0.667
Wood	$(C_6H_{10}O_5)_n$	0.444	0.062	0.494
Multi-material waste	$(C_6H_{10}O_5)_n$	0.444	0.062	0.494
Kitchen and garden waste	$(C_6H_{10}O_5)_n$	0.444	0.062	0.494
Fraction < 10 mm	$(C_6H_{10}O_5)_n$	0.444	0.062	0.494

Table 6. The share of carbon, hydrogen and oxygen in the waste

Chemical element	Share	
	dry state	operational state
C	0.381	0.245
H	0.049	0.030
O	0.352	0.201

Table 8. The share of carbon, hydrogen and oxygen in the waste calculations for the second variant

Chemical element	Share	
	dry state	operational state
C	0.376	0.218
H	0.049	0.027
O	0.379	0.204

Simulation calculations of calorific value of waste with consideration to legal requirements

For visualizing the effect of reduced share of the following fractions in waste: paper and cardboard, metal, glass, plastic, on the calorific value, the calculations (variant 2) assuming that 50% of each of these fractions is removed from the waste. Consequently, (compared to the previous calculation) the mass of these fractions in waste destined for processing will be reduced by half. Moisture content of combustible and non-combustible components of waste mass (after removal of the four waste fractions) are summarized in Table 7.

The shares of carbon, hydrogen and oxygen in the waste in the dry state and the actual state (operating) are given in Table 8.

The calorific value of the waste with moisture of 35.49% (Table 7), estimated from equation (1) is: $Q_i = 5874.34$ kJ/kg. To ensure autothermic-

ity of waste combustion with the above calorific values it will be necessary to improve the energy characteristics of waste. Improved characteristics of waste fuel can be obtained by adding external fuel e.g. alternative fuel. Such an action will increase the operating costs of waste incineration plants. The results for two variants are presented in Table 9.

An alternative method to the thermal process of processing mixed waste is the mechanical-biological treatment (MBP). The specific technical and technological solutions of the process depend on the intended use of waste fractions separated in the process. In further considerations it is assumed that as a result of mechanical processing alternative fuel will be produced, the fractions are inorganic and organic while the organic fraction is intended for composting. The calculations (3rd variant) of the calorific value of waste fraction (i.e. combustible fraction) were made for a potential use as an alternative fuel. In this vari-

Table 7. Characteristics of waste for the second variant of calculations

Group	Share [%]	Moisture [%]	Combustible components [%]	Non-combustible components [%]
I	42.40	6.92*	41.55*	51.52*
II	44.26	71.59	21.84	6.57
III	13.34	6.47	76.93	16.6
Total group	100	35.49*	37.55*	26.97*

* – Weighted average.

Table 9. The results of calculations

Item	2012	Reducing the amount of materials by 50%
Amount of recyclable waste [Mg/year]	484 997	399 988
Paper and cardboard [Mg/year]	54 069.88	27 034.94
Glass [Mg/year]	43 110.54	21 555.27
Metal [Mg/year]	11 468.19	5734.10
Plastics [Mg/year]	61 367.72	30 683.86
Moisture [%]	30.66	35.49
Combustible components [%]	40.49	37.55
Non-combustible components [%]	28.85	26.97
Calorific value [MJ/kg]	7.217	5.874
Heat energy from waste combustion [TJ/year]	2231.5	1497.9

Table 10. Summary of the results of calculations of the calorific value of the combustible fraction of waste

Item		Value
Stream of waste mass to be processed in MBP installation [Mg/year]		399 988
Combustible fraction		
Amount in the stream of waste [Mg/year]		91 110.23
Share in mass of waste to be processed [%]		22.78
Moisture [%]		12.88
Combustible components [%]		77.34
Non-combustible components [%]		9.78
C [%]	Dry state	0.525
	Operational state	0.467
H [%]	Dry state	0.058
	Operational state	0.05
O [%]	Dry state	0.416
	Operational state	0.352
Calorific value [MJ/kg]		14.747

ant, the calculation assumes that the waste from group II and group III (Table 3) are separated in a mechanical process. The combustible fraction will be constituted from mechanically separated waste classified as Group I: paper and cardboard, plastics, textiles, wood, multi-material packaging waste. The combustible mass fraction was determined by subtracting the mass of glass, metal, hazardous waste, mineral waste and fractions waste: other categories of waste from the mass of the I group waste groups in the second variant. The results of calculations are given in Table 10. The estimated calorific value of combustible waste fraction is 14747.74 kJ/kg.

SUMMARY

The amendment to the Polish law in the area of municipal waste management requires increasing the level of recycling and reprocessing waste useable fractions. The fulfilment of legal requirements will affect the properties of mixed waste to be processed.

In the case of Lubelskie Voivodeship segregation of 50% of the waste feedstock, i.e. 17.52% of waste (included in the categories I – III; Table 3) generated in 2012 will reduce the stream of waste for thermal treatment, which should be taken into account in the design of the intended incineration plant.

The decrease of each fraction by 50% of weight will lower the calorific value of the waste from 7.217 MJ/kg to 5.874 MJ/kg. As a conse-

quence, the amount of energy obtainable from the incineration of waste will reduce, which may affect the profitability of the planned project – the construction of a waste incineration plant.

To ensure autothermicity of the incineration process it will probably be necessary to use additional fuel – formed fuel or fossil fuels. Such actions will have an adverse impact on economic profitability of the waste incineration plant. This may affect the price of collecting waste to such installations.

An alternative to incineration of the mixed waste combustion processing is the mechanical-biological treatment. The calculated calorific value of the combustible fraction of the waste obtained after eliminating 50% of fractions is 14.747 MJ/kg. Alternative fuel from the combustible fraction of waste is currently used in the cement industry, and its further use in the energy sector is now debated.

REFERENCES

1. Central Statistical Office. Environment 2013. Municipal waste in 2012. ZWS Warszawa.

2. Chudzinski J. et al. 1974. Poradnik termoenergetyka. Praca zbiorowa – 2nd edition, revised and updated. WNT Warszawa.

3. Czop M. 2013. Badania podstawowych właściwości paliwowych odpadów poliolefinowych. Archiwum Gospodarki Odpadami i Ochrony Środowiska, 15(3), 71–80.

4. Jaglarz G. 2014. Symulacja zmian parametrów energetycznych odpadów komunalnych w wyniku

budowy systemu gospodarki odpadami w nowych ramach prawnych. Archiwum Gospodarki Odpadami i Ochrony Srodowiska, 16, 11–20.

5. Marczak H. 2013. Ecological and economically optimal management of waste from healthcare facilities. Journal of Ecological Engineering 14(2), 43–48.

6. Resolution of the Polish Parliament Chairman of 13 September 2013 on unified version of the *Bill on cleanness in municipalities* Dz.U. 2013 nr 0 poz. 1399.

7. Skowron H. 2006. Optymalne rozwiązania w dziedzinie termicznego przekształcania odpadów komunalnych. Katowice; www.ietu.katowice/aktual/ Debata_spoleczna/ prezentacje/Termiczne_przeksztalcanie_odpadow_optymalne_rozwiązania.pdf

8. Szyszkowski P., Kobiela K., Moczulski M. at al. 2012. Plan gospodarki odpadami dla województwa lubelskiego 2017. Załącznik do uchwały nr XXIV/396/2012 Sejmiku Województwa Lubelskiego z dnia 30 lipca 2012.

22

EFFECT OF DIFFERENT METHODS OF TREATMENT OF MUNICIPAL SEWAGE SLUDGE ON THEIR PHYSICOCHEMICAL PROPERTIES AND THEIR AGRICULTURAL UTYLIZATION

Elżbieta Malinowska[1], Kazimierz Jankowski[1], Beata Wiśniewska-Kadżajan[1], Jacek Sosnowski[1], Roman Kolczarek[1]

[1] Institute of Agronomy, Siedlce University of Natural Sciences and Humanities, Prusa, Siedlce, Poland, e-mail: malinowskae@uph.edu.pl

ABSTRACT

In the study the physicochemical properties of selected municipal sewage sludge were compared using the reports on waste generation from the years 2007–2012 in the Mazovia voivodship. The selection was done on the basis of different methods of sludge processing and the number of equivalent inhabitants (NEJ) supported by sewage. Physical and chemical properties of municipal sewage sludge were significantly dependent on the method of purification and treatment methods. These sludges were characterized by a high content of organic matter and macronutrients. The amount of heavy metals (Pb, Cd, Cr, Cu, Ni, Hg and Zn) were within acceptable standards for municipal sewage sludge used in agriculture. Municipal sewage sludge treated by biological method with higher nutrients removal did not create bacteriological danger and were used in agriculture, generally to the cultivation of all agricultural products. Sewage sludge sanitized with lime and subjected to anaerobic digestion did not meet bacteriological standards, which eliminated their use in agriculture.

Keywords: municipal sewage sludge, physicochemical properties, agricultural utilization.

INTRODUCTION

The chemical composition of sewage sludge is very varied, dependent on the season of year, the city's infrastructure and the quantity and quality of effluents discharged by the industry to municipal wastewater treatment plants. Processes in sewage directly or indirectly affect the sediment, changing its properties. The processes selection of sludge treatment has a significant influence on the subsequent their utilization [Simonetti et al. 2014]. Most often municipal wastewater in treatment plants are mechanically and biologically treated. In this process significant amounts of soluble organic substances, difficulty sedimenting suspensions and colloids, reducing the amount of viruses and bacteria are removed from the waste water. From the inorganic substances only those absorbed by microorganisms the nitrogen and phosphorus compounds are removed. The stream of municipal sewage sludge in Poland are increasing rapidly, as is the effect of dynamically running modernization and the introduction of more advanced wastewater treatment technology or the construction of new wastewater treatment [Stypka, Flaga-Maryańczyk 2013]. The approach to the sludge from the technology site of little or no waste is at the moment one of the most important aspects [Leszczyński, Brzychczyk 2007, Pawłowski et al. 2000]. Municipal sewage sludge, in accordance with the legislation in force in Poland can be used for fertilizer, as evidenced by the operating law regulations, including the Act on Waste [2012] and the Regulation of the Minister of Environmental Protection [2002, 2010]. The suitability of sewage sludge to agricultural utilization is determined bysuch factors as: the content of organic matter, nutrients for plants, content of permanently damaging (undegradable and degradable very difficult) to the environment, the presence of pathogenic organisms, consistency [Simonetti et al. 2014].

The aim of the study was to evaluate the physicochemical properties of municipal sewage sludge treated by various methods from selected wastewater treatment plants with different throughput mass of Mazovia voivodship.

MATERIAL AND METHODS

The paper presents the results of studies from the reports on waste generation in the years 2007–2012 in the Mazovia voivodship [Report of the production of waste]. Of the 598 municipal wastewater and industrial treatment plants exploited in Mazovia - as in 2012 [http://wios. warszawa.pl], four municipal sewage treatment plants using different methods of treatment were selected. On the basis of the wastewater load the treatment plant was expressed by a number of equivalent inhabitants (NEJ) tested treatments were classified to II, III, IV and V group. Group I – treatment support up to 2 thousand of NEJ, II – 2–10 thousand of NEJ, III – 10–50 thousand of NEJ, IV – 50–100 thousand of NEJ V - over 100 thousand of NEJ. Most of the selected wastewater treatment uses biological method of sludge treatment. Administrative policy of the community and the activities of exploiter can significantly affect the quality of the produced sludge, as well as its quantity. Two of the selected treatment plants belonging to II and III group purified and wastewater by biological method with increased nutrient removal (MB). Wastewater treatment by this method allows an increased reduction of nitrogen and phosphorus. Sludge from wastewater treatment plants belonging to the second group is subjected to further composting (MB-K). This process ensures the stabilization of organic compounds, natural disinfection, reducing the weight of sludge and production of a stable final product. In the largest wastewater treatment plants (group V) in Warsaw sewage sludge is disposed in the anaerobic process (fermentation) (MB-F). In the next wastewater treatment (group IV) chemical method for treatment of waste water is used by the addition of quicklime (MCh-CaO). The estimation of sewage sludge includes such parameters as pH, content of viable eggs of parasites, the dry matter, organic matter content, macronutrients, except for potassium, and the content of heavy metals. Reports

show that the sediments produced by the selected treatment in Mazovia province were not always clean in bacteriological terms (Table 1).

Table 1. Number of live eggs of parasites in municipal sewage sludge treated by various methods in Mazovia voivodship in the years 2007–2012

Sewage sludge	Year		
	2007	2009	2012
MB-K (G II)	0	0	0
MB (G III)	0	0	0
MCh-CaO (G IV)	0	198.0	0
MB-F (G V)	0	93.0	155.0

Data on the content of macronutrients and heavy metals were compared statistically using Statistica program, Version 10.0 StatSoft, for a detailed comparison of the mean values Tuke'y test at $p < 0.05$ were used.

RESULTS AND DISCUSSION

Physical and chemical properties of municipal sewage sludge are variable, depending primarily on the type of sewage, method of purification and processing methods [Maćkowiak 2000]. The analysis (table 2) shows that the lowest pH average 6.16 was characterized by sludge from sewage treatment plants applying biological treatment associated with composting, and almost twice higher pH value = 12 was recorded in the sludge from wastewater applying lime stabilization. Deacidifying activity of sewage sludge used in agriculture is confirmed by the research of many authors [Speir et al. 2003]. The dry matter content in sewage sludge ranged from 22.63 to 51.5%. The highest content of dry matter was characterized by sludge from a biological treatment plant with increased nutrient removal and the least sludge subjected to fermentation. Municipal sewage sludge contain diverse content of organic matter, which ranged from 16.7% to 51.7%. Most of organic matter was observed in the sediment subjected to composting. Maćkowiak [2000], based on eight years of research gives an average of 50.62% of organic matter in municipal sewage sludge and agri-food sewage in Polish. Sanitary characteristics of sewage sludge are variable and are based on many factors, among others, dependent on the quality of life and health status of residents in the area, the type of treated wastewater and process-

Table 2. Mass and selected parameters of municipal sewage sludge

Sewage sludge	Mass of sewage sludge (Mg / years)		Aim of application	pH			D.M. (%)			Organic material (% D.M.)		
	F.M.	D.M.		min.	max.	average	min.	max.	average	min.	max.	average
2007												
MB-K (G II)	756000	227000	1	5.30	6.80	5.80	21.5	32.0	26.0	41.3	57.8	51.7
MB (G III)	1309000	261880	3	8.27	8.07	7.90	49.9	64.3	51.5	13.3	17.9	16.7
MCh-CaO (G IV)	1843000	626900	5	11.2	11.7	11.4	28.2	38.9	34.0	17.0	26.5	20.5
MB-F (G V)	1329000	342000	5	8.50	12.0	11.0	16.5	33.8	22.6	38.1	65.5	53.4
2009												
MB-K (G II)	807000	242000	1	6.10	6.50	6.30	31.8	42.3	37.1	32.0	45.3	38.7
MB (G III)	1317000	263000	1	7.47	8.46	7.86	24.8	50.7	35.2	18.1	33.5	28.0
MCh-CaO (G IV)	3750000	1200000	5	12.0	12.5	12.2	27.8	36.6	32.0	23.3	40.5	31.8
MB-F (G V)	56810700	16313000	5	7.80	12.5	8.80	22.5	45.9	28.5	20.7	54.4	45.6
2012												
MB-K (G II)	696000	209000	1	5.70	6.76	6.37	35.1	37.8	36.7	43.2	45.8	44.6
MB (G III)	1217000	245000	1	7.00	8.25	7.82	15.3	32.9	21.6	29.0	64.4	44.9
MCh-CaO (G IV)	3160000	742000	5	12.5	12.5	12.5	18.5	28.1	23.5	36.2	50.9	43.6
MB-F (G V)	3731000	836750	5	10.1	13.2	11.9	20.4	27.2	23.3	23.4	42.7	35.0
Average of the years												
MB-K (G II)	753000	226000		5.70	6.69	6.16	29.5	37.4	33.3	38.8	49.6	45.0
MB (G III)	1281000	256629		7.58	8.26	7.86	30.0	49.3	36.1	20.1	38.6	29.9
MCh-CaO (G IV)	2917000	856300		11.9	12.2	12.0	24.8	34.5	29.8	25.5	39.3	31.9
MB-F (G V)	20623000	5830583		8.80	12.6	10.6	19.8	35.6	24.8	27.4	54.2	44.7

Table 3. Content of selected macroelements (g kg^{-1} D.M.) in municipal sewage sludge

Sewage sludge	N			P			Ca			Mg		
	min.	max.	average	min.	max.	average	min.	max.	average	min.	max.	average
2007												
MB-K (G II)	18.4	19.0	18.7	19.0	49.5	36.2	25.2	37.3	32.7	4.61	4.50	4.60
MB (G III)	48.0	84.0	70.0	8.50	9.70	9.10	1.20	20.5	10.1	0.60	3.60	2.30
MCh-CaO (G IV)	22.0	30.0	27.0	11.0	18.0	14.0	179.0	300.0	247.0	3.00	4.30	3.80
MB-F (G V)	26.0	51.0	39.3	21.0	37.5	29.8	45.2	100.7	76.4	5.10	8.26	6.50
2009												
MB-K (G II)	16.2	25.4	20.8	34.2	38.3	36.3	39.9	44.3	42.1	4.80	5.00	4.90
MB (G III)	11.1	15.7	13.3	8.50	17.3	12.4	8.60	31.2	17.4	2.70	6.50	4.90
MCh-CaO (G IV)	19.1	31.0	24.1	1.40	22.3	8.15	81.3	246.0	153.0	2.50	5.70	3.70
MB-F (G V)	11.3	40.0	30.3	13.2	27.2	21.8	9.80	19.0	14.6	3.30	5.30	4.20
2012												
MB-K (G II)	20.8	23.4	22.3	28.1	37.6	33.8	33.6	41.4	37.3	3.90	4.50	4.10
MB (G III)	30.2	42.6	36.5	25.3	51.1	41.6	38.7	59.5	46.3	6.90	24.4	16.0
MCh-CaO (G IV)	26.2	49.9	34.5	13.5	21.1	18.2	155.0	263.0	209.5	3.90	5.00	4.40
MB-F (G V)	20.3	33.0	28.2	9.80	19.0	14.6	150.0	239.0	191.3	4.78	5.93	5.47
Average of the years												
MB-K (G II)	18.5	22.6	20.6	27.1	41.8	35.4	32.9	41.0	37.4	4.44	4.67	4.53
MB (G III)	29.8	47.4	39.9	14.1	26.0	21.0	16.2	37.1	24.6	3.40	11.5	7.73
MCh-CaO (G IV)	22.4	36.9	28.5	8.63	20.5	13.5	138.4	269.7	203.2	3.13	5.00	3.97
MB-F (G V)	19.2	41.3	32.6	14.7	27.9	22.1	68.3	119.6	94.1	4.39	6.50	5.39
LSD$_{0.05}$ for: A–methods of sludge treatment	1.60	4.41	3.81	0.898	0.733	0.724	3.84	8.02	1.44	0.277	0.745	0.543
B –years	1.25	3.46	2.98	0.704	0.575	0.568	3.01	6.29	1.13	0.217	0.584	0.426
A/B- interaction	2.51	7.46	6.60	1.56	1.27	1.25	6.66	13.9	2.49	0.479	1.29	0.940
B/A - interaction	2.77	6.92	5.98	1.41	1.15	1.14	6.03	12.6	2.26	0.434	1.17	0.851

ing methods [Sahlström et al. 2004, Bagge et al. 2005, Walczak, Lalke-Porczyk 2009]. The data show that smaller wastewater treatment plants produce sludge bacteriological safe, which can be used in agriculture, in general, to the cultivation of all agricultural products (Table 1, 2).

The content of biogenic elements (essential in plant nutrition) in the presented sewage sludge was significantly varied and depends on the method of wastewater treatment (Table. 3). Sludges are very rich in nitrogen, phosphorus, magnesium. The amounts of these macronutrients are usually much higher than in manure [Rosik-Dulewska 2000, 2001, Skorbiłowicz 2002]. The nitrogen content ranged from 20.6 to 39.9 g kg[-1], with variations from 11.1 to 84 g kg[-1]. Phosphorus content was on average from 13.5 to 35.4 g kg[-1], magnesium from 3.97 to 7.73 g kg[-1]. Very large amounts of calcium in the sludge hygenized with CaO have been reported. The average content of the element was 203.2 g kg[-1], and in the remaining sewage sludges it was much lower, up to 5-fold. The contents of macronutrients in these

sludges are comparable with the results of many authors [Mazur 2000, Jakubus 2005, 2006, Malinowska, Kalembasa 2013].

In the sludge the main criterion for their eligibility for the natural use is the content of heavy metals [Czechowska-Kosacka 2007]. In these sludge did not exceed the permissible content of heavy metals [Regulation 2010] (Table 4, 5).

The average lead content in sewage sludge ranged in each year on average from 9.80 to 42.8 mg kg[-1] (Table 4). As follows from the data content of this metal in the sludges stabilized by different methods did not exceed 35 mg kg[-1] D.M. It was reported that sludges analyzed in 2012 contained much less lead than in previous years. Maćkowiak [2000] obtained similar content of lead in municipal sewage sludge in eight years of research. The average cadmium content ranged from 0.420 mg kg[-1] in sewage sludge stabilized with lime to 11.7 mg kg[-1] in sludges subjected to fermentation, from sewage treatment plant with the largest throughput mass. The chromium content in sewage sludge was significantly varied and

Table 4. Contents of selected heavy metals (mg kg[-1] D.M.) in municipal sewage sludge

Sewage sludge	Pb			Cd			Cr		
	min.	max.	average	min.	max.	average	min.	max.	average
	2007								
MB-K (G II)	25.0	53.0	42.8	1.80	2.37	2.03	209.0	392.0	293.0
MB (G III)	15.7	19.9	17.8	0.910	1.19	1.02	7.38	17.2	13.5
MCh-CaO (G IV)	6.30	50.4	23.7	0.840	2.00	1.58	0.100	46.6	28.1
MB-F (G V)	24.4	42.8	33.4	8.66	16.3	11.7	103.4	195.0	150.1
	2009								
MB-K (G II)	29.6	32.6	31.1	1.24	1.30	1.27	109.0	145.0	127.0
MB (G III)	13.6	48.8	30.5	0.990	2.79	1.81	7.17	15.8	12.8
MCh-CaO (G IV)	4.22	8.08	6.11	0.220	0.710	0.420	9.84	19.7	14.8
MB-F (G V)	21.0	71.9	43.7	0.574	4.13	1.99	74.8	153.0	112.0
	2012								
MB-K (G II)	23.2	29.2	27.2	0.990	1.16	1.09	272.0	498.0	388.0
MB (G III)	22.9	32.0	27.8	1.04	1.33	1.19	21.3	27.2	23.8
MCh-CaO (G IV)	6.68	12.7	9.80	0.720	1.55	1.05	13.6	23.1	18.7
MB-F (G V)	17.6	22.7	19.9	1.60	2.41	1.98	20.0	46.6	29.9
	Average of the years								
MB-K (G II)	25.9	38.3	33.7	1.34	1.61	1.46	196.7	345.0	269.3
MB (G III)	17.4	33.6	25.4	0.980	1.77	1.34	12.0	20.1	16.7
MCh-CaO (G IV)	5.73	23.7	13.2	0.593	1.42	1.02	7.85	29.8	20.5
MB-F (G V)	21.0	45.8	32.3	3.61	7.61	5.23	66.1	131.5	97.3
LSD[0.05] for: A – methods of sludge treatment B –years A/B- interaction B/A - interaction	0.823 0.645 1.43 1.29	1.38 1.09 2.40 2.18	0.770 0.604 1.33 1.21	0.064 0.050 0.111 0.101	0.422 0.331 0.732 0.663	0.295 0.232 0.511 0.463	1.66 1.31 2.88 2.61	3.16 2.48 5.48 4.96	2.93 2.33 5.15 4.66

Table 5. Contents of selected heavy metals (mg $\cdot$ kg^{-1} D.M.) in municipal sewage sludge

Sewge sludge	Cu			Ni			Hg			Zn		
	min.	max.	average	min.	max.	average	min.	max.	średnia	min.	max.	average
2007												
MB-K (G II)	124.0	176.0	144.3	23.5	32.3	27.6	0.880	1.28	1.04	900.0	1026.0	956.3
MB (G III)	53.4	67.3	59.9	7.40	62.7	26.6	0.210	0.420	0.330	395.0	493.0	444.3
MCh-CaO (G IV)	79.8	233.0	148.0	1.00	17.9	10.1	0.280	0.500	0.370	485.6	985.0	785.9
MB-F (G V)	201.2	274.0	244.9	23.4	53.8	34.5	3.50	4.21	3.85	1619.6	2567.0	2054.5
2009												
MB-K (G II)	122.0	133.0	127.5	20.4	31.6	26.0	0.630	0.730	0.680	763.0	818.0	790.5
MB (G III)	59.9	41.0	162.0	10.1	18.6	13.0	0.400	3.66	1.50	256.0	852.0	606.7
MCh-CaO (G IV)	11.3	79.3	49.7	8.57	14.8	11.9	0.130	0.190	0.150	335.0	381.0	359.0
MB-F (G V)	141.0	361.0	256.0	37.1	91.8	66.5	0.170	1.74	0.780	521.0	1212.0	973.0
2012												
MB-K (G II)	112.0	128.0	121.0	28.0	33.5	30.0	0.360	0.660	0.520	652.0	877.0	777.3
MB (G III)	159.0	270.0	222.3	14.2	16.5	15.6	0.140	0.660	0.450	823.0	1092.0	956.7
MCh-CaO (G IV)	97.3	258.0	155.3	5.39	10.6	8.80	0.160	1.51	0.540	589.0	819.0	658.0
MB-F (G V)	89.1	134.0	113.7	10.5	13.3	11.7	0.690	1.06	0.890	569.0	1070.0	783.0
Average of the years												
MB-K (G II)	119.3	145.7	130.9	23.9	32.5	27.9	0.623	0.890	0.746	771.7	907.0	841.4
MB (G III)	90.8	126.1	148.1	10.6	32.6	18.4	0.250	1.58	0.760	491.3	812.3	669.2
MCh-CaO (G IV)	62.8	190.1	117.7	4.99	14.4	10.3	0.190	0.733	0.353	469.9	728.3	601.0
MB-F (G V)	143.8	256.3	204.9	23.7	52.9	37.6	1.45	2.34	1.84	903.2	1616.3	1270.2
LSD$_{0.05}$ for: A – methods of sludge treatment	1.87	5.88	1.12	0.542	6.77	0.959	0.205	0.061	0.071	37.7	32.0	29.6
B –years	1.47	4.61	0.879	0.425	5.31	0.752	0.161	0.052	0.059	29.6	28.3	25.4
A/B- interaction	3.24	10.2	1.94	0.938	11.7	1.66	0.355	0.104	0.121	65.3	56.2	51.3
B/A -interaction	2.93	9.22	1.76	0.849	10.6	1.50	0.321	0.093	0.109	59.2	48.3	45.2

ranged on average from 12.8 to 388 mg kg^{-1}. The highest content of this element was observed in the biologically stabilized sludges and subjected to a process of composting.

The content of other heavy metals (copper, nickel, mercury and zinc) in the sludge was significantly dependent on the processing method (Table 5).

The copper content was on average from 49.7 to 244.9 mg kg^{-1} and was highest in sludge from wastewater plants applying fermentation, with the exception of 2012. A similar correlation was observed for the other heavy metals. The nickel content ranged from 8.80 to 66.5 mg kg^{-1}. The average content from six years did not exceed 40 mg Ni kg^{-1} D.M. of sludge and was more than 8-fold lower than that allowed for municipal sewage sludge for use in agriculture. Mercury is a highly toxic element which forms common environmental pollution [Gochefeld 2003, Wang et al. 2004]. It was observed decrease in the content of this metal in these sludge in the period considered. The average mercury content ranged from 0.330

to 1.50 mg kg^{-1} and Hg standard for sludge is 16 mg kg^{-1}. Toxicity of zinc is much lower than of mercury, cadmium or lead, the metal presents the highest toxic risk for biomass in the soil. The high content of this element in the sludge used in agriculture may pose a threat to soil microorganisms. The zinc content in these sewage sludges ranged from 359.0 to 2054 mg kg^{-1}, the average from six years did not exceed 1300 mg kg^{-1}. According to Wardas et al. [2002] the source of this metal in municipal sewage sludge can be galvanized water pipes, as well as large amounts of chlorine regulating the bacteriological state of water.

Based on the evaluated parameters was found that municipal sewage sludge produced by different methods in the Mazovia voivodship in the years 2007–2012 in terms of chemical composition are valuable waste, which can be used in agriculture for the cultivation of all agricultural products. Sanitary standards were only one parameter that limited agricultural use of sewage sludge stabilized chemically and biologically under anaerobic conditions.

CONCLUSIONS

- Physical and chemical properties of municipal sewage sludge were dependent on the method of purification and methods of their processing and load of sewage treatment plant.

- These sewage sludges contain a significant amount of organic matter and nutrients but heavy metals did not exceed acceptable standards for municipal sewage sludge used in agriculture.

- From the analysis of reports on waste generation in the years 2007–2012 in the Mazovia voivodship resulted that assessed municipal sludge treated biologically derived from small sewage treatment plants do not endanger bacteriological. Sewage sludge treated by other methods from wastewater treatment plants serving more than 50 thousand of equivalent number of residents did not meet sanitary standards.

REFERENCES

1. Bagge E., Sahlström L., Albihn A. 2005. The effect of hygienic on the microbial flora of bio waste at biogas plants. Water Research 39, 4879–4886.

2. Czechowska-Kosacka A. 2007. Influence of sewage sludge solidification on immobilisation of heavy metals. Polish J. Environ. Stud. 16, (2A), 625–628.

3. Gochefeld M. 2003. Cases of mercury exposure, bioavailability, and absorbtion. Ecotoxicology and Environmental Safety 56, 174.

4. http://wios.warszawa.pl

5. Jakubus M. 2005. Sewage sludge characteristics with regard to their agricultural and reclamation usefulness, Fol. Univ. Agric. Stetin. 244, Agricultura 99, 73–82.

6. Jakubus M. 2006. Wpływ wieloletniego stosowania osadu ściekowego na zmiany wybranych właściwości chemicznych gleby. Zesz. Probl. Post. Nauk Rol. 512, 209–219.

7. Leszczyński S., Brzychczyk B. 2007. Thermal utilization of sewage sludge and municipal organic waste towards hydrogen production. Pol. J. Environ. Studies 16, 3B, 290–294.

8. Maćkowiak Cz. 2000. Skład chemiczny osadów ściekowych i odpadów przemysłu spożywczego o znaczeniu nawozowym. Nawozy i Nawożenie R II, 3(4) 3a, 131-149.

9. Malinowska E., Kalembasa D. 2013. Contents of some selected elements in *Miscanthus sacchariflorus* (*Maxim.*) Hack biomass under the influence of sewage sludge fertilization in cultivation experiment. Ecol. Chem. Eng. A 20(2), 203–211.

10. Mazur T. 2000. Rolnicza utylizacja stałych odpadów organicznych. Zesz. Probl. Post. Nauk Rol. 472, 507–516.

11. Pawłowski L., Kotowski M., Kotowska U., Czechowska A. 2000. Utylization of sewage sludge in cement kilns. Kluwer Academic/Pulisher, Environ. Sci. Research, 58, 41–54, New York.

12. Raport dotyczący wytwarzania odpadów z lat 2007-2012 w województwie mazowieckim [online: dostęp dn. 4.11.2014, http://www.mazowia.pl]

13. Rosik-Dulewska Cz. 2000. Podstawy gospodarki odpadami. PWN, Warszawa, ss. 305.

14. Rosik-Dulewska Cz. 2001. Zawartość składników nawozowych oraz metali ciężkich i ich frakcji w kompostach z odpadów komunalnych. Zesz. Probl. Post. Nauk Rol. 477, 467–477.

15. Rozporządzenie Ministra Ochrony Środowiska w sprawie komunalnych osadów ściekowych z dnia 13.07.2010 r.(Dz.U. Nr 137, poz. 924).

16. Rozporządzenie Ministra Ochrony Środowiska w sprawie standardów jakości gleby oraz standardów jakości ziemi z dnia 9.09.2002 r. (Dz.U. Nr 165, poz. 1359).

17. Sahlström L., Aspan A., Bagge E., Danielsson-Tham M.L., Albihn A. 2004. Bacterial pathogen incidences in sludge from swedish sewage treatment plant. Wat. Res. 38, 1989–1994.

18. Simonetti M., Rossi G., Cabbai V., Goi D. 2014. Tests on the effect of ultrasonic treatment on two different activated sludge waste. Environ. Prot. Eng. 40 (1), 23–34.

19. Skorbiłowicz M. 2002. Charakterystyka osadów ściekowych z wybranych oczyszczalni z województwa podlaskiego pod względem zawartości metali ciężkich. Acta Agroph. 73, 277–283.

20. Speir T.W., Schaik A.P. Van, Percival H.J., Close M.E., Pang L.P. 2003. Heavy metals in soil, plants and grundwater following high-rate sewage sludge application to land. Water, Air and Soil Pollution 150 (1-4), 319–358.

21. Stypka T., Flaga-Maryańczyk A. 2013. Comparative analysis of municipal solid waste systems: Cracow case study. Environ. Prot. Eng. 39 (4), 135–153.

22. Ustawa o odpadach z dnia 14 grudnia 2012 roku, rozdz. 4. Komunalne osady ściekowe, art. 96.

23. Walczak M., Lalke-Porczyk E. 2009. Occurrence of bacteria Salmonella sp. in sewage sludge used in agriculture. Environ. Prot. Eng. 35 (4), 5–12.

24. Wang Q., Kim D., Dionysion D.D., Sorial G.A., Timberlake D. 2004. Sources and remediation for mercury contamination in aquatic systems – a literature review, J. Environ. Pollut. 131, 323.

25. Wardas M., Pawlikowski M., Gurda M., Idzik M., Jagła A., Janas J., Morusek J. 2002. Cynk w osadach gospodarki wodno-ściekowej miasta Krakowa. [W:] Cynk w środowisku – problemy ekologiczne i metodyczne. Zesz. Nauk. Komit. „Człowiek i Środowisko", PAN, Warszawa 33, 253–262.

23

RISK ASSESSMENT OF SURFACE WATERS ASSOCIATED WITH WATER CIRCULATION TECHNOLOGIES ON TROUT FARMS

Marcin Sidoruk[1], Józef Koc[1], Ireneusz Cymes[1], Małgorzata Rafałowska[1]
Andrzej Rochwerger[1], Katarzyna Sobczyńska-Wójcik[1], Krystyna A. Skibniewska[2]
Ewa Siemianowska[2], Janusz Guziur[3], Józef Szarek[4]

[1] Department of Land Reclamation and Environmental Management, Faculty of Environmental Management and Agriculture, University of Warmia and Mazury in Olsztyn, Plac Łódzki 2, 10-756 Olsztyn, Poland, e-mail: marcin.sidoruk@uwm.edu.pl

[2] Department of Security Rudiments, Faculty of Technical Sciences, University of Warmia and Mazury in Olsztyn, Heweliusza 10, 10-719 Olsztyn, Poland

[3] Department of Fish Biology and Breeding. Faculty of Environmental Sciences, University of Warmia and Mazury in Olsztyn, Oczapowskiego 14, 10-719 Olsztyn, Poland

[4] Department of Pathophysiology, Forensic Veterinary Medicine and Administration, Faculty of Veterinary Medicine, University of Warmia and Mazury in Olsztyn, Oczapowskiego 13, 10-719 Olsztyn, Poland

ABSTRACT

Dynamic development of aquaculture has led to an increasing impact on the status of surface waters. Fish production generates wastes that, at high concentrations, may present a serious risk to the aquatic environment. Studies on the assessment of the impact of water management technologies in trout production on the quality of surface waters were conducted in 2011. Six farms were selected for the studies and were divided into two groups based on water management solutions (n = 3): farms with a flow through system (FTS) and farms with a recirculation aquaculture system (RAS). On all farms, water measurement points were set and they depicted the quality of inflow water, the quality of water in ponds and the quality of outflow water. The studies did not demonstrate any impact of applied technology on electrolyte conductivity or calcium and magnesium concentrations in outflow water from a trout operation. In addition, it was found that the use of water for production purposes resulted in a slight increase in phosphorus and total nitrogen concentrations in waste waters.

Keywords: trout production, physical and chemical water properties, water contaminations.

INTRODUCTION

Cold, high quality inflow water which is usually taken from neighbouring rivers and streams or efficient sources is essential for maintaining a high level of trout production. The dynamic development of aquaculture has caused an increasing impact on the quality of surface waters. Fish production generates wastes (uneaten feed, excrements, chemical compounds and medicinal products supplied to ponds) that at high concentrations may present a serious risk to the aquatic environment [Koc, Sidoruk, 2013, Mehmet, Hüseyin, 2014]. Water contamination in fish production ponds influences the growth of microorganisms, including those harmful to fish and organisms inhabiting reservoirs where the water from fish ponds is transferred.

Pond management requires that breeding or rearing technologies combine the potential and economic production efficiency and determine the impact of these activities on the environment. Therefore, it is important to identify hazards and conflict situations found in the environment and caused by fish production in ponds [Prądzyńska, 2004; Sidoruk, 2012].

The objective of the studies was to determine the impact of water management in trout production on the quality of surface waters.

MATERIALS AND METHODS

The studies that consisted in assessing the impact of water management technology in trout production on the quality of surface waters were conducted in 2011. Six farms were selected and divided into two groups based on an applied water management solution, n = 3. One group consisted of farms with a flow through system (FTS) while the other group included farms with a recirculation aquaculture system (RAS).

On all farms, water measurement points were established and depicted the inflow water quality, the pond water quality and the outflow water quality. On the selected points, the analyses of water (in situ) were performed in spring (in April and May) and in autumn (in October and November) at monthly intervals. With a YSI 6600 multi-parametric probe designed for testing physical water parameters, oxygen saturation (%) and electrolytic conductivity (μS/cm) were determined. Water samples were taken for laboratory analyses from the same locations. The averaged water samples were collected into 5 dm^3 containers made of polyethylene, fixed and transported to a laboratory. The following measurements were taken: BOD$_5$ – with the respirometric method on an OXI-Top apparatus; nitrate nitrogen (V) – N-NO$_3$ colometrically with disulfanilic acid; nitrate nitrogen (III) – N-NO$_2$ colometrically with sulfanilic acid; ammonium nitrogen N-NH$_4$ – colometrically with a Nessler's reagent; Kiejdahl's nitrogen, N$_{kj}$ – with distillation after mineralization in a sulphuric acid; total phosphorus (P$_{og}$) – colometrically after mineralization with ammonium molibdenate and tin dichloride as a redactor; and Mg^{2+} with the colorimetric method with titanium yellow and Ca^{2+} with atomic mass spectrophotometry.

RESULTS AND DISCUSSION

The concentration of oxygen originating from phytoplankton photosynthesis in a pond ecosystem depends on a variety of factors, such as temperature, water transparency, solar radiation and content of nutrients. It may constitute up to 80% of the oxygen input in a pond ecosystem

[Jawecki, Krzemińska 2008]. Deviations in the content of oxygen are proportional to the fertility of a fish pond. The concentration of oxygen in the water is influenced by such factors as water temperature and transparency, content of nutrients and sunlight. It is also significantly impacted by the individual characteristics of ponds and procedures to which they are subjected [Jawecki 2008].

Inflow waters that supplied rearing ponds were very well-oxygenated with the concentration of dissolved oxygen ranging from 60.4% to 127.7%, of which 50% of the results were within the 74.1–101.1% range (Figure 1). On the FTS fish farms, there was an increase in the oxygenation level of outflow water by approx. 10.3% whereas on the RAS farms, the concentration of oxygen in water was reduced by approx. 29%. In the ponds with water recirculation, despite intensive trout production, there was a minor reduction of the oxygen content in outflow water, which was generated with artificial water aeration (the use of mechanical aerators or pure oxygen) to create optimal conditions for fish.

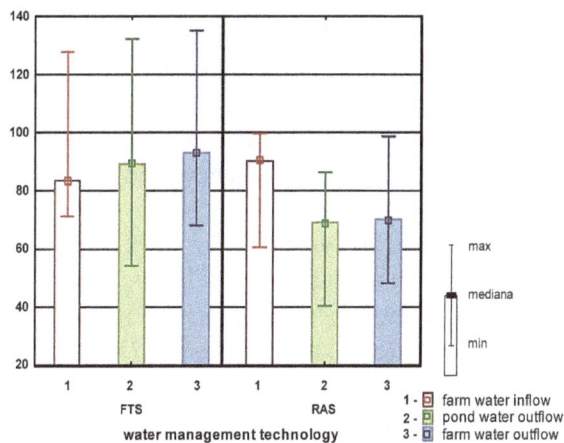

Figure 1. Oxygen saturation of inflow and outflow waters on trout production facilities (%)

Electrolytic conductivity of water corresponds to its content of mineral contaminations, a higher level of contamination was correlated with higher conductivity. Based on the value of electrolytic conductivity, it is possible to determine the degree of water salinity and the content of solutes and dry residues [Macioszczyk, Dobrzyński 2002].

Based on an analysis of electrolytic conductivity in inflow waters it was found that the median of its values was comparable on all farms and ranged between 352.5 μS/cm on the RAS farms and 372.0 μS/cm on the FTS farms (Figure 2). The maximum values of electrolytic conductivity in inflow waters on the FTS facilities did not exceed 431.0 µS/cm, whereas for the RAS farms this parameter did not exceed 402 µS/cm and, in both cases, these values were typical of flowing surface waters.

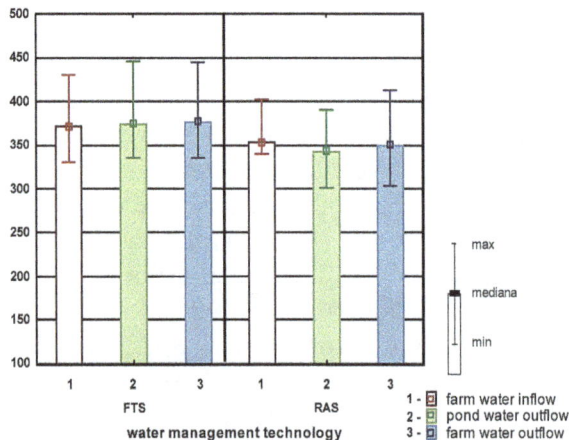

Figure 2. Electrolytic conductivity in inflow and outflow waters on trout production facilities (µS/cm)

The analysis of the results did not demonstrate any impact of applied technology on the values of electrolytic conductivity in outflow waters on trout production facilities. By assessing the changes in electrolytic in waste water outflowing from the FTS and RAS farms, it can be concluded that in the first group there was an increase in electrolytic conductivity by 5 µS/cm, while in the other group it was reduced by 2.5 µS/cm.

The water used to supply the trout production operations had a relatively low BOD_5 value and its median in the experimental period ranged from 2.07 mg/dm³ (FTS) to 2.18 mg/dm³ (RAS), of which 50% of the results with flow-through systems were in the 1.40–2.46 mg/dm³ range and in the case of RAS farms – in 1.46–4.24 mg/dm³ (Figure 3). Periodically, BOD_5 slightly exceeded the reference values for inland waters that are the habitat for salmonids (the Regulation by the Minister of Environment of October 4, 2002, Dz. U. Nr 176). Such a situation could have been caused by an increased inflow of contaminations, e.g. as a result of wash-out from fields or uncontrolled contamination of waters located above the farms.

The analysis of the results demonstrated the impact of applied water management technology on the increase of BOD_5 in outflow waters from tour production operations. By analyzing the changes in BOD_5 in the individual types of water management it was found that in post-production water outflowing from the FTS farms

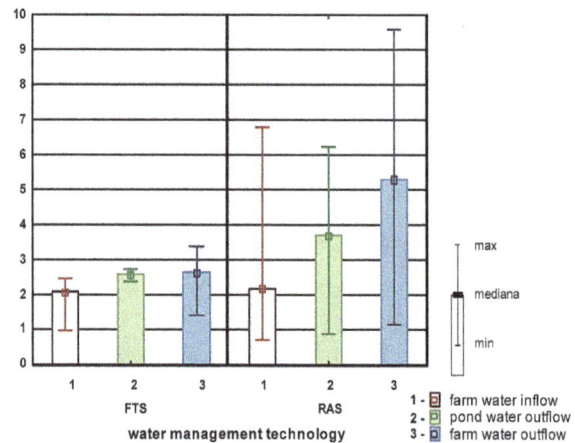

Figure 3. BOD_5 values in inflow and outflow water on trout production facilities (mg/dm³)

there was an increase in BOD_5 by approx. 0.56 mg/dm³, whereas on the RAS farms this increase was more significant amounting to, on average, 3.11 mg/dm³.

In pond water, phosphorus is found as phosphate ions, being a product of orthophosphoric acid dissociation and as dissolved organic phosphorus. In aquatic ecosystems, its main function is to regulate biological production as the basic biogene for the synthesis of organic compounds, thereby influencing the fertility of waters [Raczyńska, Machula, 2006, Sindilariu et al. 2009].

In inflow waters that supplied the farms, phosphorus was mainly found in an organic form and constituted on average 57% P_{og} at inflow on the FTS farms and 52% at inflow on the RAS farms. The median of total phosphorus concentration in inflow waters that supplied the trout production operations was comparable for both groups of farms: 0.07 mg/dm³ ranging from 0.01–0.18 mg/dm³ on the FTS farms and at 0.09 mg/dm³ (0.04 – 0.16 mg/dm³) on the RAS farms (Figure 4). After using water for rearing purposes, there was a slight increase in the concentration of P_{og} in the outflow of the examined operations. By analyzing the increments of phosphorus concentration of both farms groups, it was found that the increase of P_{og} in waste waters was smaller on the FTS farms (on average by 0.03 mg/dm³) than on the RAS farms (0.06 mg/dm³).

The slight increase in the concentration of phosphorus compounds in outflow waters could have been caused by that fact that fishery production is targeted at intensive weight gains, which results in intensive fish feeding. It is estimated that only a minor part of the compounds found in feed is incorporated into fish biomass, with the majority

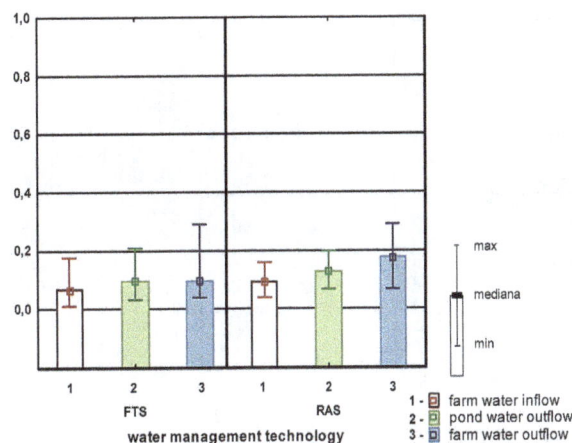

Figure 4. Content of P_{og} in inflow and outflow waters on trout production facilities (mg/dm^3)

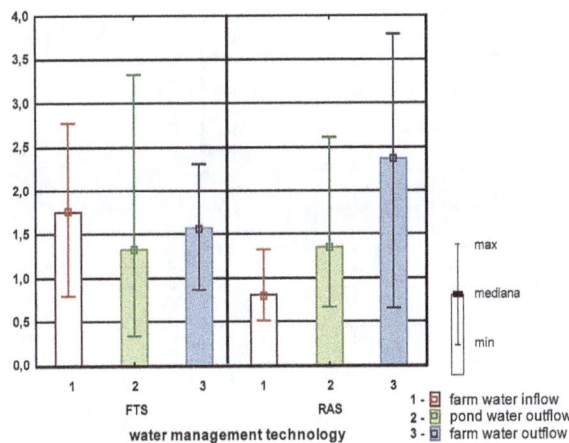

Figure 5. Content of N_{og} in inflow and outflow waters on trout production facilities (mg/dm^3)

being left in water or excreted by fish (Pulatsu et al. 2004; Tucholski, Sidoruk, 2013). It is estimated that of the feed put into fish ponds, only 5–20% of the matter is incorporated in the fish body and the rest is left in water contributing to its contamination [Goryczko 1999; Sidoruk et al., 2013].

Nitrogen accesses water mainly as mineral compounds originating from of organic nitrogen compound decomposition processes, precipitation and soils. In water, it is found in forms with a different degree of oxidation, in organic and inorganic complexes and as free dissolved nitrogen. Microbiological conversions of nitrogen in the aquatic environment are analogous as in soil. In water, organic nitrogen is most commonly found as protein, amino acids and non-protein organic compounds, i.e. urea, amines, pyridines and purine. In natural waters, it is produced from dead animal organisms and plants as well as feed residues [Koc, Sidoruk, 2013].

In inflow waters that supplied the FTS farms, nitrogen was mainly found in a mineral form that constituted 57% N_{og} whereas on the RAS farms there was an advantage of the organic form that amounted to 52% of total nitrogen. The median of total nitrogen concentration at inflow on the FTS farms was 1.76 mg/dm^3, ranging from 0.80–2.78 mg/dm^3. On the RAS farms, the median of N_{og} in inflow waters was lower, amounting to 0.80 mg/dm^3 in the range of 0.51–1.33 mg/dm^3 (Figure 5).

After using waters in trout production, it was found that there was a reduction of N_{og} concentration by 0.19 mg/dm^3 in waste waters on the FTS farms. The situation was opposite on the RAS farms where there was an increase of N_{og} by approx. 1.56 mg/dm^3, which did not generate

any unfavourable conditions for organisms in the waste water reservoirs.

Magnesium compounds in water mainly originate from the process of dissolving minerals such as dolomites or lodestones. A change in the concentration of magnesium in surface waters is associated, among others, with humic compounds found in water. These substances may be present in a dissolved or colloidal form forming magnesium-humic complexes. The ability of humic substances to bind magnesium cations largely depends on reaction (pH) and, consequently, also on the degree of dissociation of functional groups [Wezel et al. 2013]. In general, a much higher concentration of calcium is detected in comparison to magnesium, most probably due to intensive uptake of this element by plants and its concentration in precipitation [Skorbiłowicz 2013].

The content of magnesium in inflow waters for both groups of farms was low and on the FTS farms the median was 7.5 mg/dm^3, ranging from 0.5–13.1 mg/dm^3 whereas in inflow waters on the RAS farms it was slightly higher and amounted on average to 8.6 mg/dm^3 ranging within 5.3–15.0 mg/dm^3 (Figure 6).

A slightly higher concentration of magnesium in inflow waters on the RAS farms could have been impacted by different geological and soil conditions over the areas of drainage basins of watercourses that supplied the operations. Based on the analysis of the results, no impact of applied technology on the concentration of magnesium in outflow waters was found. By assessing the changes of Mg^{2+} concentration in outflow water on the FTS farms, it became evident that there was a slight increase in magnesium concentration by approx. 2.6% (0.2 mg/dm^3) whereas on the

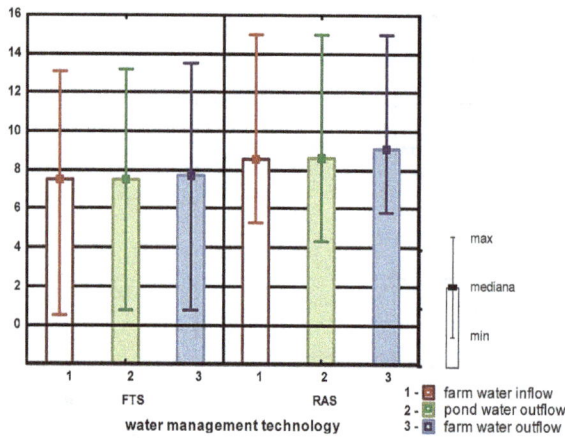

Figure 6. Content of Mg^{2+} in inflow and outflow waters on trout production facilities (mg/dm^3)

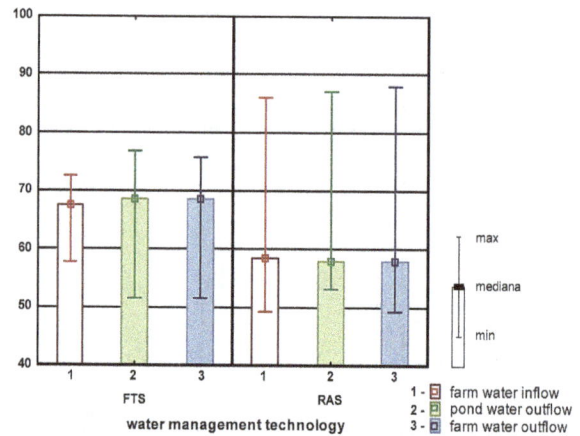

Figure 7. Content of Ca^{2+} in inflow and outflow waters on trout production facilities (mg/dm^3)

RAS farms the increase in magnesium concentration was slightly higher, amounting to approx. 0.55 mg/dm^3 (6.0%).

In surface waters, calcium is found as dissolved calcium carbonate and its content is determined by the presence of carbon dioxide in water [Sidoruk Skwierawski 2006; Degefu et al. 2011]. Calcium is an element which, together with magnesium, carbonates and sulphates, determines the hydrochemical type of most circulating waters in the drainage basins of the early post-glacial zone and its high content in surface waters mainly results from its intensive wash-out from soils [Sidoruk Skwierawski 2006]. In a moderate climate, it is intensively washed out of soil and this process is prompted by the acidification of precipitation. An appropriately high concentration of calcium in water is important due to its buffering capacity and for primary production, ensuring a sufficient concentration of CO_2 for photosynthesis [Degefu et al. 2011].

The results of the determination of calcium content in inflow waters demonstrated that its concentration was low and comparable for all operations: on the FTS farms it was 67.4 mg/dm^3 ranging from 57.8 to 72.6 mg/dm^3 whereas on the RAS farms it amounted to 58.4 mg/dm^3 (49.2–86.0 mg/dm^3). Throughout the entire experimental period, waters used to supply the ponds did not exceed the reference values for quality class No 1 (the Regulation by the Minister of Environment of November 9, 2011; Dziennik Ustaw Nr 257).

In waters used for breeding purposes, no impact of the activities on the content of calcium in water was observed. At the outflow from the FTS farms, there was a slight increase of Ca^{2+} concen-

tration by 1.1 mg/dm^3 whereas on the RAS farms a reduction of 0.5 mg/dm^3 was recorded.

CONCLUSIONS

1. The applied water management technology did not impact electrolytic conductivity in inflow waters on the trout production operations. On the farms with a flow-through system (FTS), there was a slight increase of electrolytic conductivity (by approx. 5 µS/cm in waste water), whereas on the farms with recirculation aquaculture systems (RAS) it was reduced by 2.5 µS/cm.

2. In outflow waters on the FTS farms, there was a slight increase in BOD_5 in relation to inflow waters (0.56 mg/dm^3), while on the 4 RAS farms the increase of BOD_5 was more pronounced – amounting to 3.11 mg/dm^3.

3. As a result of using water for rearing purposes, there was a minor increase in the concentration of phosphorus in outflow waters on both types of farms. In waste waters outflowing from the FTS farms, the increase of P_{og} was 0.03 mg/dm^3, whereas on the RAS farms it was slightly higher at 0.06 mg/dm^3.

4. The impact of the type of water management technology on the concentration of total nitrogen in post-rearing waters was recorded. On the FTS farms, there was a reduction of nitrogen in waste waters in relation to inflow waters by 0.19 mg/dm^3, while on the RAS farms, there was an increase in the concentration of N_{og} by approx. 1.56 mg/dm^3, which did not create unfavourable conditions for organisms inhabiting receiving water reservoirs.

Acknowledgements

The studies were funded by the EU and the Ministry of Agriculture and Rural Development within the framework of the Operational Programme "Sustainable development of fisheries and coastal fishing zones, "OP Fish 2007–2013", Contract No 00001-61724-OR1400002/10.

REFERENCES

1. Degefu F., Mengistu S., Schagerl M. 2011. Influence of fish cage farming on water quality and plankton in fish ponds: A case study in the Rift Valley and North Shoa reservoirs, Ethiopia. Aquaculture, 316, 129-135.

2. Goryczko K. 1999. Pstrągi, chów i hodowla, Poradnik hodowcy. p. 139.

3. Jawecki B. 2008. Dobowe ekstrema tlenowe w stawie rybnym. Zeszyty Problemowe Postępów Nauk Rolniczych, 528, 373-379.

4. Jawecki B., Krzemińska A., 2008. Wpływ temperatury wody na natlenienie strefy eutroficznej stawu karpiowego. Zeszyty Problemowe Postępów Nauk Rolniczych, 528, 381-387.

5. Koc J., Sidoruk M. 2013. Effect of trout aquaculture technology on quality of waters. W: Szarek J., Skibniewska K.A., Zakrzewski J., Guziur J., The quality of rainbow trout (*Oncorhynchus mykiss*, Walbaum 1792) from technologies applied in Poland, 101-120.

6. Longgen G., Zhongjie L. 2003. Effects of nitrogen and phosphorus from fish cage-culture on the communities of a shallow lake in middle Yangtze River basin of China. Aquaculture, 226, 201-212.

7. Macioszczyk A., Dobrzyński D. 2002. Hydrogeochemia strefy aktywnej wymiany wód podziemnych. Wydawnictwo Naukowe PWN, Warszawa, p. 448.

8. Mehmet A. T. K., Hüseyin S. 2014. Parameters selection for water quality index in the assessment of the environmental impacts of land-based trout farms. Ecological Indicators, 36, 672-681.

9. Prądzyńska D. 2004. Próba oceny oddziaływania stawów hodowlanych na środowisko przyrodnicze (na przykładzie gminy Malechowo). Studia Ekologiczno-Krajobrazowe w Prognozowaniu Rozwoju Zrównoważonego. Problemy Ekologii Krajobrazu, 13, 221-226.

10. Pulatsu S., Rad F., Köksal G., Aydın F., Benli A.C.K., Topçu A. 2004. The Impact of Rainbow Trout Farm Effluents on Water Quality of Karasu Stream, Turkey. Turkish Journal of Fisheries and Aquatic Sciences, 4, 09-15.

11. Raczyńska M., Machula S. 2006. Oddziaływanie stawów karpiowych na jakość wód rzeki Kąpiel (Pomorze Zachodnie). Infrastruktura i Ekologia Terenów Wiejskich, 4(2), PAN, Oddział w Krakowie, 141-149.

12. Sidoruk M. 2012. Wpływ chowu pstrąga w stawach ziemnych na właściwości fizyczne i chemiczne wód powierzchniowych. Inżynieria Ekologiczna, 31, 101-110.

13. Sidoruk M., Koc J., Szarek J., Skibniewska K., Guziur J., Zakrzewski J. 2013. Wpływ produkcji pstrąga w stawach betonowych z kaskadowym przepływem wody na właściwości fizyczne i chemiczne wód powierzchniowych. Inżynieria Ekologiczna, 34, 206-213.

14. Sidoruk M., Skwierawski A. 2006. Effect of land use on the calcium, sodium, potassium and magnesium contents in water flowing into the Bukwałd lake. Ecological Chemistry and Engineering, S1 vol. 13, 337-343.

15. Sindilariu P-D., Brinker A., Reiter R. 2009. Waste and particle management in a commercial, partially recirculating trout farm. Aquacultural Engineering, 41, 127-135.

16. Skorbiłowicz M. 2013. The sources of nutrients in waters of rivers in the wetland areas of Narew National Park in north-eastern Poland. Journal of Ecological Engineering, 14 (3), 1-7.

17. Tucholski S., Sidoruk M. 2013. The effect of feeding fishponds with biologically treated wastewater on pond water quality, Ecological Chemistry and Engineering A, 20(3), 391-399.

18. Wezel A., Robin J., Guerin M., Arthaud F., Vallod D. 2013. Management effects on water quality, sediments and fish production in extensive fish ponds in the Dombes region, France. Limnologica 43, 210-218.

24

WATER MANAGEMENT PROBLEMS AT THE BUKÓWKA DRINKING WATER RESERVOIR'S CROSS-BORDER BASIN AREA IN TERMS OF ITS ESTABLISHED FUNCTIONS

Mirosław Wiatkowski[1], Czesława Rosik-Dulewska[2]

[1] Institute of Environmental Engineering, Wrocław University of Environmental and Life Sciences, Plac Grunwaldzki 24, 50-363 Wrocław, Poland, e-mail: miroslaw.wiatkowski@up.wroc.pl

[2] Department of Land Protection, Opole University, Oleska Str. 22, 45-052 Opole, Poland, e-mail: czeslawa.rosik-dulewska@ipis.zabrze.pl

ABSTRACT

The paper covers the analysis of water management problems in the cross-border reservoir basin of Bukówka, located at the Bóbr river, at 271+540 km of its course, below the Czech-Polish border, in Dolnośląskie Voivodeship. The problems of water management in the context of the reservoir's functions have been analyzed; these are: flood control, the provision of water during low water level periods for the Water Treatment Plant in Marciszów, the provision of drinking water for the city of Wałbrzych as well as the provision of proper flow in the Bóbr river bed downstream from the reservoir. Due to its localization right below the border with the Czech Republic, the reservoir is exposed to a number of problems. The study has found that the main problems of water management in the basin area included unsatisfactory state of water and wastewater management in the basin, unsatisfactory state of the surface water quality in the basin area of the reservoir, poor condition of streams and drainage ditches, improper arrangement of arable lands and large downslopes and a lack of monitoring stations on tributaries of the reservoir. It has been found that the hydrochemical conditions in the Bukówka reservoir's section are unfavorable for it. From the eutrophication point of view, the water flowing into the tank is characterized by a large content of nutrients, especially nitrates, phosphates and BOD_5. In order to counteract eutrophication it is necessary to lower the concentration of nutrients in the water flowing into the tank. One of the basic ways to do so is to restore the water and wastewater management in the reservoir's basin. Studies in the Bukówka reservoir provide important information concerning the state of the purity of the water supplying the reservoir. In order to obtain accurate information on the state of purity, a monitoring of hydrological and water quality should be continued. The studies are a part of the strategy of protection of the quality of transboundary waters, proposed in the Convention on the Protection and Use of Transboundary Watercourses and International Lakes of 1992.

Keywords: drinking water reservoir, water management, cross-border basin area, water quality.

INTRODUCTION

Proper water management in retention reservoirs' basins is enormously important due to the fact that these reservoirs are places of accumulation of nutrients and various pollutants carried by a river. As a result of these processes the quality of water may deteriorate or its eutrophication and silting may occur [Balcerzak 2006, Czamara at al. 2008, Pawełek at al. 2014, Pütz, Paul 2008, Wiatkowski 2011]. These phenomena may often hinder the fulfillment of a reservoir's functions as well as reduce water utility and aesthetic values of reservoirs themselves [Kanownik at al. 2011, Wiatkowski et al. 2010]. Because of that, a reservoir's functions may not be fulfilled [Pütz,

Benndorf 1981; Wiatkowski, Paul 2009]. It is important to maintain water management of basins of multi-purpose reservoirs, including those containing drinking water, located at border areas, which due to their numerous functions are particularly vulnerable to pollutants from their basins. One of such multi-purpose reservoirs existing on the Odra is the Bukówka reservoir. It is located on the Bóbr river at 271+540 km, in Lower Silesia. Its basic functions are flood control, water provision during low water periods for the Water Treatment Plant in Marciszów and provision of drinking water for the city of Wałbrzych, although this last function is not fulfilled and the water from the reservoir is used for consumption only for the citizens of Bukówka village, and in order to ensure the proper flow into the riverbed of the Bóbr river, downstream from the reservoir [Operat 2007]. Due to the fact that the Bukówka reservoir is located just below the border with the Czech Republic, it is exposed to a number of problems, among which the main is the deterioration of the water quality in the reservoir. Due to its cross-border location and economic importance, it is crucial to monitor the reservoir's water quality as well as the water quality of its tributaries. In such a situation, the regional water administration is supposed to organize and stimulate the protection and rational use of water, under the integrated water policy of the EC. These issues are particularly important in case of transboundary waters, which according to the Art. 1 of the Convention [Konwencja 1992] are any surface or ground waters which mark and cross a border between two or more countries or are located on such borders. Tasks for such type of waters are presented in the Directive 97/11/EC [Dyrektywa 1997]. It obliges parties managing transboundary waters to analyze the state of the environment in these areas and to carry out strategic environmental impact assessments, from the plans and policies to individual investments. The Directive 97/11/EC recognizes the need to implement the following postulates:

- The obligation to remove any environmental damage at its source as well as the "polluter pays" principle,
- The achievement of good water status as a strategic goal, postulating a reduction of dangerous substances emission into the water at the same time,
- Performance of analysis of the water status in particular basins, evaluation of anthropogenic impacts including water economy (it is recommended to implement water service charges in the amount covering water management costs, including expenditures on water protection and restoration of ecological of a water-related environment).

Therefore, it is very important to characterize thoroughly selected water management problems in the basin area of a cross-border retention reservoir, to indicate results of water management interactions and to present proposals of how to improve this condition. Because transboundary impact means any impact resulting in significant adverse effects on the environment in the area under the jurisdiction of one party, being a result of changes in transboundary waters caused by human activity, in which the physical beginning is situated entirely or partly in the area under the jurisdiction of another party; such effects on the environment include, among others, impact on water and other physical structures or interactions between these factors [Konwencja 1992].

The purpose of this article is to present some water management problems in the Bukówka reservoir basin located on the Bóbr river, below the Polish-Czech border. Particular attention was paid to the water quality of the Bukówka reservoir tributaries.

METHODOLOGY

The study included an inventory of the basin between January 2006 and December 2007 concerning the operation of the Bukówka drinking water reservoir on the Bóbr river. Particular attention was paid to the assessment of the Bóbr river water quality, the Bukówka reservoir tributary. In addition, the results of the water quality of four tributaries were presented: Opawa, Złotna, Bachorzyna and Paprotki. In order to create water quality characteristics, the results of our own research were used (from the period of November 2006-December 2007), as well as the results from the Regional Inspectorate for Environmental Protection in Wrocław (period 1993–2005) [Raport 2005] and from the study "Vodohospodářské bilance for 2011, období 2006–2011 and výhledu k 2021" (period 2009–2011) [Havranek et al. 2011]. The following indicators of water quality were analyzed: N_{tot}, NO_3^-, NO_2^-, NH_4^+, PO_4^{3-}, BOD_5, water pH, electrolytic conductivity, the number of fecal coliform bacteria, water temperature, and total suspended solids.

The quality of the reservoir's tributaries was assessed in accordance with the applicable Decree on the classification of surface water bodies [Rozporządzenie 2011]; an assessment of water eutrophication was presented on the basis of the Decree of the Minister of the Environment of 2002 on the criteria for indicating waters vulnerable to pollution by nitrogen compounds from agriculture [Rozporządzenie 2002]. The basin and the reservoir exploitation were analyzed, as well as water management at the reservoir. Particular attention was paid to the quality of the water supplying the reservoir, the water of the reservoir itself, as well as the water outflow from it; proposals of actions to limit the impact of improper water management were presented.

THE OBJECT CHARACTERISTICS

The Bukówka water reservoir is located on the Bóbr river at 271+540 km (left-hand tributary of the Odra river), in the Lubawka municipality, Kamienna district. The Bukówka reservoir is one of the highest above-sea-level dam reservoirs in Poland. Its main components are the reservoir's bowl (Figure 1a, b): an earth dam with drain devices. The central dam is located between two mountains: Zameczek and Zadzierna (Figure 1c). Moreover, the reservoir consists of bottom release, surface spillway, intake of tap water for Wałbrzych Water and Sewage Enterprise, which is currently not in operation, hydroelectric power plant and side dam Miszkowice (Figure 1d, e). Almost the

Figure 1. The Bukówka reservoir's bowl (a, b), ground dam (c), the Miszkowice reservoir (d), the side dam Miszkowice and the polder ourflow (s), The Bóbr riverbed from the Bukówka reservoir (f)

entire reservoir is limited by natural banks, except for the front and side dams (Figure 1f). The side dam is located in the Miszkowice village area in the western part of the reservoir. The function of the side dam is to protect the Miszkowice village from the reservoir's backflow. Since Miszkowice polder is not capable of performing gravitational drainage, drainage pumping is done using three pumps. Water management in the polder is automated [Manual 2007]. The parameters of the tank: 12.01 million m³ and 167.0 ha at a normal level and 16.790 million m³ and 199.0 ha at maximum flood level [Operat 2007].

The outflow from the reservoir is done through bottom release and surface spillway (Figure 1f). On the wall of the bottom release a gauge

patch is placed (Figure 1a, b), and also the reservoir is equipped with a limnigraph. An operator discharges water from a modern control room located in the tower of inflow and outflow (Figure 1a, b). The dam is also equipped with piezometers and a drainage trench located on the air side of the dam.

RESULTS AND DISCUSSION

Exploitation of the basin and the reservoirs

The area of the basin in the Bukówka reservoir dam section is 58.5 km², of which the Bóbr river flows 4.6 km² into Poland. The Bóbr river flows into the Czech Republic for about two kilo-

a)   b)

c)   d)

e)   f)

Figure 2. The Bukówka reservoir's tributaries: Bóbr riverbed (a, b), Złotna (c), Opawa (d), Bachorzyna (e), Paprotki (f)

meters, then more than four kilometers in Poland, and flows into the Bukówka reservoir [Ocena 2007; Report 2005]. The longitudinal slope of the Bukówka reservoir's basin is approx. 8.8%. The basin area is used for agriculture (approx. 50%) and covered with woods (approx. 45%). There is also a visible slope of the basin, mostly from west to east, and in the southern part of the basin from south west to north east [Operat 2007]. The Bukówka reservoir is a recipient of several watercourses (Figure 2): rivers Bóbr, Złotna, Opawa and Bachorzyna; it also receives waters from several drainage ditches: Paprotki and pumped water from the Miszkowice polder.

In the basin area of the reservoir are located villages: Miszkowice, Opawa, Jarkowice, Niedamirów and Paprotki. The average population in the basin area is approx. 25 persons/km^2. At the moment of the research on the basin area of the reservoir, there was no overall sewage system. The municipal sewage formed in the reservoir's basin was channeled mainly to the Złotna river, and Paprotki ditch, which are above the tank.

Agriculture, water and wastewater management in the reservoirs basins, both direct and indirect, are not without significance to the quality of water bodies. According to Pütz and Paul [2008] and Pütz and Benndorf [1981] a kind of

reservoir basin, in case of which the issue of water management should be thoroughly considered, is a drinking water reservoir basin. In addition, in a case where slopes in the near basin area of the reservoir are large (in case of the Bukówka reservoir longitudinal north-east declines from Zadzierna mountain range from 10 to 60% and at the southern, Szczepanów side from 5–12%), and along the slope there is agriculture (Figure 3a), with surface runoff waters and various pollutions, including fertilizing substances, get into a reservoir.

During the study period the Bukówka reservoir was also used by anglers. According to the rules any other possibility of recreational use of the reservoir during the study period was not allowed. However, as it was noted, the reservoir was used by the public for recreational purposes and bathing (Figure 3b, c, d). There is no doubt that the Bukówka reservoir due to its large area of various coastlines, perfectly blended into the surrounding mountains, is a recreational and landscape attraction. Therefore, attempts were made to obtain permission for recreational use of the reservoir. It is worth noting, however, that in this case it is necessary to perform the assessment of the possibilities of adapting it for recreational purposes. The most important is the tourism capacity, not only because of the perfect leisure condi-

Figure 3. The usage of the Bukówka reservoir

tions, but mostly due to the need to protect the tank from excessive tourism pressure and preservation of the principles of balanced development and protection. There is no doubt that in the present conditions of declining water resources such measures become necessary, though they must be related to the proper water and wastewater management in the reservoir's basin.

Water management at the reservoir

We may distinguish four periods of water management at the Bukówka reservoir. It is water management under normal operating conditions, the conditions of flood, drought and the period of ice phenomena. As is stated in the Manual of water management [Instrukcja, 2007], water management at the reservoir during the normal operating period is a management within the usable capacity, ranging between the minimum level of damming – 521.30 meters above sea level and the normal level of 534.30 meters above sea level. Water management in these circumstances should ensure drainage of the reservoir's proper flow downstream (0.10 m^3·s^{-1}) and provide water to the intake in Marciszów (0.56 m^3·s^{-1}). Outflow instruction is given once a day at 7.00 am (also at other times when the situation requires a change in the size of an outflow). During the period of study on the reservoir as well as on its inlet the gauge patches were installed.

Flood conditions of the flood management at the Bukówka reservoir are managed when impoundment at the reservoir is equal or bigger than normally and when inflow to the reservoir is larger than the permitted Q_{doz} flow of 8.00 m^3·s^{-1}. The reservoir's flood management applies to the constant reserve management, i.e. the tank capacity between the normal (534.30 meters above sea level) and the maximum level of impoundments (536.30 meters above sea level). Moreover it applies to the forced flood reserve, i.e. the tank capacity between the maximum and the extraordinary levels of impoundments – 537.10 meters above sea level, meant to be filled only in exceptional cases of catastrophic surge or failure of release and spillway mechanisms, as well as during renovation closures. In case of a flood, water is passed through with one or two holes of bottom release, or through both spillway spans. Biological water release is closed. Water management during drought is reduced to maintain a minimum level of impoundment through maintaining Q_{dys}

overflow from the reservoir equal to the Q inflow. If the flow is below its needs, a limitation of water supply for users downstream of the reservoir occurs. The period of drought is defined as a period when the following conditions occur together: the level of impoundment on the reservoir is less than or equal to the minimum level of impoundment, inflow to the reservoir is lower than the proper flow downstream (0.10 m^3·s^{-1}), hydro-meteorological conditions do not indicate the occurrence of rainfall in the reservoir area which results in further deepening a state of low water (no flow increase in the Bóbr river). In contrast, during periods of ice cover (Figure 1b), it is advisable to maintain reservoir impoundment at normal levels, so the ice is not drawn into the spillway (dropping ice through spillways is forbidden). Dropping water in winter should be done by bottom release. Spillway trapdoor, niches, and gauge patches should be cleaned [Instrukcja 2007].

During the exploitation of reservoirs many problems appear, which may hinder or prevent its proper use. The Bukówka reservoir, just like other reservoirs, faces numerous problems. One of the main ones is a lack of constant hydrological and water quality monitoring at the tributaries and the reservoir itself, which would contribute to the proper use of water in the reservoir.

Water quality

Particularly important for the proper functioning of reservoirs, especially those containing drinking water, are water management issues in their basin areas. Dam reservoirs, due to their location at the lowest point of a basin are receivers of pollutants from the whole basin area, which make them very vulnerable to the processes occurring in a basin [Benndorf, Pütz 1987; Granlund et al. 2005; Skonieczek et al. 2013; Wegner et al. 1975]. A significant threat to the reservoirs is the process of eutrophication. Often the ability to use water in reservoirs depends on its quality, which is influenced, inter alia, by water and wastewater management in the basin of a reservoir [Wiatkowski, Paul 2009]. It was found that the sources of water pollution in the reservoirs' basins are point, dispersed and territorial sources [Koc et al. 2008]. The main causes of pollution of surface waters in the discussed area are: domestic waste waters containing organic compounds and nutrients channeled into watercourses without treatment, pollution from agricultural production, pollution flowing through surface watercourses from the Czech Republic, illegal landfills, area flows and linear pollution [Informacja 2011]. Studies showed that the main sources of the Bóbr river pollution are domestic and industrial waste water from urban centers and domestic waste water from rural community centers [Raport 2005]. In the basin area the villages of Paprotki and Miszkowice have sewage systems. Water is taken from intakes: Paprotki – drainage intake and Miszkowice – bank intake at the Złotna creek. Villages completely or partially void of sewage systems in the basin area are places like: Jarkowice. The villages of Opawa and Niedamirów were undergoing the process of sewage system construction [Informacja 2011]. During the study period in the basin area a fragmentary sewage system functioned, a small part of Miszkowice and Jarkowice had their own sewerage. Lubawka municipality has two waste water treatment plants which receive waste water from Bukówka, Miszkowice and Jarkowice. Waste waters from other villages of the municipality are transported to the treatment plant by a sanitation fleet or after domestic wastewater pretreatment channeled through the water or ground. During the study period the Lubawka municipality carried out or attempted to carry out tasks involving construction of a sanitary sewage system in villages: Jarkowice, Miszkowice, Opawa and Niedamirów.

At the Bukówka reservoir's basin there is no continuous monitoring of the water quality flowing into the tank. Periodic assessment of the water quality of the Bóbr river above the reservoir the Regional Inspectorate for Environmental Protection in Wroclaw carried out in 1993–2005. Table 1 shows the results of the quality of the water flowing into the Bukówka reservoir.

The Bóbr river water quality at the border showed considerable variability. An unsatisfacto-

ry state of water was registered. Water flowing to Poland from the Czech Republic was characterized by unsatisfactory quality. A large number of faecal coliform bacteria, very high concentrations of phosphorus and increased color and phenolic index determined this classification. High pollution by phosphorus and nitrogen compounds in 2005 was much lower than in the previous year though. A significant annual fluctuation in the concentrations of nutrients for the period 1993–2005 in the water flowing to Polish was observed (Table 2) [Raport 2005]. As authors of the report concluded [2005], high concentrations of biogenic compounds pose a threat for the Bukówka dam reservoir eutrophication. Water quality assessment of the Bóbr river considering parameters characterizing the process of eutrophication, demonstrated exceeding annual average concentrations of phosphates and nitrates. The analysis of the least favorable indicators of water quality at the border (Table 1) showed that the Bóbr river water, due to its conductivity, may be classified into class I of water quality. In contrast, concentrations of nitrates, phosphates and BOD_5 exceeded the water quality limits relating to surface water bodies in natural watercourses, such as a river, appropriate for class II [Rozporządzenie 2011].

Similar results, indicating high pollution of the Bóbr river flowing from the Czech Republic are presented Havranek [2011]. It concluded that the Bóbr river water flowing out from the Czech Republic should be included into the class V of water quality, due to the high content of total phosphorus.

On the other hand, our own water quality testing at the reservoir, conducted in 2006–2007, confirmed that from the point of view of eutrophication, hydrochemical conditions occurring in the Bukówka reservoir's basin are unfavorable for it, in terms of functions performed by the reservoir

Table 1. Average annual water quality indicators of the Bóbr river above the Bukówka reservoir at the border (km 269.6) from the period between 1993 and 2005, from WIOŚ Wroclaw

Indicator	WIOS investigations (1993–2005)	
	Bóbr river above the Bukówka reservoir, cross border	
	min. (year)	max. (year)
N tot [mg N·dm⁻³]	2.8 (1997)	>20 (1994)
Nitrates [mg NO₃⁻·dm⁻³]	12 (1998)	>50 (1993, 1994, 1995, 1996, 2000, 2004)
Phosphates [mg PO₄³⁻·dm⁻³]	0.20 (1997)	>2.00 (1993, 1994, 2000, 2003, 2004)
BOD₅ [mg O₂·dm⁻³]	5.0 (1997, 1999,2003)	9.0 (1995)
Electrolytic conductivity [µS·cm⁻¹]	200 (1997, 1998, 1999)	470 (1994)
The number of faecal coliform bacteria	7000 (1997)	400 000 (1993, 1994)

[Wiatkowski et al. 2015]. In the waters flowing into the Bukówka tank concentrations of: nitrates ranged from 0.88 mg $NO_3^-\cdot dm^{-3}$ (Złotna and Bachorzyna) to 17.70 mg $NO_3^-\cdot dm^{-3}$ (Opawa); phosphates ranged from 0.02 mg $PO_4^{3-}\cdot dm^3$ (Złotna) to 1.28 mg $PO_4^{3-}\ dm^3$ (Paprotki); ammonium ions ranged from 0.020 mg $NH_4^+\cdot dm^{-3}$ (Złotna) to 0.640 mg $NH_4^+\cdot dm^{-3}$ (Opawa); nitrites ranged from 0.003 mg $NO_2^-\cdot dm^{-3}$ (Złotna) to 0.368 mg $NO_2^-\cdot dm^{-3}$ (Ferns); BOD_5 ranged from 1.0 to 11.00 mg $O_2\cdot dm^{-3}$ (Opawa) and electrolytic conductivity ranged from 74 (Bachorzyna) to 714 µS/cm (Bóbr). The temperature of the water streams flowing into the reservoir during the analyzed period ranged from 2.1 °C (Bachorzyna and Opawa) to 22.5 °C (Bachorzyna). The highest concentration of total suspended solids – 180 mg·dm^{-3} – was observed in the waters flowing into the tank (Bachorzyna) Wiatkowski at al. 2015]. This is confirmed by tests of the water quality of the Bóbr river presented in [Operat 2007; Ocena 2007]. The tested waters from the Bukówka reservoir's basin are not vulnerable to pollution by nitrogen compounds from agricultural sources because concentrations of nitrate are lower than those recommended (50 mg $NO_3^-\cdot dm^{-3}$) [Rozporządzenie 2002].

In view of the fact that during the study period at the Bukówka reservoir a surface water intake protective zone "Bukówka" was established according to the permission OS-6210/8/97 for the planned Water Treatment Plant and water transfer to Wałbrzych [Operat 2007], the research should continue. At the time of the research only direct water intake for the citizens of the village located below the reservoir took place.

The basic water management problems at the Bukówka reservoir's basin include:
- The unsatisfactory state of water and waste water management in the catchment area (illegal channeling of sewage into the river Złotna and Paprotki ditch) (Figure 1),
- The unsatisfactory state of surface water quality in the basin of the reservoir (Figure 1),
- The poor condition of streams and drainage ditches (Figure 2),
- Improper arrangement of arable land and large slopes (which cause a rapid outflow of water and pollutants from the reservoir) (Figure 3a).
- A lack of water monitoring stations on the reservoir's tributaries (Bóbr, Opawa, Złotna, Bachorzyna, Paprotki). Currently, the gauge patch is installed downstream from the reservoir. This means that the reservoir does not have any hydrological protection. Water monitoring devices should be installed in order to measure water levels on the reservoir's tributaries, even more so considering the fact that in the past such devices were present on Bóbr and Złotna.

Implementation of water management tasks should be assessed on the basis of its effectiveness, e.g. improvement of water quality, flood losses reduction etc. On the other hand, an identification of the water management problems, particularly in reservoirs' basin areas, is necessary to implement recovery measures.

CONCLUSIONS

Proper water management in the reservoir's basins should aim at improving water quality and seek to create conditions for its retention especially when it comes to reservoirs of water for human consumption. Eutrophication processes intensification in aquatic environments causes the decrease in water utility and aesthetic values of reservoirs. This may prevent a fulfillment of the reservoir's tasks.

Hydrochemical conditions in the Bukówka reservoir section are unfavorable for it. From the eutrophication perspective, water flowing into the reservoir is characterized by a significant content of nutrients. Levels of nitrates, phosphates and BOD_5 have exceeded the water quality limits relating to surface water bodies in natural watercourses, such as a river, appropriate for class II.

In order to counteract the eutrophication it is necessary to lower the concentration of nutrients in the water inflow. One of the basic solutions is to organize proper water and wastewater management in the reservoir's basin.

Proposals for changes in water management in the reservoir's basin area, which start with the construction of sewage systems in towns located in the reservoir's basin, are (during the study period the measures were taken to provide sewage systems to the villages in the reservoir's basin area) the introduction of a protection zone around the reservoir limiting the use of land, a creation of monitoring stations on the tributaries of the reservoir and finally a restoration of clear flow in drainage ditches.

For a proper implementation of recovery actions it is necessary to identify water management problems, particularly at the drinking water reservoirs' basin areas.

Studies carried out at the Bukówka reservoir provide valuable information on the state of waters supplying the reservoir. In order to obtain accurate information on the state of these hydrological and water quality monitoring should be continued.

REFERENCES

1. Balcerzak W. 2006. The protection of reservoir water against the eutrophication process. Polish J. of Environ. Stud. Vol. 15, No. 6, 837–844.

2. Benndorf J., Pütz K. 1987. Control of eutrophication of lakes and reservoirs by means of pre-dams – I. Mode of operation and calculation of the nutrient elimination capacity. Wat. Res. 21, 829–838.

3. Czamara W., Czamara A., Wiatkowski M. 2008. The use of pre-dams with plant filters to improve water quality in storage reservoirs, Archives of Environ. Protection, vol. 34, 79–89.

4. Dyrektywa Rady 97/11/WE z dnia 3 marca 1997 r. zmieniająca dyrektywę 85/337/EWG w sprawie oceny wpływu wywieranego przez niektóre publiczne i prywatne przedsięwzięcia na środowisko.

5. Granlund K., Räike A., Ekholm P., Rankinen K., Rekolainen S. 2005. Assessment of water protection targets for agricultural nutrient loading in Finland. Journal of Hydrology 304, 251–260.

6. Havránek L., Kovář A., Zapletal T. 2011. Zpráva o hodnocení jakosti povrchových vod pro území ve správě Povodí Labe, státní podnik Odbor péče o vodní zdroje. Hradec Králové, wrzesień 2011, pp. 77.

7. Informacja o stanie środowiska w powiecie kamiennogórskim, Kamienna Góra, maj 2011 r. pp. 27.

8. Instrukcja gospodarowania wodą dla zbiornika wodnego Bukówka. Powiat kamiennogórski, woj. dolnośląskie, Integrated Engineering Sp. z o.o. na zlecenie RZGW Wrocławiu, Raszyn 2007, pp. 22.

9. Kanownik W., Kowalik T., Bogdał A., Ostrowski K., Rajda W. Jakość i walory użytkowe wód odpływających ze zlewni zbiorników małej retencji planowanych w rejonie Krakowa. Wyd. UR w Krakowie, Kraków 2011, pp. 110.

10. Koc J., Duda M., Tucholski S. 2008. Znaczenie zbiornika retencyjnego dla ochrony jeziora przed spływami fosforu ze zlewni rolniczej. Acta Sci. Pol., Formatio Circumiectus 7 (1), 13–24.

11. Konwencja o ochronie i użytkowaniu cieków transgranicznych i jezior międzynarodowych, Sporządzona w Helsinkach, dnia 17 marca 1992 roku, Narody Zjednoczone 1992.

12. Ocena stanu jakości rzek województwa dolnośląskiego w 2007 roku. Wojewódzki Inspektorat Ochrony Środowiska We Wrocławiu, pp. 54.

13. Operat wodnoprawny dla zbiornika wodnego Bukówka. Powiat kamiennogórski, woj. dolnośląskie, Integrated Engineering Sp. z o.o. na zlecenie RZGW we Wrocławiu, Raszyn 2007, pp. 22.

14. Pütz, K., Benndorf J. 1981. Die zielgerichtete Wassergütebewirtschaftung von Talsperren und Speichern – Information zum Fachbereichsstandard TGL 27885/03, Acta hydrochim. et hydrobiol., 9, 25–36.

15. Pütz K., Paul L. 2008. Suspended matter elimination in a pre-dam with discharge dependent storage level regulation. Limnologica, 38, 388–399.

16. Przybyła Cz., Kozdrój P., Sojka M. Ocena Jakości wód w lateralnych zbiornikach Jutrosin i Pakosław w pierwszych latach funkcjonowania. Inżynieria Ekologiczna, Vol. 39, 2014, 123–135.

17. Raport o stanie środowiska województwa dolnośląskiego w 2005 r. WIOŚ, Wrocław 2005.

18. Skonieczek P., Koc J., Duda M. 2013. Wpływ zbiorników retencyjnych w ochronie jezior przed zanieczyszczeniami spływającymi z obszarów wiejskich. DOI: 10.2429/Proc.2013.7(1)033.

19. Rozporządzenie Ministra Środowiska z 23 grudnia 2002 roku w sprawie kryteriów wyznaczania wód wrażliwych na zanieczyszczenie związkami azotu ze źródeł rolniczych. Dz. U. Nr 241, poz. 2093.

20. Rozporządzenie Ministra Środowiska z 9 listopada 2011 roku w sprawie sposobu klasyfikacji stanu jednolitych części wód powierzchniowych oraz środowiskowych norm jakości dla substancji priorytetowych. Dz. U. Nr 257, poz. 1545.

21. Wegener U., Dörter K., Beuschold E. 1975. Der Einfluss der landwirtschaftlichen Nutzung von Talsperreneinzugsgebieten auf den Nährstoffeintrag in Trinkwassertalsperren. Acta hydrochim. hydrobiol. 13, pp. 553–561.

22. Wiatkowski M., Paul L. 2009. Surface water quality assessment in the Troja river catchment in the context of Włodzienin reservoir construction. Polish J. of Environ. Stud., Vol. 18, No. 5, 923–929.

23. Wiatkowski M., Rosik-Dulewska Cz., Wiatkowska B. 2010. Charakterystyka stanu użytkowania małego zbiornika zaporowego Nowaki na Korzkwi. Rocz Ochr Środow., T. 12, 351–364.

24. Wiatkowski M. 2011. Influence of Słup dam reservoir on flow and quality of water in the Nysa Szalona river. Polish J. of Environ. Stud. Vol. 20, No. 2, 467–476.

25. Wiatkowski M., Rosik-Dulewska Cz., Kasperek R. 2015. Analysis of the impurity supply to Bukówka reservoir from the transboundary Bóbr river basin. Roczn. Ochr. Środow. (Ann. Set The Environ. Prot.), in print.

25

THE ANALYSIS OF THE MANUFACTURING AND USING ALTERNATIVE FUEL – A MIXTURE OF RAPESEED OIL AND ALCOHOL

Piotr Kardasz[1], Radosław Wróbel[2]

[1] Wrocław University of Technology, 27 Wyspiańskiego Str., 50-370 Wrocław, Poland, e-mail: piotr.kardasz@pwr.wroc.pl

[2] Wrocław University of Technology, 27 Wyspiańskiego Str., 50-370 Wrocław, Poland, e-mail: radoslaw.wrobel@pwr.wroc.pl

ABSTRACT

The following article is an analysis of designed process of manufacturing a mixture of 50% rapeseed oil and 50% alcohol and using it as a fuel. The analyzed eco-fuel is completely based on renewable sources, and can be a good alternative to diesel fuel. The analysis was made according to the assumptions of Life Cycle Assessment, which is a method that divides the whole life cycle of the product into the unit processes. It is used especially for measuring the environmental impact of the product. The life cycle of fuel mixture in an amount of 10 000 l was divided into six unit processes: the production of oilseed and biomass on the farm, transport of rapeseed to oil extraction works, oil production, the production of alcohol from biomass, the transport of mixture into a transport company and the use of total fuel delivered by the company. The use of energy and the amount of pollutants emitted were particularly important in the analysis. Fuel mixture, the same as the analyzed, was used during the whole designed process. In the production of rape on a farm the tractor and the harvester were used, and caused highest emissions of pollutants during all steps involved in the production of fuel. Alcohol, the component of the mixture, was produced through the fermentation of biomass that cames from waste from rapeseed processing, which caused no energy consumption or emissions. The analysis shows that total emissions of harmful gases is lower than that of conventional diesel, which proves that the tested fuel mixture is more environmentally friendly.

Keywords: alternative fuel, mixture, rapeseed oil, alcohol, unit process, biomass.

INTRODUCTION

Traditional fuel for powering diesel engines is diesel oil. As a product derived from petroleum it belongs to non-renewable energy sources. It's price, alike the price of petroleum, has been rising for years, and is highly dependent on political issues. In addition, its combustion emits a number of compounds hazardous to the environment.

Alternative fuel that can replace diesel oil, should be primarily produced with renewable sources, and also should be characterized by lower price for similar performance and lower emissions of harmful substances during combustion. The fuel that could be a mixture of rapeseed oil and alcohol, which is still under development. The fuel is the subject of the following analysis.

SUBJECT AND METHOD OF THE ANALYSIS

The analyzed alternative fuel consists of 50% rapeseed oil and 50% alcohol produced in biomass fermentation. The analysis is carried out for 10 000 liters of fuel.

The materials used for the designed production process are from fully renewable sources, it is mainly rape, grown in farms. Rapeseeds are converted into the oil, and the waste (e.g. straw) is used as biomass to obtain alcohol.

Basing on previous studies it can be stated that, compared to conventional diesel, the mixture of rapeseed oil and the alcohol will be characterized by only a few percent lower energy value, which results in a little lower efficiency [2].

The study was based on assumptions of Life Cycle Assessment, which are to divide the whole life cycle of product info unit processes and to analyze each one of them. This method is mainly used to assess the environmental risks posed by the product. Unit processes may be associated with the production, transport, use or disposal, and for each of these, the use of raw materials, energy, emissions or waste volume is determined, depending on the purpose of analysis [3].

The following analysis focuses primarily on emissions of pollutants and energy consumption in each of the phases of the whole designed process, which led to information on total emissions and energy consumption for the life cycle of the fuel mixture.

UNIT PROCESSES IN THE MIXTURE FUEL LIFE CYCLE

The following analysis includes all unit processes making up the finished product, including the emissions they cause. The process of using the fuel is also included.

Figure 1 illustrates successive processes making up the life cycle of a mixture of vegetable oil and alcohol.

Single processes making up the whole designed process of production and use of tested alternative fuel are characterized below.

Production of oilseed and biomass

In order to obtain raw materials which are necessary to produce elements of the mixture, it is needed to cultivate appropriate intermediates. Vegetable oil is produced from rapeseed, and alcohol is formed in the fermentation of biomass.

It was assumed that the production of the two components is carried out on the farm located in Lower Silesia.

For the production of 5000 liters of rapeseed oil the area of 5 hectares required. This value was estimated basing on average yields and the amount of oil obtained from one tone

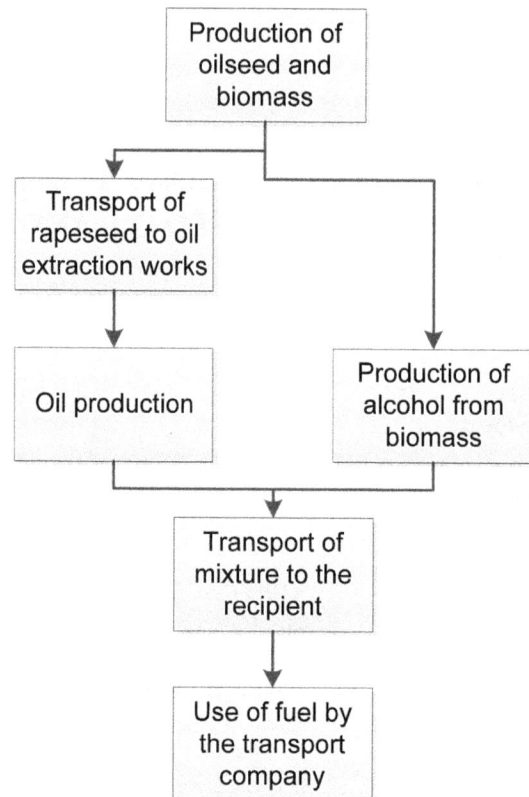

Figure 1. Processes of the production of life cycle of fuel mixture

of rapeseed. It can be gained on average 2.5 tones of rapeseed from one hectare, while one tone can be processed into around 400 liters of oil [1].

Biomass intended to be used in the fermentation process to produce alcohol is derived from wastes from the production of rapeseed (straw, oil cake). Therefore, the production of components for the production of rapeseed oil and the alcohol is treated as a single process. During the fermentation of one ton of biomass 220 liters of alcohol are produced, and one hectare can produces 5 tons of biomass. Therefore, to produce 5000 liters of alcohol 4.55 ha is required, so cultivation area will remain the same as in the case of rape crops – 5 ha.

It is assumed that an agricultural machinery (tractor and harvester) is used in the production process of these materials, also powered by a mixture of rapeseed oil and alcohol. Previous experiments showed that this fuel is characterized by 10% higher consumption than conventional diesel fuel. For the production of necessary raw materials (12.5 t rapeseed and 22.75 t rape straw) the estimated total fuel consumption is 494 liters (Table 1). It takes into account such processes as skimming, plowing, tillage, seeding, fertilization, protective treatments, seed harvest and commuting.

Table 1. Fuel consumption during rapeseed and straw production

Type of activity	Fuel mixture consumption [dm³]
Skimming	63
Plowing	110
Tillage	104
Seeding	29
Fertilization	42
Protective treatments	42
Seed harvest	104
Commuting	33
Sum	494

Because the energy value of the fuel used is similar to that of diesel oil, it is assumed that it is 36 MJ/L. Therefore, the energy value of whole fuel used were as follows:

$$E_{zon} = 36\ MJ/l \times 494\ l = 17784\ MJ \qquad (1)$$

When cultivating, it was also necessary to use some fertilizers and pesticides, which is shown in Table 2.

Energy value of fertilizers and pesticides was calculated 37 390 MJ. Therefore, total energy value is:

$$E_{pr} = 17784\ MJ + 37390\ MJ = 55174\ MJ \quad (2)$$

The combustion of fuel by a tractor and combine harvester, issues a certain amount of air pollution. It was assumed that a tractor (Zetor 12245) workes on average power 63 kW. Operating time of the tractor on 5 ha of land in the whole process of cultivation of rape is 43.75 h. Therefore, the tractor uses 2756.25 kWh.

The combine harvester (Bison Rekor workes for 18.75 hours, and its average power is 75 kW. Thus, the energy consumption is 1406.25 kWh. Therefore, the total energy consumption in the production of 12.5 tonnes of rape is 4162.5 kWh. Emission level for the combustion of a mixture of rapeseed oil and the alcohol is shown in Table 3.

Therefore, a total production of rapeseed and biomass to produce 5000 liters and 5000 liters of alcohol requires 55 174 MJ. It uses 494 liters of fuel, 1400 kg of nitrogen fertilizer, 570 kg of phosphate fertilizer, 630 kg of potassium fertilizers and 15 kg of pesticides. 3330 g of carbon monoxide and 29 137.5 g of nitrogen oxides are emitted.

Transport of rapeseed to oil extraction works

The next step is to transport the produced rapeseed oil to an oil mill, located at a distance of 50 km from the farm. It is assumed that transportation is done by already used Zetor 12 245 with two trailers with total capacity of 20 000 kg. During transport to the oil mill and return to the farm the tractor uses 841 liters of fuel mixture, which energy value is 36 MJ/l. So total energy consumption of the process is 3042 MJ.

During the transportation, the average tractor power is 70 kW, and the transportation time is 6 h. Therefore, the energy consumption is 420 kWh. In total, in this process 841 liters of fuel are used, which causes an emission of 336 g carbon monoxide and 2940 g nitrogen oxides (Table 4).

Table 2. Consumption and energy value of fertilizers and pesticides

Fertilizer / pesticide	Consumption [kg]	Conversion coefficient [MJ/kg]	Energy value [MJ]
Nitrogen fertilizers	1400	20	28 000
Phosphate fertilizers	570	7	3990
Potassium fertilizers	630	5	3150
Pesticides	15	150	2250
Sum	2615	182	37 390

Table 3. Emission of pollutants during rape production

Pollutants	Average emission [g/kWh]	Emission for 4162.5 kWh [g]
CO	0.8	3330
NO_x	7	29 137.5
SO_x	0	0

Table 4. Emission of pollutants during the transportation of rapeseed

Pollutants	Average emission [g/kWh]	Emission for 420 kWh [g]
CO	0.8	336
NO_x	7	2940
SO_x	0	0

Oil production

The next step is the extraction of 5000 liters of oil from the rapeseed provided. For this purpose, the oil press TLS-30 may be used. Total power of this machine is 120 kW (including auxiliary drives) and it processes 25 tons of rapeseed per day.

Therefore, the press running time to process 12.5 tons of rape is 12 hours. So the energy consumed is 2106 MJ. The device uses 1440 kWh during the process (Table 5).

Table 5. Emission of pollutans during rapeseed oil production

Pollutants	Average emission [g/kWh]	Emission for 1440 kWh [g]
CO	0.014	20.16
NO$_X$	0.088	126.72
SO$_X$	0.338	486.72

Source: own elaboration, based on user's manual of the oil press.

Therefore, the process of producing 5000 liters of oil consumes 12.5 tons of rapeseed and issues 20.16 g of carbon monoxide, 126.72 g of nitrogen oxides and 486.72 g of sulfur oxides.

Alcohol production from biomass

For the manufacturing of the fuel mixture, it is also necessary to produce 5000 liters of alcohol. Alcohol is produced by the fermentation of biomass acquired earlier, involving bacterium Clostridium acetobutylicum.

In order to produce the required amount of alcohol, 22.75 tons of biomass has to be used. The process does not require any energy and does not cause the emission of harmful substances into the environment.

Transport of fuel mixture to recipient

After the combination of the components, 10 000 l of alternative fuel is made. Then, it is transported to the transport company, established in Wrocław, which is located at a distance of 10 km from the factory.

It is assumed that in the process of transport, the tractor unit Mercedes-Benz Actros with tank-type trailer NPA-33 is used. The vehicle is supplied with a mixture of rapeseed oil and alcohol. The entire vehicle, including the filled trailer has a weight of 21.2 tons.

The total process energy consumption is 127.57 MJ. Fuel consumption over a distance of 10 km is 3.5 liters. The average power of the tractor is about 250 kW, therefore, the journey lasts 0.166 hours (assuming average vehicle speed of 60 kph) and the energy consumption is 42 kWh.

Table 6. Emission of pollutans during the transportation do the recipient

Pollutants	Average emission [g/kWh]	Emission for 42 kWh [g]
CO	0.8	33.6
NO$_X$	7	294
SO$_X$	0	0

In the process of transport of 10 000 liters fuel for the company, 3.5 liters of fuel is consumed. The process generates 33.6 g of carbon monoxide and 294 g of nitrogen oxides (Table 6).

Use of fuel by the transport company

The transport company uses the tractor Mercedes-Benz Actros with trailers. The energy value of the whole fuel delivered is 360 GJ.

$$E_{wemor} = 10\ 000\ l \times 36\ MJ/l = 360\ 000\ MJ = 360\ GJ \ (3)$$

Assuming an average speed of 75 kph and consumption of 35 liters per 100 km it can be calculated that 10 000 liters of fuel enable to drive 28 571 km during 381 h. Therefore, if the average engine power is 335 kW, then power consumption is 127 635 kWh.

Table 7. Emission of pollutants during the use of fuel mixture

Pollutants	Average emission [g/kWh]	Emission for 127 635 kWh [g]
CO	0.8	102 108
NO$_X$	7	893 445
SO$_X$	0	0

Table 8. Summary of the energy consumption in the production of a mixture of rapeseed oil and an alcohol

Process	Energy consumption [MJ]
Rape production	55 174
Rapeseed transportation	3024
Rapeseed oil production	2106
Transport of fuel mixture to the recipient	127.57
Sum	60 431.57

Table 9. Summary of emissions of pollutants for the production and use of a mixture of rapeseed oil and alcohol as a fuel

Process	Emission [g]		
	CO	NO$_x$	SO$_x$
Rape production	3330	29 137.7	0
Rapeseed transportation	336	2940	0
Rapeseed oil production	20.08	126.72	486.72
Biomass fermentation to produce alcohol	0	0	0
Transport of fuel mixture to the recipient	33.6	294	0
Use of fuel mixture	102 108	893 445	0
Sum	105 827.68	925 943.42	486.72

In total, 10 000 liters of rapeseed oil mixture with alcohol are used by the transport company, generating the 102 108 g of carbon monoxide and 893 445 g of nitrogen oxides (Table 7).

CONSLUSIONS

In the designed process, the manufactured fuel delivered 360 GJ of energy. Table 8 shows the power consumption in processes of fuel production and total energy. In conclusion, with the effort of 60 GJ, there was nearly six times more power provided – 360 GJ.

Although the designed process of production of mixture of rapeseed oil and alcohol used environmentally friendly methods, it is inevitable to emit some pollutants, mainly associated with the burning fuel.

In the life cycle of fuel mixture it more than 100 000 g of carbon monoxide, more than 900 000 g of nitrogen oxides and almost 500 g of sulfur oxides were emitted (Table 9). This emission appears to be high, but it is lower than the production and use of conventional diesel, tested previously according to the same method (carbon monoxide 235 376 g, nitrogen oxides 944 921 g, sulfur oxides 83 287 g).

The unit process that caused the highest emission was the production of rapeseed on the farm. This may be an indication for further improvement of the manufacturing process.

The fuel consisting of rapeseed oil and alcohol produced from biomass has similar energy to that of diesel oil, but it is more environmentally friendly and made from entirely renewable sources.

REFERENCES

1. Bieniek J., Molendowski F., Kopa D. 2010. Analiza opłacalności produkcji estrów metylowych oleju rzepakowego na przykładzie wytwórni rolniczej W-400. Instytut Inżynierii Rolniczej, Uniwersytet Przyrodniczy we Wrocławiu, Inżynieria Rolnicza, 1(119).

2. Bocheńska A., Bocheński C. 2008. Olej rzepakowy paliwem do silników Diesla. Czasopismo Techniczne, Wydawn. Politechniki Krakowskiej, z. 8-M.

3. Grzesik K. 2006. Wprowadzenie do oceny cyklu życia (LCA) – nowej techniki w ochronie środowiska. Inżynieria Środowiska, 11(1).

26

ORGANIC WASTE USED IN AGRICULTURAL BIOGAS PLANTS

Joanna Kazimierowicz[1]

[1] Department of Environmental Systems Engineering, Bialystok University of Technology, Wiejska 45A, 15-351 Białystok, Poland, e-mail: j.kazimierowicz@pb.edu.pl

ABSTRACT

Treatment of organic waste is an ecological and economical problem. Searching method for disposal of these wastes, interest is methane fermentation. The use of this process in agricultural biogas plants allows disposal of hazardous waste, obtaining valuable fertilizer, while the production of ecologically clean fuel – biogas. The article presents the characteristics of organic waste from various industries, which make them suitable for use as substrates in agricultural biogas plants.

Keywords: agricultural biogas plants, organic waste, anaerobic digestion, biogas.

INTRODUCTION

An indispensable element of existence and human activity is the production of waste. With the development of civilization and the associated lifestyle followed the evolution of consumption growth and change in the model, as a consequence of the emergence of more and more waste and its becoming more diverse in terms of composition [Bliht 1999, Yamamura 1983].

Waste, regardless of their type, are a serious problem in both environmental and economic perspective [Kucharczyk at al. 2010].

Disposal of the waste consists of subjecting biological, physical or chemical processes, in order to bring them into a state that does not pose a threat to human life or health and the environment.

Proper waste management is of great importance in achieving sustainable development. If conditions permit, the type of building, it is important to promote local solutions, and even individual, such as backyard composting. Such activities contribute to the maintenance of local circulation of matter and shape the environmental awareness of residents. The recommended method, among others, for large farms may be the use of methane fermentation. It ensures to obtain positive results, both in terms of waste disposal, as well as a valuable source of renewable energy.

OPPORTUNITIES TO DEAL WITH BIODEGRADABLE WASTE

Municipalities must meet a number of obligations in the field of biodegradable waste. This will include: weight reduction of biodegradable municipal waste, transferred to storage, organization of their selective collection, providing construction, maintenance and operation of regional plant for processing waste.

Biodegradable waste can be subjected to composting, mechanical-biological treatment, anaerobic digestion, incineration, or can be stored. Storage of waste poses a number of dangers to the environment. These include the pollution of groundwater, surface water, soil and air.

Therefore, it is necessary to limit the amount of biodegradable municipal waste going to landfills and enhance the use of alternative methods of disposal.

Composting is one of the methods of biological waste, leading to the formation of compost. It occurs in the biochemical processes involving the decomposition of organic matter. Due to the low cost of this method it attracted a lot of interest in many countries. Using specialized measuring apparatus, it is possible to maintain optimal process parameters: temperature and humidity, as well as oxygen necessary for the growth of microorganisms. It also requires the selection of the appropriate fractions and pre-selected waste [Kucharczyk at al. 2010]. The development of the main product of the composting process, i.e. compost, creates more and more problems. Protection of soil, including agricultural products, pollution has forced the introduction of very strict requirements regarding the content of harmful substances in all materials applied to the soil. This mainly concerns the content of heavy metals and some organic pollutants. The material must also be safe in terms of sanitary-epidemiological [Manczarski 2012].

Mechanical-biological waste treatment is a process of "other" municipal waste, unsorted or any other bio-waste unsuitable for composting or anaerobic decomposition in order to stabilize and reduce their volume [Manczarski 2012].

Incineration, as one of the main ways of waste management can be implemented both in waste incineration plants, as well as power boilers [Kotlicki, Wawszczak 2010].

The most technologically advanced incinerators may be too expensive for many countries, but the costs are just one of the major factors limiting the use of this technology. It is difficult to achieve complete combustion. The big problem is the formation of trace amounts of products of incomplete combustion [Marchwińska, Budka 2014]. During the incineration of waste produced highly toxic chemicals mainly dioxins and furans, which if spilled into the environment, pose a danger. It is also high dust emissions into the atmosphere and hydrocarbons that are threats. The removal of the exhaust gases can be very difficult and costly.

Fermentation is a biochemical process involving microorganisms, wherein the organic substances are transformed into several phases to methane and carbon dioxide. In order to hinder the processes microbes and their enzymes are used [Pilarski, Adamski 2009].The fermentation process can be divided into four phases. In phase I, insoluble organic compounds (proteins, fats and carbohydrates) are processed by hydrolyzing bacteria that produce suitable enzymes-hydrolases. Proteases break down proteins, glycosidases-carbohydrates, lipase-fat. The action of these enzymes leads to the formation of soluble monomers or dimers. These processes are responsible for the speed of fermentation process, which in turn determines the phase distribution of the remaining methane-generating formation. Not all organic matter is decomposed in the process of hydrolysis. About 40–50% is not biodegradable because of lack of suitable enzyme-degrading polymers or monomers dimers. In phase II (acydogenezie) monomers and dimers, produced in phase I, are metabolized to short organic acids, having from one to six carbon atoms in a molecule. The most commonly produced acids include: formic, acetic, propionic, butyric, valeric, caproic. Alcohols are also generated: methyl and ethyl esters, aldehydes formic acid and acetic acid. The by-products of the reaction are carbon dioxide and hydrogen gas. Conversion of the products to acetic acid authors occurs only with external energy supply. However, it may occur freely (exothermic reaction) while hydrogen is continuously discharged, that is, in an environment where the partial pressure is suitably low, which takes place during the reduction of carbon dioxide to methane [Schlesinger 1997]. Lower hydrogen partial pressure, and the formation of reduced products more – the more desirable. In the system of methane to obtain a stabilized primarily by lead acetate, hydrogen and carbon dioxide, and the remainder of the acids and aldehydes correspond to a marginal role. This provides a fermentation process to produce more energy and the possibility of direct use of the substrates for the production of methane by the methanogenic bacteria. If the fermentation process produce large amounts of acids containing more than two carbon atoms, they cannot be used by methanogenic miroorganisms and are converted in the next step [Schink 1997]. In octanogenezie (phase III) organic acids, typically containing from three to six carbon atoms, are converted by the action of suitable bacterial strains to acetic acid, hydrogen and carbon dioxide. They can serve as substrates metanogennym bacteria for the production of methane. This phase is very energy-intensive and should, therefore, be most difficult. If the reactions were to occur spontaneously hydrogen should be removed from the system, and the partial pressure cannot exceed 400 Pa. In order to obtain octanogenesis, one must seek a synthrophy of actagenes with hydrogen absorbing methagenes. In phase IV (methanogenesis), which is the last phase, methane is produced. Stoichiometric calculations indicate that about 65–70% of the methane produced in the reduction process acetates [Smith at al. 1980]. Acetates are, therefore, one of the key intermediate formation of methane-generating substrates [Pilarski, Adamski 2009].

Fermentation has a triple role [Ledakowicz, Krzystek 2005]:

- allows you to convert the energy contained in the waste into useful fuel (biogas) that can be stored and transported,
- ensures recycling of organic waste into stable soil improvers, valuable liquid fertilizer and energy,
- allows inerting of waste, which aims to reduce the adverse impact on the environment.

AGRICULTURAL BIOGAS PLANTS

Agricultural biogas plants are gaining more and more supporters among agricultural producers. The reasons for this phenomenon must be sought in more widely available information on agricultural biogas applications, pressure emerging delivery companies, changes in energy law and necessity, resulting from directives for EU, proper handling of slurry and manure. The biogas plant, taking into account rising production costs and declining profitability of animal and plant production, is the perfect means of income sources diversification [Gniazdowski 2009].

The biogas plant is an installation in which a controlled biomass methane fermentation process, is biogas. Biomass, as defined by the European Union, means the biodegradable fraction of products, waste and residues from agro-industry, forestry and related industries, as well as the biodegradable fraction of industrial and municipal waste [Pilarski, Adamski 2009, Dyrektywa 2001/77/WE].

Biogas is a mixture of methane CH_4, carbon dioxide CO_2, and trace amounts of hydrogen sulphide, nitrogen, oxygen, hydrogen and other substances. Percentage of individual components is presented in Table 1.

The elements found in most biogas plants are pre-dam biomass, fermentation tank, covered airtight membrane, post-fermentation tank or lagoon CHP system generating electricity and heat, plumbing, safety, electrical, including control systems that integrate all the elements in functional unit, connection to the grid and heat. The technological process and the use of the substrates affect the composition of the resulting biogas, including methane content, which provides a gross calorific value of biogas. The larger the percentage of methane, the higher the calorific value of biogas. The ability to adjust the contribution of each waste for the generation of biogas of high.

The functioning of biogas carries a number of benefits. The use of organic waste in an unfermented form emits significant amounts of methane. Obtaining methane in biogas by the controlled fermentation and used for energy production can reduce emissions of methane and other greenhouse gases from the decomposition of animal manure. Wastes subjected to anaerobic digestion are a better fertilizer than unfermented manure. Combustion of the biogas is characterized by the significantly lower emissions of sulfur dioxide and nitrogen oxides, as compared to the combustion of fossil fuels. In addition, the formation of biogas plants provide additional jobs, the creation of a local source of energy, especially heat energy, which can be used to heat public buildings.

Technical possibilities of utilization of the energy contained in the biogas include the following variants [Dudek, Zaleska-Bartosz 2010]:
- direct combustion of gas boilers, thermal devices,
- production of electricity in gas engines with power generator,
- cogeneration or trigeneration (production of heat and electricity in combination)
- the production of biomethane, which can be injected into the natural gas distribution networks, used in industrial processes or as a transport fuel.

Table 1. The chemical composition of biogas from agricultural biogas plant [Steppa 1988]

Component of biogas	Concentration [%]
Methane (CH_4)	52 – 85
Carbon dioxide (CO_2)	14 – 48
Hydrogen sulphide (H_2S)	0.08 – 5.50
Hydrogen (H_2)	0.0 – 5.5
Carbon monoxide (CO)	0.0 – 2.1
Nitrogen (N_2)	0.6 – 7.5
Oxygen (O_2)	0.0 – 1.0

ORGANIC WASTE USED IN BIOGAS PLANTS

The primary sector, of which organic waste can be used in biogas plants include the [Curkowski at al. 2013]:
- agriculture,
- meat industry,
- breweries,
- distilleries,

- dairies,
- fruit and vegetable processing.

The main waste generated by agriculture are natural fertilizers such as manure, urine and manure from pig farms and cattle. The possibility of their agricultural use are limited by periods of fertilization and the requirement is not exceeded the limit dose. Is meaningful for biogas installations due to the fact that they are good stabilizers of the fermentation process and to reduce the need for dilution water because the increase hydration of the batch.

The meat waste can include, among others, the contents of the gastrointestinal tract and blood. Depending on the detailed classification in terms of epidemics, the feed can be subjected to thermal utilization. An alternative might be to use some of the waste in biogas plants. They have a high energy value and improve the dynamics of fermentation.

Brewers Brewing, represent about 77% of waste organic matter in the production of beer. Their recovery as a result of drainage or compaction is expensive. They can, however, be used in biogas plants as rich in nutrients, such as fiber, protein and fatty Bevelled, can be used for fodder purposes, and are characterized by a high yield of methane [Kasprzak 2012].

Distillers due to the short suitability for storage and a low dry matter content make it difficult to dispose. Dried decoction process of feed utilization requires the use of energy intensive processes, such as: filtration, drying, evaporation, phase separation). Distillers can replace manure biogas plants in the process because of its nutritional value (rich in minerals, fats and vitamins) and high hydration [Hanczakowska 2005].

Whey represents about 80–90% of the volume of waste from the production of milk and cheese. It is necessary to fractionation solid and liquid components, for example by the use of membrane filtration for industrial development. High hydration whey, rich in proteins, fats and lactose and the ability to produce high-biogas makes it a suitable substrate for use in a biogas plant [Jodłowski 2008].

Fruit and vegetable waste generated mainly such as bagasse, primarily grapes, apples, carrots, potato pulp and beet pulp. They are impermanent and unstable material, requiring immediate processing. This poses a threat to the rapid growth of microbiological contaminants. Therefore, storage requires fixation (ensiling or drying). Up to 80%

of organic matter from bagasse can be converted into biogas which is rich in nutrients: fiber, carbon and nitrogen compounds, pectins and polysaccharides [Tarko et al. 2012, Misiura 2013, Kuczyńska at al. 2011].

CONCLUSIONS

1. Waste that cause difficulties in rendering can be used successfully in agricultural biogas plants.
2. The use of biomass is the best way to produce energy from renewable sources.
3. The development of biogas enables the structural reconstruction of power based on the sustainable development of agriculture energy, the flow of private capital to rural areas, as well as the stabilization of energy supply.
4. Disposal of waste in agricultural production and food processing helps preserve ecological safety.
5. A reduction in CO_2 emissions in the production of electricity and heat.
6. The development of biogas plants, not only in agriculture, is in line with the implementation of the commitments to the European Union.

Acknowledgments

This article is part of the work of the registered S/WBiIŚ/2/2011.

REFERENCES

1. Bliht G.E. at al.: The effect of waste composition on leachate an gas quality: a study in South Africa. Waste Management & Research, Vol. 17, 1999, 124-140.
2. Curkowski A., Oniszk-Popławka A., Haładyj A.: Biogas – a deliberate choice. Foundation Institute for Sustainable Development. Warsaw 2013.
3. Dudek J., Zaleska-Bartosz J.: Acquisition and use of biogas for energy purposes. Problems of Ecology, Vol. 14, No. 1, 2010.
4. Directive 2001/77/WE.
5. Gniazdowski J.: Performance evaluation of biogas for the planned biogas plant at the dairy farm. Problems of Agricultural Engineering 3, 2009, 67-73,.
6. Hanczakowska E.: Dried Distillers' Grains (DDGS) in swine nutrition. Department of Animal Nutrition and Paszoznawstwa, Institute of Animal Production – National Research Institute in Cracow, Gauteng. Publisher Equipment Sp. z o.o., 2005.

7. Jodłowski P.J., Jodłowski G.S.: Whey as a starting material for the biogas in the methane fermentation process. Conference of Young Scientists, Krakow 2008.

8. Kasprzak J.: Environmental determinants of economic recirculation in the brewing industry. Engineering and Chemical Equipment No. 5, 2012.

9. Kotlicki T., Wawszczak A.: Waste incineration in power boilers. Mining and Geoengineering, 35(3), 2010.

10. Kucharczyk K., Stępień W., Gworek B.: Composting of municipal waste as a method of recycling organic matter. Environment and Natural Resources, Vol. 42, 2010, 240-254.

11. Kuczyńska I., Nogaj A., Pomykała R.: Waste in biogas production. Thu. II. Recycling 10(130), 2011.

12. Ledakowicz S., Krzystek L.: The use of methane fermentation in waste agri-food industry. Biotechnology 3(70), 2005, 165-183,.

13. Manczarski P.: Mechanical-biological treatment and disposal of waste in light of the new legislation in force. [In:] Waste Management New Regulations, Ch. 1, Polish Association of Sanitary Engineers and Technicians Poznań 2012, 117-144.

14. Marchwińska E., Budka D.: The problem of waste in terms of public health. Access on: 11.01.2014: http://www.srodowiskoazdrowie.pl/wpr/Aktualnosci/Czestochowa/Referaty/Marchwinska.pdf?f2 7ba39e183cc4811d3754669e5fce7a=96a08867e4 09ac927ff0b619a555c326.

15. Misiura A.: By-products of fruit and vegetable industry and its use for fodder purposes. University of Life Sciences in Lublin, Faculty of Production Engineering, Cattle Breeder 3, 2013.

16. Pilarski K., Adamski M.: Perspectives of biogas production with taking into consideration reaction mechanism in the range of quantitative and qualitative analyses of fermentation processes. J. Res. and Applic. in Agric. Engin., 54(2), 2009, 81-86,.

17. The Biogas Invest 2012 Renewable Energy Institute, Warsaw 2012.

18. Schink B.: Energetics of Syntrophic Cooperation in Methanogenic Degradation Microbiology and Molecular, Biology reviews, 61(2), 1997, 262-280.

19. Schlesinger W.H.: Biogeochemistry. An analysis of global change. Academic Press, San Diego, 1997, 231-238.

20. Smith M.R., Zinder S.H., Mah R.A.: Microbial methanogenesis from acetate, Proc. Biochem., 15, 1980, 34-39.

21. Steppa M.: Agricultural biogas plants. IBMER, Warsaw 1988.

22. Tarko T., Duda-Chodak A., Bebak A.: The biological activity of selected fruit and vegetable pomace, Food. Learning. Technology. Quality, 4(83), 2012.

23. Yamamura K.: Current status of waste management. [In:] Japa. Waste Management & Research, Vol. 1, 1983, 1-15.

27

DECOMPOSITION OF TARS IN MICROWAVE PLASMA – PRELIMINARY RESULTS

Mateusz Wnukowski[1]

[1] Institute of Heat Engineering and Fluid Mechanics, Wrocław University of Technology, Wybrzeże Wyspiań-skiego 27, 50-370 Wrocław, Poland, e-mail: mateusz.wnukowski@pwr.wroc.pl

ABSTRACT

The paper refers to the main problem connected with biomass gasification - a presence of tar in a product gas. This paper presents preliminary results of tar decomposition in a microwave plasma reactor. It gives a basic insight into the construction and work of the plasma reactor. During the experiment, researches were carried out on toluene as a tar surrogate. As a carrier gas for toluene and as a plasma agent, nitrogen was used. Flow rates of the gases and the microwave generator's power were constant during the whole experiment. Results of the experiment showed that the decomposition process of toluene was effective because the decomposition efficiency attained above 95%. The main products of tar decomposition were light hydrocarbons and soot. The article also gives plans for further research in a matter of tar removal from the product gas.

Keywords: tar, gasification, microwave plasma.

INTRODUCTION

With an increasing demand for fuels and their decreasing resources at the same time, a much higher pressure is put on alternative, renewable energy sources. Biomass is definitely one of them and nowadays it is one of the most important new energy resources. Significance of biomass comes from the fact that it is easy to obtain, widely and relatively evenly available all over the world and it is relevantly cheap. In addition, biomass is considered to be neutral with respect to CO_2 emission.

There are a few ways of biomass utilization in a field of energy production i.e. fermentation, combustion, pyrolysis and gasification. Gasification is a process that has recently attracted high interest, mainly because of the fact, that it allows to transform solid biomass into much more useful (in the meaning of transport and diversity of application) products. As a result of gasification mostly gaseous products are obtained. Their concentration depends on the type of gasifier, a sort of biomass and process parameters. The produced gas may be used for a production of heat, electricity or as a raw chemical material for synthesis of liquid fuels or other chemicals.

The major drawback related to the biomass gasification is a presence of solid particulates and tar in a produced gas. While the solid particles can be easily separated, the main problem is tar, which at a high temperature, is in a vapor state – what makes it difficult to separate. Upon condensation, tar blocks downstream pipelines and foul engines and turbines (1). Tar is a product of high temperature reactions of a cellulose, hemicellulose and lignin depolymerization. Tars are characterized and classified by few, similar classification's systems (1). All of them state that tar is a mixture of heterocyclic, aromatic and poly-aromatic hydrocarbons.

The amount of produced tar strongly depends on a gasification process (type of gasifier and work parameters) and it is in the range of 0.5 to 100 g/Nm^3 (2). At the same time, the allowed concentration of tars for a gas used in ICE (internal combustion engines) is stated on a level of 50 to 100 mg/Nm^3 and below 5 mg/m^3 for gas turbines (1). As a result, the presence of tar in the product gas precludes one of the most important applications of biomass gasification.

Although it is clear that the gasication process is promising, its application in larger and wider scale requires a reliable and efficient way of removal/conversion of tar from a produced gas.

Many attempts have been made in order to develop methods which allow the purification of gasification products from tar. These include primary methods, which are based on a gasifier design and biomass properties (2) and secondary methods i.e. mechanical (3), thermal (4), catalytic (5), and plasma methods (6), (7), (8). The last two methods show the highest efficiency of tar removal (9), (10). It is also worth to mention that these two methods can be matched together, showing even a higher efficiency in consequence (11) – that is a further goal for the authors of this article.

Plasma is a high temperature, strongly ionized medium with high concentration of electrons, ions, radicals and excited molecules and atoms. Such a composition makes plasma a highly reactive medium. Therefore, plasma may be considered as a coupling of a temperature source and a catalytic medium. The chemical reactions taking place during the high temperature tar decomposition can be described as follows (10):

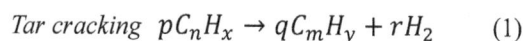

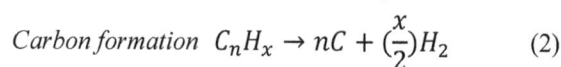

$$Tar\ cracking\quad pC_nH_x \rightarrow qC_mH_y + rH_2 \qquad (1)$$

$$Carbon\ formation\quad C_nH_x \rightarrow nC + (\frac{x}{2})H_2 \qquad (2)$$

where: C_nH_x represents tar and C_mH_y represents a hydrocarbon with a carbon number smaller than that of C_nH_x.

The lab-scale investigations were carried out on the microwave plasma application for decomposition of toluene as a tar surrogate. The preliminary results were to show the efficiency of the applied method and to identify products of the decomposition process.

THE LABORATORY SET-UP

Figure 1 presents a scheme of the laboratory research installation. The most important part of the installation is a microwave plasma reactor of the tubular type (4) presented in Figure 2. The main two reactor's elements are microwave generators (1) and quartz tube. Three microwave generators (Promis), each of 2 kW power, provided microwaves of 2.45 GHz frequency. Microwaves radiation was absorbed by gas providing energy into it and as a consequence, exciting and

ionizing gas molecules and creating plasma. Plasma was generated at the top of the quartz reactor and transported down the reactor by the carrier gas flow. A length of the quartz tube was 1200 mm and depending on the gas flow, the plasma can reach up to one third of the tube's length. The inner diameter of the tube was 60 mm. It is worth to notice that this type of reactor, in opposite to plasma arc for example, it was hardly ever used in a tar removal process in other research and, therefore, its impact on tar decomposition is not known very well. Its advantage is a simple and compact construction, what might be a great convenience in coupling it with a catalyst bed.

In this study, nitrogen was used as a plasma agent. It is not without a reason if we consider that nitrogen is the main ingredient in a generator gas obtained through gasification with air. It is planned to use a mixture of gases simulating syngas instead of pure nitrogen – this will prevent syngas dilution. Nitrogen was supplied from a steel bottle with pressured gas (1). Nitrogen was separated on three streams – all flow rates were controlled with a use of mass flow meters (2) (Aalborg GFM 67 and XFM 47). Two of the streams were used only for the reactor feeding purpose – one of them was to provide molecules for plasma creating (it had axial flow) and the purpose of the second one is to protect the quarts tube from a high temperature (it had swirl flow). The third stream was used as a carrier gas for toluene stored in a glass tank. The temperature of toluene was measured due to the fact that it has an influence on toluene vapors pressure.

The products of the decomposition process of toluene were analyzed with a use of two devices: gas chromatograph (6) and stationary gas analyzer (7). The chromatograph used in the research was HP 6890 with a HP-5 (Crosslinked 5% PH ME Siloxane) column and the flame ionization detector (FID). The GC analysis required sampling with a syringe (5) at the outlet of the reactor. The gas analyzer was GAS 3000 (GEIT Europe) which evaluated concentrations of CO, CO_2, CH_4 and H_2. The analyzer was connected with a computer and worked online through the whole experiment.

EXPERIMENTAL

Toluene was used in many studies on tar decomposition as a tar surrogate (11), (12). This de-

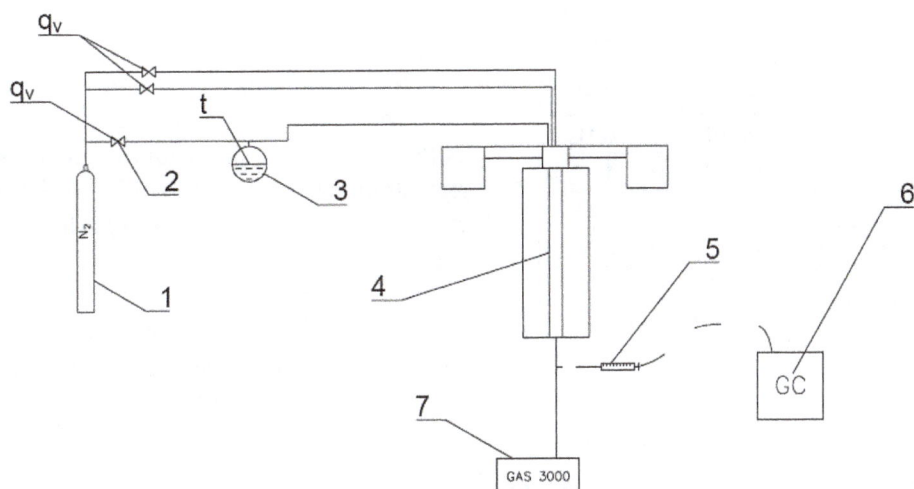

Figure 1. Scheme of the installation for plasma tar decomposition

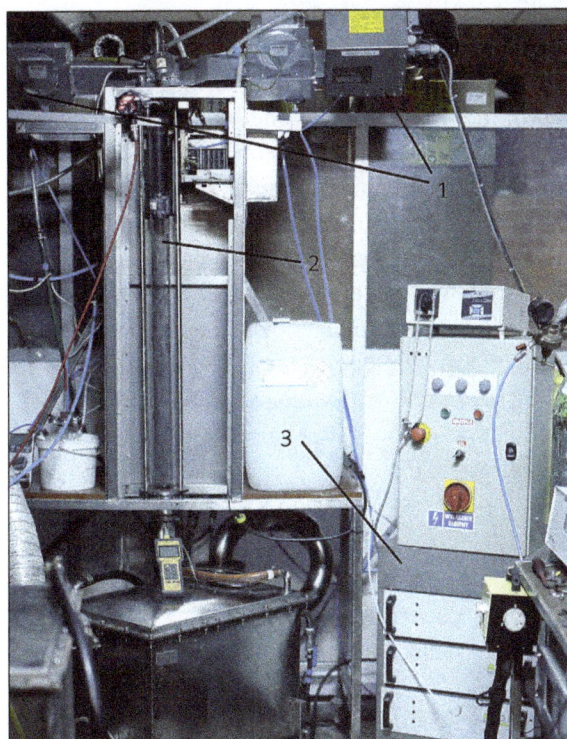

Figure 2. The microwave plasma reactor: 1 – microwave generators, 2 – quartz tube, 3 – power supply

cision was mainly justified by a significant concentration of toluene in a tar from biomass gasification process – it is one of the main compounds that create tar (13). Approach like this, also simplifies analyses and clarifies measurements allowing for a more precise interpretation of results and comparison of them with other studies. Therefore, in our experiment toluene was used as a model compound.

The first step in research was to find out what concentration of toluene could be achieved with a different flow of carrier gas. During scaling, as well as in the main experiment, the plasma agent and protective gas flows were on a level of 15 l/min. With this flow the plasma work was stable, the toluene concentration was not diluted and nitrogen consumption was on a low level. Flow of the carrier gas was regulated on the following levels: 3, 5, 10, 15 and 20 l/min. The results achieved for different carrier gas flow are given in Table 1. As it can be seen in Figure 3, the relationship between carrier gas flow and toluene concentration/amount is not linear. It is assumed that at the beginning of increasing gas flow, velocity of gas is too high to be saturated with toluene vapors. With further flow increase, however, the entrainment mechanism may have a greater influence on toluene transport - but at the same time it also had an impact on toluene dilution. Thus, although the amount of toluene may have slightly increased, its concentration kept decreasing. Despite that, the results show clearly that the highest concentration/amount of toluene was achieved for the lowest gas flow – that is 3 l/min. During the scaling samples were taken from an inlet port at the top of the reactor through which toluene is introduced into the reactor. During the scaling the toluene had an ambient temperature of 22 °C.

For the decomposition experiment, a carrier gas flow rate of 5 l/min was chosen. This choice was dictated by two reasons: a demand of testing the installation in conditions of high toluene concentration and a problem with stabilizing the gas flow on a lower level where every deviation may have an impact on the results. During decomposition investigations toluene stored in the bottle had an ambient temperature of 20 °C. Only two

Table 1. Parameters obtained for toluene feeder scaling

q_v l/min	Toluene concentration		F g/min
	ppm	g/Nm³	
3	49338	185.07	6.11
5	37092	139.13	4.87
10	31704	118.92	4.76
15	31291	116.31	5.28
20	27369	102.66	5.13

Explainations: q_v – volumetric flow of carried gas, F – mass flow of toluene.

microwaves generators were used in these experiments. Their power was set up on a maximum, which was 2 kW for each one.

In both, scaling and decomposition research, from three to five samples of gas were taken, depending on a repeatability of measurements.

For the determination of toluene decomposition efficiency the samples were taken at the outlet of the reactor. The temperature at the outlet of the plasma reactor was about 22 °C. Just before the principal toluene removal process, the measurements similar to those for feeder scaling were carried out in a purpose of evaluating inlet toluene concentration (C_0) – it was 131 g/m³.

Additionally, during the toluene decomposition process the GAS 3000 analyzer was measuring concentrations of CO, CO_2, CH_4 and H_2. The chromatograph program was set up on 70 °C (the column temperature) and 3 minutes (previous research shown that after the toluene's peak none other were shown up to 15 minutes). Identification of decomposition products was done with a use of GC (HP6890) with MS (HP 5973) by an external laboratory (Laboratory of Gas Chromatography, Department of Polymer and Carbon Materials, Wroclaw University of Technology).

RESULTS AND DISCUSSION

Preliminary investigations were performed to recognize a potential of the method. The carrier gas (5 l/min) was doped with toluene to bring its concentration to approx. $C_0 = 130$ g/m³. The products of toluene decomposition were identified and the effectiveness of destruction was calculated for the selected experiment.

Figure 4 shows a chromatogram obtained from the analysis of toluene decomposition products. There are four main products and marginal amounts of other compounds. Qualitative analyzes allowed to identify three products: methane, benzene and toluene. The fourth product of toluene decomposition is some light compound (with retention time of 1.507 and peak's area of 12.07 pA·s) that is hard to define because its identification is disturbed by the presence of other light compounds such as methane, nitrogen and carbon oxide. Some deeper and extended further analyses all required for identification of that compound, however, according to other observations (14), (15) it is suggested that it might be C_2 compound like ethane or ethylene.

The retention times, average concentrations of toluene, organic products of its decomposition and conversion efficiencies are given in Table 2.

The effectiveness of toluene conversion in the decomposition process was calculated with a use of the following formula:

$$X = \left(1 - \frac{C_m}{C_0}\right) \cdot 100\% \qquad (3)$$

where: X – toluene conversion efficiency, %
C_m – concentration of toluene after decomposition, g/m³
C_0 – concentration of toluene before decomposition – 131 g/m³

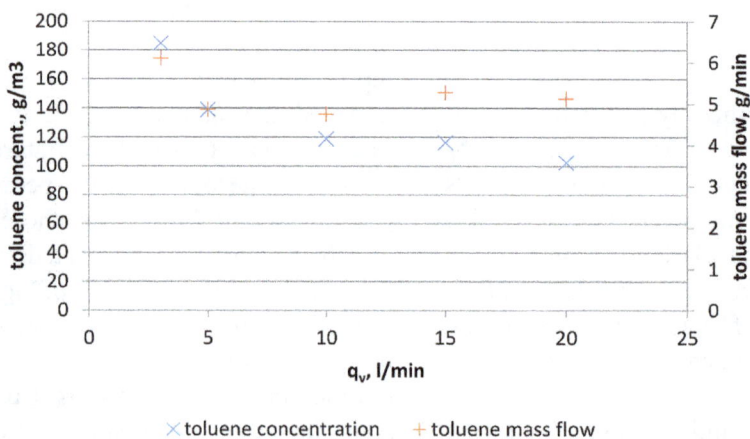

Figure 3. Relationships between carried gas flow and toluene mass flow and concentration

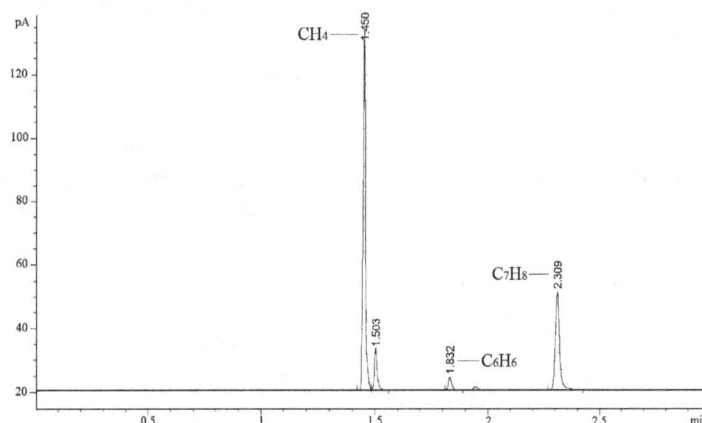

Figure 4. Chromatogram of the toluene decomposition process

Table 2. Concentrations of toluene and selected products of decomposition and efficiency of conversion

Compound	Retention time [min]	Peak's area [pA·s]	Concentration C_m [g/m³]	Conversion [%]
Toluene	2.309	42.75	5.89	95.50
Benzene	1.832	5.96	0.08	0.06
Methane	1.450	97.43	2.03	1.55

The conversion degree of toluene conversion into benzene and methane was calculated as a ratio of obtained product and introduced toluene:

$$X_{CH_4, C_6H_6} = \left(\frac{C_m}{C_0}\right) \cdot 100\% \qquad (4)$$

No hydrogen or carbon dioxide were detected by the Gas 3000 analyzer in the produce gas in the reactor outlet. Concentration of CO was on a level of 0.06%. A presence of CO is intrigue since no oxygen was introduced – this might be explained with a small, unavoidable leaks and air sucking. Also a lack of hydrogen might be surprising. This may be explained in a few, not excluding ways. Firstly, hydrogen might have been adsorbed on soot. Secondly, considering air leaks, some of hydrogen might have been burned into water. Finally, it is possible that hydrogen was used in a process of hydrocracking of toluene and benzene.

Beside those products, it is also important to mention that noticeable amounts of soot were produced during the toluene decomposition, that is typical for that kind of process and was mentioned in other publications (16). Matching those concentrations of inorganic compounds with those for organic, shows that most of toluene was in fact transformed into soot.

If the structure of the toluene is considered, it seems reasonable and intuitive that such a compound would convert into methane and benzene. However, the results show that the amount of benzene is far smaller than methane. This may suggest that benzene was converted into methane, soot and probably some other lighter hydrocarbons. Those reactions, explaining high concentration of soot, lack of hydrogen and disproportion between benzene and methane, might be presented by the following simplified chemical formulas:

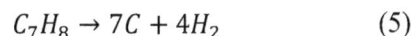

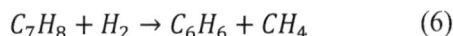

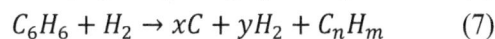

$$C_7H_8 \rightarrow 7C + 4H_2 \qquad (5)$$
$$C_7H_8 + H_2 \rightarrow C_6H_6 + CH_4 \qquad (6)$$
$$C_6H_6 + H_2 \rightarrow xC + yH_2 + C_nH_m \qquad (7)$$

where C_nH_m represents a hydrocarbon with a carbon number smaller than benzene and which might be an unidentified compound that was previously mentioned. More precise chemical mechanism requires further researches including identification of more compounds.

CONCLUSION

The presented results of the initial lab-scale experiment leads to the following conclusions:
1. The microwave plasma is highly effective in toluene decomposition (> 95%), which indicates that this type of installation might be used for tar removal from a syngas.
2. The main product of toluene conversion is soot, which can be easily separated from a gas with a use of mechanical devices such as cyclones of fabric filters.

3. In conversion process some small amounts of hydrocarbons, lighter thn toluene were also obtained, which can increase caloric value of syngas.

Next steps in research will include water steam introduction into the reactor, what should allowed to transform soot into hydrogen and carbon monoxide, and use of a catalyst to increase the conversion even more. Besides that, the impact of plasma generator's power and plasma agent gas composition on the decomposition process will be investigated.

REFERENCES

1. Anis S., Zainala Z.A. Tar reduction in biomass producer gas via mechanical, catalytic and thermal methods: A review. Renewable and Sustainable Energy Reviews, 15, 2011, 2355-2377.

2. Devi L., Ptasinski K.J., Janssen F.J.J.G. A review of the primary measures for tar elimination in biomass gasi cation processes. Biomass and Bioenergy, 24, 2003, 125-140.

3. Unal W.J.O., J. Andries, Hein K.R.G. Biomass and fossil fuel conversion by pressurised fluidised bed gasification using hot gas ceramic filters as gas cleaning. Biomass and Bioenergy, 25, 2003, 59-83.

4. Bridgwater A.V. The technical and economic feasibility of biomass gasification for power generation. Fuel, Vol. 74, 1995, 631-653.

5. Robert R.Z., Brown C., Suby A., Cummer K. Catalytic destruction of tar in biomass derived producer gas. Energy Conversion and Management, 45, 2004, 995-1014.

6. Nair S.A., Pemen A.J.M., Yana K., van Gompel F.M., van Leuken H.E.M., van Heesch E.J.M., Ptasinski K.J., Drinkenburg A.A.H. Tar removal from biomass-derived fuel gas by pulsed corona discharges. Fuel Processing Technology, 84, 2003, 161-173.

7. Pikoń K., Czekalska Z., Stelmach S., Ścierski W.

Zastosowanie metod plazmowych do oczyszczania gazu procesowego ze zgazowania biomasy. Archiwum Gospodarki Odpadami i Ochrony Srodowiska, 12, 2010, 61-72.

8. Wacławiak K. Research area for reactors with electric spark discharge, producing low-temperature plasma for cleaning of gas. Archives of Waste Management and Environmental Protection, 16, 2014, 69-76.

9. Tiejun Wang Jie Chang, Xiaoqin Cui, Qi Zhang, Yan Fu. Reforming of raw fuel gas from biomass gasification to syngas over highly stable nickel–magnesium solid solution catalysts. Fuel Processing Technology, 87, 2006, 421-428.

10. Tippayawong N., Inthasan P. Investigation of light tar cracking in a gliding arc plasma system. International Journal of Chemical Reactor Engineering, 8, 2010, 1-16.

11. Kai Tao Naoko Ohta, Guiqing Liu, Yoshiharu Yoneyama, Tiejun Wang, Noritatsu Tsubaki. Plasma enhanced catalytic reforming of biomass tar model compound to syngas. Fuel, 104, 2013, 53-57.

12. Baofeng Zhao Xiaodong Zhang, Lei Chen, Rongbo Qu, Guangfan Meng, Xiaolu Yi, Li Sun. Steam reforming of toluene as model compound of biomass pyrolysis tar for hydrogen. Biomass and Bioenergy, 34, 2010, 140-144.

13. Salvado R.C.J., Farriol X., Montane D. Steam reforming model compounds of biomass gasification tars: conversion at different operating conditions and tendency towards coke formation. Fuel Processing Technology, 74, 2001, 19-31.

14. Andreas J. Mechanisms and kinetics of thermal reactions of aromatic hydrocarbons from pyrolysis of solid fuels. Fuel, 75, 1996, 1441-1448.

15. Young Nam Chun Seong Cheon Kim, Kunio Yoshikawa. Removal characteristics of tar benzene using the externally oscillated plasma reformer. Chemical Engineering and Processing: Process Intensification, 57-58, 2012, 65-74.

16. Nogueira R.M.E., Sobrinho A.S.S., Couto B.A.P., Maciel H.S., Lacava P.T. Tar Reforming under a Microwave Plasma Torch, 27, 2013, 1174-1181.

28

POTENTIAL APPLICATIONS OF SOS-GFP BIOSENSOR TO *IN VITRO* RAPID SCREENING OF CYTOTOXIC AND GENOTOXIC EFFECT OF ANTICANCER AND ANTIDIABETIC PHARMACIST RESIDUES IN SURFACE WATER

Marzena Matejczyk[1], Stanisław Józef Rosochacki[1]

[1] Department of Sanitary Biology and Biotechnology, Faculty of Civil Engineering and Environmental Engineering, Bialystok University of Technology, Wiejska 45E, 15-351 Bialystok, Poland, e-mail: m.matejczyk@pb.edu.pl

ABSTRACT

Escherichia coli K-12 GFP-based bacterial biosensors allowed the detection of cytotoxic and genotoxic effect of anticancer drug– cyclophosphamide and antidiabetic drug – metformin in PBS buffer and surface water. Experimental data indicated that *recA::gfpmut2* genetic system was sensitive to drugs and drugs mixture applied in experiment. *RecA* promoter was a good bioindicator in cytotoxic and genotoxic effect screening of cyclophosphamide, metformin and the mixture of the both drugs in PBS buffer and surface water. The results indicated that *E. coli* K-12 *recA::gfp mut2* strain could be potentially useful for first-step screening of cytotoxic and genotoxic effect of anticancer and antidiabetic pharmacist residues in water. Next steps in research will include more experimental analysis to validate *recA::gfpmut2* genetic system in *E. coli* K-12 on different anticancer drugs.

Keywords: SOS-gfp biosensor, cytotoxicity, genotoxicity, cyclophosphamide, metformin.

INTRODUCTION

Cyclophosphamide, ifosfamide, methotrexate, 5-fluorouracil, taxol, vinca alkaloids and platinum compounds are commonly used as chemotherapeutic agents on cancer. Cytostatic compounds have a generally polar structure and have been detected in hospital wastewaters, the influents and effluents of WWTPs and surface waters [Besse et al. 2012, Zhang et al. 2013, Yu-Chen Lin et al. 2014]. According to a review by Kosjek and Heath [Kosjek and Heath 2011], last studies have been primarily focused on hospital effluents, and only a few of them have focused on environmental samples and their fate. Most cytostatic compounds are not likely to undergo biodegradation or volatilization processes, and a limited number of studies has reported their degradation by sunlight photolysis [Kosjek and Heath 2011].

Cyclophosphamide (CP) – a cytotoxic agent that alkylates DNA, has a wide spectrum of clinical uses in the chemotherapy treatment of many neoplastic diseases. The acute toxicities of CP are associated with its genotoxicity. CP was detected worldwide at ng/l to µg/l levels in surface water (<30–64,8 ng/l) [Kosjek and Heath 2011, Besse et al. 2012, Zhang et al. 2013, Yu-Chen Lin et al. 2014].

The antidiabetic drug metformin is among the pharmaceuticals with the highest production amounts world-wide. Measurements performed in sewage and surface waters, showed an almost ubiquitous presence of metformin in the aquatic environment. Metformin was found in all investigated river waters. Concentration levels depend on the sewage fraction of the analyzed waters and in most rivers they are in the range of several to 100 ng/l, i.e. in the same order of magnitude or even higher than for other relevant pharmaceutical residues [Scheurer et al. 2009, Quinn et al. 2013].

The pharmaceutical residues of cyclophosphamide and metformin were detected worldwide at ng/l to µg/l levels in environmental samples (influents and effluents, surface water). Due to their highly potent mechanism of action (they directly or indirectly act with structure and function of DNA) these specific groups of drugs are conceived to be hazardous to living organisms and human health. There is a need to target them with environmental significance, quantify them, and assess their cytotoxic and genotoxic risk to living organisms [Scheurer et al. 2009, Kosjek and Heath 2011, Besse et al. 2012, Quinn et al. 2013, Zhang et al. 2013, Yu-Chen Lin et al. 2014].

In case of cytotoxicity and genotoxicity assessment bioassays are valuable bacterial assays based on genetically modified bacteria carrying a SOS-regulon, DNA damage-inducible *recA* promoter upstream of enhanced-mutant variant of *gfp* gene that is expressed when DNA repair is induced by chemical agents. So far, several constructs were tested, the SOS-*gfp* biosensor with *recA* promoter was found to be highly sensitive to the detection of carcinogens, cyto- and genotoxins. In such living cell systems, bacteria are especially attractive due to their rapid growth rate, low cost, and easy handling [Ptitsyn et al., 1997, Kostrzyńska et al. 2002, Zaslaver et al. 2004, Matejczyk 2010, Alhandrami and Paton 2013, Park et al. 2013].

The aim of the present study was to evaluate the potential applications of SOS-*gfp* biosensor for *in vitro* rapid screening of cytotoxic and genotoxic effect of residues of anticancer and antidiabetic pharmacist in water. In experiment reporter strains of *Escherichia coli* K-12 *recA::gfpmut2* with a plasmid-borne transcriptional fusion between DNA-damage, genotoxin inducible *recA* promoter involved in the SOS regulon response and fast folding GFP variant reporter gene-*gfpmut2* was used.

In the presented data more stable and fast folding mutant of *gfp* gene – *gfpmut2* with excitation and emission wavelengths of 485 and 507 nm was used [Zaslaver et al. 2004].

MATERIALS AND METHODS

Bacteria strain and plasmid. In the experiment genetically modified *Escherichia coli* K-12 MG1655 logarythmic phase cells: *Escherichia coli* K-12 *recA::gfpmut2* and *Escherichia coli* K-12 *promoterless::gfpmut2* were used (a gift from Prof. Uri Alon, Department of Molecular Cell Biology & Department of Physics of Complex Systems, Weizmann Institute of Science Rehovot, Israel). They contained a pUA66 plasmid-borne transcriptional fusion between DNA-damage, genotoxin-sensitive *recA* promoter involved in the SOS regulon response and fast folding GFP variant reporter gene-*gfpmut2* [Zaslaver et al. 2004].

Bacteria growth condition. *Escherichia coli* K-12 MG1655 strains: *Escherichia coli* K-12 *recA::gfpmut2* and *Escherichia coli* K-12 *promoterless::gfpmut2* were cultured overnight in LB agar medium (Merck, Germany) at 30 °C supplemented with 100 µg/ml of kanamycin (Sigma-Aldrich, Germany). Colonies were carried to LB broth medium (10 g NaCl, 10 g tryptone and 5 g yeast extract per 1000 ml of distilled water) with 100 µg/ml of kanamycin and incubated overnight at 30 °C. After that, bacteria cultures were refreshed in LB broth medium with 100 µg/ml of kanamycin and were cultivated to logarithmic phase of growth (2 hour cultivation). Following that, the cells were washed with PBS buffer (1.44 g Na_2HPO_4, 0.24 g KH_2PO_4, 0.2 g KCl, 8 g NaCl per 1000 ml of distilled water, pH=7).

Monitoring of bacteria growth and bacteria concentration. At the start of the experiment the initial bacteria cells density was standardized to OD = 0.2 (Optical Density) value by the use of spectrophotometer (Perkin Elmer Enspire 2300) at wavelength of 600 nm. The concentration of bacteria cells per ml of PBS was assessed by series dilutions system and expressed as Colony Forming Units per ml (CFU/ml) values. The growth dynamic of bacteria strains treated with CP (Sigma Aldrich, USA Company) and metformin (Bialystok pharmacy) was monitored with the use of standard spectrophotometer analysis of Optical Density values at wavelength of 600 nm. The values of bacteria growth inhibition (GI) during the treatment with drugs at the start of bacteria incubation with drugs - time 0 and after 3 and 24 hours were calculated according to the formula: GI (%) = OD_{CS} (%) – OD_{DS} (%), Where: OD_{CS} (%) – Optical Density of control sample =100%, OD_{DS} (%) – Optical Density of bacteria samples treated with drugs.

Bacteria cells treatment with cyclophosphamide and metformin: 1 ml of stationary phase bacteria cells (1×10^8 CFU/ml; OD=0.2) were suspended in 4 ml of PBS buffer and the following drugs were used in testing: cyclophosphamide (CP), metformin (M) and CP+metformin (CP+M) in five different concentrations, for CP: 0,0001; 0.001; 0.01; 0.1 and 1 mg/ml; for metformin: 0.3; 0.7 and 1 mg/ml. For CP+metformin three different combinations were used: (1) 0.0001+0.3; 0.001+0.3; 0.01+0.3; 0.1+0.3 and 1+0.3 mg/ml; (2) 0.001+0.7; 0.001+0.7; 0.01+0.7; 0,1+0,7 and 1+0.7 mg/ml and (3) 0.001+1; 0.001+1; 0.01+1; 0,1+1 and 1+1 mg/ml. Bacteria strains were incubated with drugs in 3 and 24 hours at 30 °C. Drugs concentrations were selected experimentally and after reviewing the reference recommendation [Rhizos and Elisaf 2013]. The time of bacteria incubation with drugs (3h and 24 h) was estimated for monitoring the sensitivity of recA::gfp genetic construct for quickly (3 h) and later (24 h) response. The control sample – Escherichia coli K-12 recA::gfpmut2 strain in PBS buffer was not treated with drugs. To verify the correct activity of recA promoter, Escherichia coli K-12 strain containing pUA66 plasmid without promoter – Escherichia coli K-12 promoterless::gfpmut2 was used as a control. Additionally, to assess genotoxic sensitivity of recA::gfpmut2 construct, 4% acetone was used as a negative control and 50 μM methylnitronitrosoguanidine (MNNG, known genotoxin) as positive control [Ptitsyn et. al. 1997, Kostrzyńska et. al. 2002].

Bacteria cells treatment with cyclophosphamide and metformin in surface water. Surface water samples were collected in sterile flasks from Białka river. Samples were sterilized by filtration. 1 ml of logarythmic phase bacteria cells (2×10^8 CFU/ml; OD=0.2) was suspended in 4 ml of surface water at combination of CP (0.1 mg/ml) and metformin (1 mg/ml) used in genotoxicity testing. Drug concentrations were selected to the highest stimulation of *gfp* gene expression in PBS buffer (for IF= 10.42). The conditions of bacteria incubations and the control protocols were the same as above.

Analytical method for the intensity of *gfp* gene fluorescence (FI) analysis. After exposition of bacteria cultures to tested drugs strains were washed with PBS buffer and the intensity of fluorescence of *gfp* gene in the volume of 1 ml of bacteria cells suspension (1×10^4 CFU/ml) in PBS buffer was measured with spectrofluorometer (Perkin Elmer Enspire 2300). The measurements were taken at excitation and emission wavelengths of 485 and 507 nm.

Assessment of SFI values. The specific fluorescence intensity (SFI) value which is defined as the fluorescence intensity (FI) divided by the optical density (OD) measured at each time point at 600 nm was calculated according to the following formula to detect the level of genotoxic activity of drugs: $SFI = \dfrac{FI}{OD}$, where: *SFI* – Specific Fluorescence Intensity, *FI* – the Fluorescence Intensity of the strains at excitation and emission wavelengths

of 485 and 507 nm, *OD* – Optical Density at 600 nm of the strains.

Detection of $S_{gfpexp.}$ and $I_{gfpexp.}$ values. For each concentration of the tested drugs the levels of stimulation of *gfp* ($S_{gfpexp.}$) or inhibition ($I_{gfpexp.}$) were calculated, according to the formulas:

- for the SFI values with an increase with the level of *gfp* expression (CP and CP+metformin) in comparison with the control sample: $S_{gfpexp.}$ (%) = SFI_{DS} (%) – SFI_{CS} (%), where SFI_{DS} (%) – SFI values for tested drugs sample, SFI_{CS} (%) – SFI for control sample, =100%,

- for the SFI values with a decrease with the level of *gfp* expression (metformin) in comparison to the control sample: $I_{gfpexp.}$ (%) = SFI_{CS} (%) – SFI_{DS} (%), SFI_{CS} (%) – SFI for control sample, =100%, SFI_{DS} (%) – SFI values for tested drugs sample.

Assessment of F_I values. For each concentration of the tested drugs induction factors (F_I) were calculated. $F_{I=}$ (FI_I/OD_0)/(FI_0/OD_I), where FI_I is the fluorescence intensity of the culture treated with DNA – damaging compound; FI_0 is the fluorescence intensity of the control sample without genotoxin; OD_I is the optical density at 600 nm of treated culture and OD_0 is the optical density of the control sample. The F_I, $S_{gfpexp.}$ and $I_{gfpexp.}$ values express the potency of genotoxic activity of both drugs.

Classification of tested drugs as genotoxins. The F_I values were calculated for classification of the tested drugs as genotoxins. According to Ptitsyn et. al. [1997] and Kostrzyńska et. al. [2002], genotoxin was identified as a chemical if its induction factor was 2 or more ($F_I \geq 2$).

Statistical analysis. Experiments were conducted in three independent series. Statistical data obtained in this study are expressed as mean ± standard deviation (SD) for n = 6. The data were analyzed with the use of standard statistical analyses, including one-way Student`s test for multiple comparisons to determine the significance between different groups. The values of *P*<0.05 were considered as significant.

RESULTS

Logarithmic phase *Escherichia coli* K-12 MG1655 *recA::gfpmut2* strain treatment with cyclophosphamide (CP) and metformin (M) showed that separate administration of both drugs caused a significant dose- and time-dependent decreased ($P<0.05$) in SFI value and increased the inhibition of *recA* promoter activity and intensified $I_{gfpexp.}$ value compared to non-treated cells in PBS buffer (control sample) (Table 1). A sustained decrease in SFI values was observed after 24 hours incubation of bacteria cells with both drugs.

3 h and 24 h simultaneous co-administration of CP+M to a logarithmic phase the bacteria (especially up to 24 h) significantly modulated CP activity and intensified the sensitivity of *recA* promoter and *gfp* gene expression and stimulated SFI, F_I ($F_I \geq 2$) and $S_{gfpexp.}$ values compared to control sample and samples treated separately with CP and M. Progressive significant stimulation of SFI, F_I and $S_{gfpexp.}$ values were obtained for higher concentration of CP (1; 0.1; 0.01; 0.001 mg/ml) and M (1; 0.7 mg/ml) during 3 h and 24 h incubation comparing to control sample. The maximum point for *recA* promoter stimulation was observed for co-adminiastration of 0.1 mg/ml CP and 1 mg/ml M in 24 h ($S_{gfpexp.} = 838\%$).

Table 1. SFI values for logarithmic phase E. coli K-12 recA::gfp mut2 treated with cyclophosphamide (CP), metformin (M) and combination of CP and metformin (CP+M) in three different metformin concentrations (0.3; 0.7; 1 mg/ml) in comparison with the control sample (bacteria strain in PBS buffer)

CP (mg/ml)	M (mg/ml)	t	Control sample SFI±SD	M SFI±SD	$I_{gfp\,exp}$ (%)	CP SFI±SD	$I_{gfp\,exp}$ (%)	CP+M SFI±SD	F_I	$S_{gfp\,exp}$ (%)
1	0.3	3	18.45±3.43	15.54±2.84*	15.8	16.23±2.3**	12	22.30±2.45*bc	–	–
		24	38.90±6.56	31.12±4.20a	20	19.34±2.73ab	50	103±8.20abc	2.65	165
	0.7	3	18.45±3.43	14.14±3.84*	15.8	16.23±2.3**	12	41.0±4.40abc	2.22	122
		24	38.90±6.56	27.65±4.43a	20	19.34±2.73ab	50	163±8.23abc	4.20	320
	1	3	18.45±3.43	13.70±2.03*	15.8	16.23±2.3*b	12	89±6.40abc	4.82	382
		24	38.90±6.56	25.30±3.43a	20	19.34±2.73ab	50	270±10.75abc	6.94	594
0.1	0.3	3	18.45±3.43	15.54±2.84*	15.8	17.14±2.22**	7	21.14±2.54*b*	–	–
		24	38.90±6.56	31.12±4.20a	20	21.14±2.73ab	55.66	130±8.43abc	3.34	134
	0.7	3	18.45±3.43	14.14±3.84*	15.8	17.14±2.22**	7	38±3.10abc	2.06	106
		24	38.90±6.56	27.65±4.43a	20	21.14±2.73ab	55.66	240±9.20abc	6.17	517
	1	3	18.45±3.43	13.70±2.03*	15.8	17.14±2.22*b	7	150±7.20abc	8.13	713
		24	38.90±6.56	25.30±3.43a	20	21.14±2.73a*	55.66	365±11.30abc	9.38	838
0.01	0.3	3	18.45±3.43	15.54±2.84*	15.8	17.93±1.92**	2.8	20.12±1.13*b*	–	–
		24	38.90±6.56	31.12±4.20a	20	22.30±2.45ab	42.67	98±5.45abc	2.52	152
	0.7	3	18.45±3.43	14.14±3.84*	15.8	17.93±1.92*b	2.8	32±2.22abc	–	–
		24	38.90±6.56	27.65±4.43a	20	22.30±2.45ab	42.67	160±7.24abc	4.11	311
	1	3	18.45±3.43	13.70±2.03*	15.8	17.93±1.92*b	2.8	121±6.23abc	6.56	556
		24	38.90±6.56	25.30±3.43a	20	22.30±2.45a*	42.67	200±9.17abc	5.14	414
0.001	0.3	3	18.45±3.43	15.54±2.84*	15.8	18.02±2.43**	2.33	24.22±2.46*b*	–	–
		24	38.90±6.56	31.12±4.20a	20	29.34±2.93**	24.58	120±6.56abc	3.08	208
	0.7	3	18.45±3.43	14.14±3.84*	15.8	18.02±2.43*b	2.33	36±3.74abc	–	–
		24	38.90±6.56	27.65±4.43a	20	29.34±2.93a*	24.58	180±8.40abc	4.63	363
	1	3	18.45±3.43	13.70±2.03*	15.8	18.02±2.43*b	2.33	110±6.10abc	5.96	496
		24	38.90±6.56	25.30±3.43a	20	29.34±2.93ab	24.58	130±7.23abc	3.34	234
0.0001	0.3	3	18.45±3.43	15.54±2.84*	15.8	18.28±2.87*b	0.9	20±1.76*b*	–	–
		24	38.90±6.56	31.12±4.20a	20	37.34±3.83*b	4	40±4.84*b*	–	–
	0.7	3	18.45±3.43	14.14±3.84*	15.8	18.28±2.87*b	0.9	26±2.24*b*	–	–
		24	38.90±6.56	27.65±4.43a	20	37.34±3.83*b	4	42±4.32*b*	–	–
	1	3	18.45±3.43	13.70±2.03*	15.8	18.28±2.87*b	0.9	23.3±3.16*b*	–	–
		24	38.90±6.56	25.30±3.43a	20	37.34±3.83*b	4	48.30±5.03*b*	–	–

Comments: T – time of bacteria strain incubation with drugs, FI – induction factor values, Igfpexp. (%) – the percent of inhibition of gfp expression after treatment of bacteria cells with M and CP in comparison with the control sample (100%), Sgfpexp. (%) – the percent of stimulation of gfp expression after treatment of bacteria cells with CP and CP+M in comparison with the control sample (100%). Data points represent mean values ± SD; n=6; a - significantly different from control ($p<0.05$); b - significantly different from metformin (M) group ($p<0.05$); c - significantly different from cyclophosphamide (CP) group ($p<0.05$); * - no significantly different.

In case of bacteria incubation bacteria incubation with CP and M administrated separately there was no increase in F_I values ($F_I \geq 2$) for two phases of bacteria cells. F_I values ≥ 2 were obtained for the highest concentration of CP and M in simultaneous co-administration of both drugs. The treatment of bacteria with the smallest concentration of CP (0.0001 mg/ml) resulted in a progressive decrease in F_I (below 2) values.

The monitoring of bacteria cultures growth (OD) at the start of bacteria incubation (time 0) and after 3 and 24 h with drugs, indicated significant increase in GI (growth inhibition) values for all tested concentration of M and CP in 24 h treatment. Simultaneous action of both drugs on bacteria cells significantly enhanced the growth inhibition values in 24 h incubation in comparison to separate administration of the drugs and comparable to control sample. In shorter time (3 h) of drugs influence on bacteria cells, there were no statistically important differences for OD values.

Metformin, in relation to CP caused statistically significant increase in the cytotoxic and genotoxic activity of CP. Logarithmic phase prolonged treatment (up to 24 h) of bacteria cells with metformin at concentrations of 0.7 and 1 mg/ml significantly influenced the growth inhibition of bacteria. After 3 h of incubation there were no significant changes in OD values. Bacteria incubated with PBS buffer (control sample) without any drug, resulted in no statistical differences in OD value from 0 to 24 hours of continuous cultivation.

The treatment of *gfp* biosensor bacteria strain in surface water (n=6) enhanced the sensitivity of *recA::gfpmut2* genotoxic system and increased the stimulation of *gfp* expression and SFI value in comparison to incubation in PBS buffer. Prolonged treatment (up to 24 h) of bacteria cells with combination of CP (0.1 mg/ml) and metformin (1 mg/ml) in surface water significantly influenced *gfp* expression with the maximum values of IF=13.20 and 1220% of S_{gfpexp} values comparable to control sample.

DISCUSSION

Previous studies showed that *recA* promoter was induced by known genotoxins and selected anticancer drugs [Ptitsyn et al. 1997, Kostrzyńska et al. 2002, Zaslaver et al. 2004, Matejczyk 2010, Alhandrami and Paton 2013, Park et al. 2013]. According to the results obtained in our experiment M in simultaneous co-administration with CP was shown to modulate and dramatically increase the reactivity of *recA* promoter in relation to separate bacteria treatment with CP and M.

Amador et. al. [2012] indicated, that chronic metformin exposure may be potentially genotoxic *in vitro*. The results of the above experiment provided the confirmation of the possible influence of metformin on the genes, especially in quickly dividing cells, because in 50% of cases there were significant differences (comparable to the control sample) in the level of *recA* promoter sensitivity and *gfp* expression after logarithmic phase bacteria treatment with the whole applied concentrations of metformin and in longer time of incubation (up to 24 h).

Our results indicated that prolonged exposure of bacteria to metformin (up to 24 h) in co-administration with CP resulted in a progressive stimulation of *recA* promoter reactivity and *gfp* gene expression. F_I values ≥ 2 were obtained for CIS+M after 24 hours and for 3 h of incubation. The strongest stimulation of *recA* promoter and *gfp* expression was noticed after addition of the higher CP concentration – 1, 0.1 and 0.01 mg/ml than for lower concentrations of antidiabetic drug. The investigated concentrations of M and CP inhibited SFI values of bacteria growth in the logarithmic phase. It can be suggested the possible repression of *recA::gfp* genetic construct on transcription by both drugs. Our results showed that CP treatment significantly inhibited *E. coli* K-12 longer (up to 24 h) bacteria cells growth. Metformin could has cytotoxic effect by inhibition of bacteria cells growth, for highest applied concentrations and 24 h treatment. CP is a very active drug, especially for dividing cells. The above data confirmed that coadministration of CP+M importantly intensified cytotoxic effect and *recA* promoter activity. The obtained results are in agreement with earlier empirical studies of other authors (with the use of cisplatin) who demonstrated that co-administration of metformin with chemotherapeutic agents intensified the inhibition of cancer cells proliferation and significantly improved cisplatin-induced cytotoxicity [Quinn et al. 2013]. DNA damage can initiate a cascade of cellular biological effects including cell death. The direct and indirect metformin influence on DNA could be the main biological mechanism of enhancement the cytotoxic and genotoxic activity of CP, especially in simultaneous co-administration of both drugs.

Preliminary results indicated stronger reactivity of *recA::gfpmut2* genetic system in surface water for drugs treated samples (n=6) (data not shown). The influence of surface water on genetic system, which was used, could be a consequence of the presence of different chemicals (e.g. compounds from hospitals) in surface water than we used in our experiment. These unknown water compounds could increase the *gfp* expression in bacteria strain, similarly to the drugs (cyclophosphamide and metformin) used in our experiments. The maximum point for *recA* promoter stimulation was observed for co-adminiastration of 0,1 mg/ml CP and 1mg/ml M in 24 h being $S_{gfpexp.} = 1220\%$ (as compared to $S_{gfpexp.} = 838\%$ for CP and M in PBS).

To assess genotoxic sensitivity of *recA::gfp* genetic biosensing system 4% acetone was tested as a negative control. For this chemical F_I values did not increase during 3 h and 24 h of incubation. Methylnitronitrosoguanidine (MNNG) at concentration of 50 μM was used as a positive control. For this chemical F_I=8.4 during 24 h incubation and F_I=2.8 during 3h were obtained (data not shown). These results showed stronger sensitivity of *recA::gfp* biosensing system for MNNG than acetone stressor.

CONCLUSIONS

1. The results of the presented study indicated that *recA::gfpmut2* genetic system was sensitive to drugs applied in experiment and drugs mixture.

2. *RecA* promoter was a good bioindicator for cytotoxic and genotoxic effect screening of cyclophosphamide, metformin and the mixture of the both drugs in PBS buffer and surface water.

3. The results indicated that *E. coli* K-12 *recA::gfp mut2* strain could be potentially useful for first-step screening of cytotoxic and genotoxic effect of anticancer and antidiabetic pharmacist residues in water.

4. Next steps in research will include more experimental analysis to validate *recA::gfpmut2* genetic system in *E. coli* K-12 on different anticancer drugs.

Acknowledgements

Authors are very grateful to Prof. Uri Alon, Department of Molecular Cell Biology & Department of Physics of Complex Systems, Weizmann Institute of Science Rehovot, Israel for providing bacteria strains. This work was financially supported by research project number S/WBiIŚ/3/2011.

REFERENCES

1. Besse J.P., Latour J.-F., Garric J. 2012. Anticancer drugs in surface waters. What can we say about the occurrence and environmental significance of cytotoxic, cytostatic and endocrine therapy drugs? Environ. Intern., 39, 73–86.

2. Zhang J., Chang V.W.C., Giannis A., Wang A.–Y. 2013. Removal of cytostatic drugs from aquatic environment: A review., Scien. Tot. Environ., 445–446, 281–298.

3. Yu-Chen Lin A., Lin Y.C., Lee W.N. 2014. Prevalence and sunlight photolysis of controlled and chemotherapeutic drugs in aqueous environments. Environm. Poll., 187, 170–181.

4. Kosjek T., Heath E. 2011. Occurrence, fate and determination of cytostatic pharmaceuticals in the environment, Tr. AC. Trends Anal. Chem., 30, 1065–1087.

5. Quinn B.J., Kitagawa H., Memmott R.M., Gills J.J., Dennis P.A. 2013. Repositioning metformin for cancer prevention and treatment, Trends. Endocrinol. Metabol., 24, 9, 469–480.

6. Scheurer M., Sacher F., Brauch H.J. 2009. Occurrence of the antidiabetic drug metformin in sewage and surface waters in Germany, J Environ. Monit., 11, 9, 1608–1613.

7. Kostrzyńska M, Leung K.T., Lee H., Trevors J.T. 2002. Green fluorescence protein based biosensor for detecting SOS-inducing activity of genotoxic compounds. J Microbiol. Meth., 48, 43–51.

8. Matejczyk M. 2010. The Potency of application of microbial biosensors. Advances in Microbiology (In Polish). 49, 4, 297–304.

9. Alhadrami H.A., Paton G.I. 2013. The potential applications of SOS-lux biosensors for rapid screening of mutagenic chemicals. FEMS Microbiol. Lett., 344, 1, 69–76.

10. Park M., Tsai S.L., Chen W. 2013. Microbial Biosensors: Engineered microorganisms as the sensing machinery. Sensors, 13, 5777–5795.

11. Zaslaver A., Mayo A.E., Rosemberg R., Bashkin P., Sberro H., Tsalyuk M., Surette M.G., Alon U. 2004. Just-in-time transcription program in metabolic pathways. Nat. Genet., 36, 5, 486–491.

12. Ptitsyn L.R., Horneck G., Komova O., Kozubek S., Krasavin E.A., Bonev M., Rettberg P. 1997. A biosensor for environmental genotoxin screening based on an SOSlux assay in recombinant Escherichia coli cells. Appl. Environm. Microbiol., 63, 4377–4384.

13. Amador R.R., Longo J.P.F., Lacava Z.G., Dorea J.G. 2012. Almeida Santos M. de F.M. Metformin (dimethyl-biquanide) induced DNA damage in mammalian cells. Gen. Mol. Biol., 35, 1, 153–158.

Permissions

All chapters in this book were first published in JEE, by Polish Society of Ecological Engineering (PTIE); hereby published with permission under the Creative Commons Attribution License or equivalent. Every chapter published in this book has been scrutinized by our experts. Their significance has been extensively debated. The topics covered herein carry significant findings which will fuel the growth of the discipline. They may even be implemented as practical applications or may be referred to as a beginning point for another development.

The contributors of this book come from diverse backgrounds, making this book a truly international effort. This book will bring forth new frontiers with its revolutionizing research information and detailed analysis of the nascent developments around the world.

We would like to thank all the contributing authors for lending their expertise to make the book truly unique. They have played a crucial role in the development of this book. Without their invaluable contributions this book wouldn't have been possible. They have made vital efforts to compile up to date information on the varied aspects of this subject to make this book a valuable addition to the collection of many professionals and students.

This book was conceptualized with the vision of imparting up-to-date information and advanced data in this field. To ensure the same, a matchless editorial board was set up. Every individual on the board went through rigorous rounds of assessment to prove their worth. After which they invested a large part of their time researching and compiling the most relevant data for our readers.

The editorial board has been involved in producing this book since its inception. They have spent rigorous hours researching and exploring the diverse topics which have resulted in the successful publishing of this book. They have passed on their knowledge of decades through this book. To expedite this challenging task, the publisher supported the team at every step. A small team of assistant editors was also appointed to further simplify the editing procedure and attain best results for the readers.

Apart from the editorial board, the designing team has also invested a significant amount of their time in understanding the subject and creating the most relevant covers. They scrutinized every image to scout for the most suitable representation of the subject and create an appropriate cover for the book.

The publishing team has been an ardent support to the editorial, designing and production team. Their endless efforts to recruit the best for this project, has resulted in the accomplishment of this book. They are a veteran in the field of academics and their pool of knowledge is as vast as their experience in printing. Their expertise and guidance has proved useful at every step. Their uncompromising quality standards have made this book an exceptional effort. Their encouragement from time to time has been an inspiration for everyone.

The publisher and the editorial board hope that this book will prove to be a valuable piece of knowledge for researchers, students, practitioners and scholars across the globe.

List of Contributors

Dorota Krzemińska
Institute of Environmental Engineering, Czestochowa University of Technology, Brzeznicka 60a, 42-200 Czestochowa, Poland

Ewa Neczaj
Institute of Environmental Engineering, Czestochowa University of Technology, Brzeznicka 60a, 42-200 Czestochowa, Poland

Gabriel Borowski
Faculty of Fundamentals of Technology, Lublin University of Technology, Nadbystrzycka 38, 20-618 Lublin, Poland

Czesława Rosik-Dulewska
Institute of Environmental Engineering of the Polish Academy of Sciences, Skłodowskiej-Curie Str. 34, 41-819 Zabrze, Poland

Tomasz Ciesielczuk
Opole University, Department of Land Protection, Oleska Str. 22, 45-052 Opole, Poland

Urszula Karwaczyńska
Opole University, Department of Land Protection, Oleska Str. 22, 45-052 Opole, Poland

Hanna Gabriel

Monika Puchlik
Białystok University of Technology, 45A Wiejska Str., 15-351 Białystok, Poland

Katarzyna Ignatowicz
Białystok University of Technology, 45A Wiejska Str., 15-351 Białystok, Poland

Wojciech Dąbrowski
Białystok University of Technology, 45A Wiejska Str., 15-351 Białystok, Poland

Robert Rosa
Department of Vegetable Crop, Siedlce University of Natural Sciences and Humanities, Bolesława Prusa 14, 08-110 Siedlce. Poland

Roman Kolczarek
Institute of Agronomy, University of Natural Sciences and Humanities, B. Prusa 14, 08-110 Siedlce, Poland

Agnieszka Kamińska
Department of Applied Mathematics and Computer Science, University of Life Sciences in Lublin, Poland

Antoni Grzywna
Department of Environmental Engineering and Geodesy, University of Life Sciences in Lublin, Leszczyń- skiego 7, 20-950 Lublin, Poland

Cezary Tkaczuk
Department of Plant Protection, Siedlce University of Natural Sciences and Humanities, B. Prusa 14, 08-110 Siedlce, Poland

Anna Król
Department of Plant Protection, Siedlce University of Natural Sciences and Humanities, B. Prusa 14, 08-110 Siedlce, Poland

Anna Majchrowska-Safaryan
Department of Plant Protection, Siedlce University of Natural Sciences and Humanities, B. Prusa 14, 08-110 Siedlce, Poland

Łukasz Nicewicz
Department of Plant Protection, Siedlce University of Natural Sciences and Humanities, B. Prusa 14, 08-110 Siedlce, Poland

Aleksandra Steinhoff-Wrześniewska
Institute of Technology and Life Sciences, Lower Silesian Research Centre in Wrocław, Zygmunta Berlinga 7, 51-209 Wrocław, Poland

Joanna Szczykowska
Faculty of Civil and Environmental Engineering, Bialystok University of Technology, Wiejska 45b, 15-351 Białystok, Poland

Anna Siemieniuk
Faculty of Civil and Environmental Engineering, Bialystok University of Technology, Wiejska 45b, 15-351 Białystok, Poland

Jolanta Dąbrowska
Institute of Environmental Engineering, Wrocław University of Environmental and Life Sciences, Plac Grun-waldzki 24, 50-363 Wrocław, Poland

Olgierd Kempa
Department of Spatial Economy, Wrocław University of Environmental and Life Sciences, Plac Grunwaldzki 24, 50-363 Wrocław, Poland

Joanna Markowska
Institute of Environmental Engineering, Wrocław University of Environmental and Life Sciences, Plac Grun-waldzki 24, 50-363 Wrocław, Poland

Jerzy Sobota
Institute of Building, Wrocław University of Environmental and Life Sciences, Plac Grunwaldzki 24, 50-363 Wrocław, Poland

Samuel Agarry
Ladoke Akintola University of Technology, Ogbomoso, Nigeria

Ganiyu K. Latinwo
Ladoke Akintola University of Technology, Ogbomoso, Nigeria

Kazimierz H. Dyguś
Faculty of Ecology, University of Ecology and Management, 12 Olszewska Str., 00-792 Warsaw, Poland

Anna Siemieniuk
Department of Technology in Engineering and Environmental Protection, Faculty of Civil and Environmental Engineering, Białystok University of Technology, Wiejska 45A, 15-351 Białystok, Poland

Joannna Szczykowska
Department of Technology in Engineering and Environmental Protection, Faculty of Civil and Environmental Engineering, Białystok University of Technology, Wiejska 45A, 15-351 Białystok, Poland

Józefa Wiater
Department of Technology in Engineering and Environmental Protection, Faculty of Civil and Environmental Engineering, Białystok University of Technology, Wiejska 45A, 15-351 Białystok, Poland

Piotr Salachna
Department of Horticulture, Faculty of Environmental, Management and Agriculture, West Pomeranian University of Technology in Szczecin, Papieża Pawła VI 3, 71-459 Szczecin, Poland

Agnieszka Zawadzińska
Department of Horticulture, Faculty of Environmental, Management and Agriculture, West Pomeranian University of Technology in Szczecin, Papieża Pawła VI 3, 71-459 Szczecin, Poland

Agnieszka Wysocka-Czubaszek
Department of Environmental Protection and Management, Białystok University of Technology, Wiejska 45A, 15-351 Białystok, Poland

Robert Czubaszek
Department of Environmental Protection and Management, Białystok University of Technology, Wiejska 45A, 15-351 Białystok, Poland

Irena Niedźwiecka-Filipiak
Institute of Landscape Architecture, Wrocław Universiy of Environmental and Life Sciences, pl. Grunwaldzki 24a, 50-363 Wrocław, Poland

Liliana Serafin
Institute of Landscape Architecture, Wrocław Universiy of Environmental and Life Sciences, pl. Grunwaldzki 24a, 50-363 Wrocław, Poland

Agnieszka Wysocka-Czubaszek
Department of Environmental Protection and Management, Białystok University of Technology, Wiejska 45A, 15-351 Białystok, Poland

Robert Czubaszek
Department of Environmental Protection and Management, Białystok University of Technology, Wiejska 45A, 15-351 Białystok, Poland

Agnieszka Kisło
Department of Technology in Engineering and Environmental Protection, Faculty of Civil and Environmental Engineering, University of Technology in Bialystok, 45A Wiejska Str., 15-351 Bialystok, Poland

Iwona Skoczko
Department of Technology in Engineering and Environmental Protection, Faculty of Civil and Environmental Engineering, University of Technology in Bialystok, 45A Wiejska Str., 15-351 Bialystok, Poland

Tomasz Kowalik
Department of Land Reclamation and Environmental Development, University of Agriculture in Krakow, Mickiewicza 24-28, 30-050 Kraków, Poland

Włodzimierz Rajda
Department of Land Reclamation and Environmental Development, University of Agriculture in Krakow, Mickiewicza 24-28, 30-050 Kraków, Poland

Elżbieta Halina Grygorczuk-Petersons
Department of Technology in Engineering and Environmental Protection, Bialystok University of Technology, Wiejska 45, 15-351 Bialystok, Poland

Józefa Wiater
Department of Technology in Engineering and Environmental Protection, Bialystok University of Technology, Wiejska 45, 15-351 Bialystok, Poland

Marek Gugała
Siedlce University of Natural Sciences and Humanities, B. Prusa 14, 08-110 Siedlce, Poland

Krystyna Zarzecka
Siedlce University of Natural Sciences and Humanities, B. Prusa 14, 08-110 Siedlce, Poland

Anna Sikorska
Siedlce University of Natural Sciences and Humanities,
B. Prusa 14, 08-110 Siedlce, Poland

Halina Marczak
Mechanical Engineering Faculty, Lublin Technology
University, Nadbystrzycka 36, 20-618 Lublin, Poland

Elżbieta Malinowska
Institute of Agronomy, Siedlce University of Natural
Sciences and Humanities, Prusa, Siedlce, Poland

Kazimierz Jankowski
Institute of Agronomy, Siedlce University of Natural
Sciences and Humanities, Prusa, Siedlce, Poland

Beata Wiśniewska-Kadżajan
Institute of Agronomy, Siedlce University of Natural
Sciences and Humanities, Prusa, Siedlce, Poland

Jacek Sosnowski
Institute of Agronomy, Siedlce University of Natural
Sciences and Humanities, Prusa, Siedlce, Poland

Roman Kolczarek
Institute of Agronomy, Siedlce University of Natural
Sciences and Humanities, Prusa, Siedlce, Poland

Marcin Sidoruk
Department of Land Reclamation and Environmental
Management, Faculty of Environmental Management
and Agriculture, University of Warmia and Mazury in
Olsztyn, Plac Łódzki 2, 10-756 Olsztyn, Poland

Józef Koc
Department of Land Reclamation and Environmental
Management, Faculty of Environmental Management
and Agriculture, University of Warmia and Mazury in
Olsztyn, Plac Łódzki 2, 10-756 Olsztyn, Poland

Ireneusz Cymes
Department of Land Reclamation and Environmental
Management, Faculty of Environmental Management
and Agriculture, University of Warmia and Mazury in
Olsztyn, Plac Łódzki 2, 10-756 Olsztyn, Poland

Małgorzata Rafałowska
Department of Land Reclamation and Environmental
Management, Faculty of Environmental Management
and Agriculture, University of Warmia and Mazury in
Olsztyn, Plac Łódzki 2, 10-756 Olsztyn, Poland

Andrzej Rochwerger
Department of Land Reclamation and Environmental
Management, Faculty of Environmental Management
and Agriculture, University of Warmia and Mazury in
Olsztyn, Plac Łódzki 2, 10-756 Olsztyn, Poland

Katarzyna Sobczyńska-Wójcik
Department of Land Reclamation and Environmental
Management, Faculty of Environmental Management
and Agriculture, University of Warmia and Mazury in
Olsztyn, Plac Łódzki 2, 10-756 Olsztyn, Poland

Krystyna A. Skibniewska
Department of Security Rudiments, Faculty of Technical
Sciences, University of Warmia and Mazury in Olsztyn,
Heweliusza 10, 10-719 Olsztyn, Poland

Ewa Siemianowska
Department of Security Rudiments, Faculty of Technical
Sciences, University of Warmia and Mazury in Olsztyn,
Heweliusza 10, 10-719 Olsztyn, Poland

Janusz Guziur
Department of Fish Biology and Breeding. Faculty of
Environmental Sciences, University of Warmia and
Mazury in Olsztyn, Oczapowskiego 14, 10-719 Olsztyn,
Poland

Józef Szarek
Department of Pathophysiology, Forensic Veterinary
Medicine and Administration, Faculty of Veterinary
Medicine, University of Warmia and Mazury in Olsztyn,
Oczapowskiego 13, 10-719 Olsztyn, Poland

Mirosław Wiatkowski
Institute of Environmental Engineering, Wrocław
University of Environmental and Life Sciences, Plac
Grunwaldzki 24, 50-363 Wrocław, Poland

Czesława Rosik-Dulewska
Department of Land Protection, Opole University, Oleska
Str. 22, 45-052 Opole, Poland

Piotr Kardasz
Wrocław University of Technology, 27 Wyspiańskiego Str.,
50-370 Wrocław, Poland

Radosław Wróbel
Wrocław University of Technology, 27 Wyspiańskiego Str.,
50-370 Wrocław, Poland

Joanna Kazimierowicz
Department of Environmental Systems Engineering,
Bialystok University of Technology, Wiejska 45A, 15-351
Białystok, Poland

Mateusz Wnukowski
Institute of Heat Engineering and Fluid Mechanics,
Wrocław University of Technology, Wybrzeże Wyspiań-
skiego 27, 50-370 Wrocław, Poland

Marzena Matejczyk
Department of Sanitary Biology and Biotechnology, Faculty of Civil Engineering and Environmental Engineering, Bialystok University of Technology, Wiejska 45E, 15-351 Bialystok, Poland

Stanisław Józef Rosochacki
Department of Sanitary Biology and Biotechnology, Faculty of Civil Engineering and Environmental Engineering, Bialystok University of Technology, Wiejska 45E, 15-351 Bialystok, Poland